GLOBALIZATION

Yesterday, Today, and Tomorrow

GLOBALIZATION

Yesterday, Today, and Tomorrow

Edited by

Jim Sheffield,
Victoria University of Wellington

Andrey Korotayev
Russian State University for the Humanities

Leonid Grinin
Volgograd Center for Social Research

EMERGENT™
PUBLICATIONS
3810 N 188th Ave
Litchfield Park, AZ 85340

Globalization: Yesterday, Today, and Tomorrow

Edited by: Jim Sheffield, Andrey Korotayev, & Leonid Grinin

Library of Congress Control Number: 2013936973

ISBN: 978-1-938158-08-7

Copyright © 2013 Emergent Publications
3810 N 188th Ave, Litchfield Park, AZ 85340, USA

Printed in the United States of America

ABOUT THE EDITORS

Jim Sheffield is a systems theorist at the School of Management at Victoria University of Wellington, New Zealand. In this capacity and his previous role as Director of the Decision Support Centre, University of Auckland he has designed, implemented and evaluated more than 100 action research initiatives. Major projects focussed on the facilitation of national policy in response to globalization. Jim has published widely in the systems perspectives that underpin aspects of globalization, especially those related to decision making, knowledge management, systemic development and ethical inquiry. He has over 150 scholarly publications and is coeditor of the *Journal of Globalization Studies*. He serves on the editorial board of journals and is active in professional societies, including the *International Society for the Systems Sciences (ISSS)*. He is the author of several books, and editor of *Systemic Development* and *My Decisive Moment*.

Andrey V. Korotayev is Senior Research Professor of the Oriental Institute and Institute for African Studies, Russian Academy of Sciences, Professor and the Head of the Department of Modern Asian and African Studies, Russian State University for the Humanities, Full Professor of the Faculty of Global Studies of the Moscow State University, and the Head of Laboratory of the Monitoring of Sociopolitical Destabilization Risks as the National Research University Higher School of Economics, Moscow. Together with Leonid Grinin he edits the *Journal of Globalization Studies* and the *Social Evolution and History*. He is the author of over 300 scholarly publications, including such monographs as *Ancient Yemen* (1995); *World Religions and Social Evolution of the Old World Oikumene Civilizations: A Cross-Cultural Perspective* (2004); *Introduction to Social Macrodynamics: Compact Macromodels of the World System Growth* (2006); and *Modeling the World Dynamics* (2012).

Leonid E. Grinin is a Russian sociologist, political anthropologist, and a scholar of historical trends and future studies. He has PhD and is Senior Research Professor at the Institute for Oriental Studies of the Russian Academy of Sciences in Moscow and serves as Deputy Director of the Eurasian Center for Big History & System Forecasting (Russian Academy of Sciences). He is also coeditor of the *Journal of Globalization Studies* and Editor-in-Chief of the journal *Age of Globalization* (in Russian). His academic interests are connected with the analysis of problems of globalization and modernization. His published research includes topics such as forecasting world political change, social-economic development and social evolution. He has also published on the theory of historical process; and the evolution of statehood. He is the author of more than 360 scholarly publications in Russian and English, including 25 monographs.

ACKNOWLEDGEMENTS

The editors would like to express their thanks to the many authors who contributed their research to this monograph. We would also like to thank our technical editors, Elena Emanova, Kseniya Uhova, and Elena Nikiforova, who were responsible for the preparation of the papers and the cover. Finally we express our gratitude to Kurt Richardson and all the others at Emergent Publications for their guidance and support throughout the publishing process.

CONTENTS

GLOBALIZATION AS A LINK BETWEEN THE PAST AND THE FUTURE

Jim Sheffield, Andrey Korotayev, & Leonid Grinin

CHAPTER 1—THE ORIGINS OF GLOBALIZATION

Leonid Grinin & Andrey Korotayev

CHAPTER 2—THE LEAD ECONOMY SEQUENCE IN WORLD POLITICS (FROM SUNG CHINA TO THE UNITED STATES): SELECTED COUNTERFACTUALS

William R. Thompson

CHAPTER 3—CONTINUITIES AND TRANSFORMATIONS IN THE EVOLUTION OF WORLD-SYSTEMS

Christopher Chase-Dunn

CHAPTER 4—GEOPOLITICAL CONDITIONS OF INTERNATIONALISM, HUMAN RIGHTS, AND WORLD LAW

Randall Collins

CHAPTER 5—CONTEMPORARY GLOBALIZATION AND NEW CIVILIZATIONAL FORMATIONS

Shmuel N. Eisenstadt

CHAPTER 6—THE 'RETURN' OF RELIGION AND THE CONFLICTED CONDITION OF WORLD ORDER

Roland Robertson

CHAPTER 7—CULTURE AND THE SUSTAINABILITY OF THE GLOBAL SYSTEM

Ervin Laszlo

CHAPTER 8—MEASURING GLOBALIZATION—OPENING THE BLACK BOX: A CRITICAL ANALYSIS OF GLOBALIZATION INDICES

Axel Dreher, Noel Gaston, Pim Martens, and Lotte Van Boxem

CHAPTER 9—ON FREE TRADE, CLIMATE CHANGE, AND THE W.T.O.

Rafael Reuveny

CHAPTER 10—THE E-WASTE STREAM IN THE WORLD-SYSTEM

R. Scott Frey

CHAPTER 11—GREAT POWER POLITICS FOR AFRICA'S DEVELOPMENT: AN OVERVIEW ANALYSIS OF IMPACT OF THE EU'S AND CHINA'S COOPERATION WITH THE CONTINENT

Zinsê Mawunou and Chunmei Zhao

CHAPTER 12—CONNECTING LOGISTICS NETWORKS GLOBALLY VIA THE U.N. SINGLE WINDOW CONCEPT

Michael Linke

CHAPTER 13—THE RECENT GLOBAL CRISIS UNDER THE LIGHT OF THE LONG WAVE THEORY

Tessaleno C. Devezas

CHAPTER 14—LOCAL SOLUTIONS IN A GLOBAL ENVIRONMENT: FACILITATING NATIONAL STRATEGIES IN NEW ZEALAND

Jim Sheffield

CHAPTER 15—GLOBAL BIFURCATION: THE DECISION WINDOW

Ervin Laszlo

CHAPTER 16—TOMORROW'S TOURIST: FLUID AND SIMPLE IDENTITIES

Ian Yeoman

CHAPTER 17—WORLD ENERGY AND CLIMATE IN THE TWENTY-FIRST CENTURY IN THE CONTEXT OF HISTORICAL TRENDS: CLEAR CONSTRAINTS TO FUTURE GROWTH

Vladimir V. Klimenko and Alexey G. Tereshin

CHAPTER 18—WILL THE GLOBAL CRISIS LEAD TO GLOBAL TRANSFORMATIONS?

Leonid Grinin & Andrey Korotayev

Introduction

GLOBALIZATION AS A LINK BETWEEN THE PAST AND THE FUTURE

Jim Sheffield, Andrey Korotayev, & Leonid Grinin

We see globalization as the growth of the sizes of social systems and the increase in the complexity of intersocietal links. Thus, in certain respects, globalization may be regarded as a process connecting the past, the present, and the future—as a sort of bridge between the past and the future. The title and the composition of the present volume reflect this idea.

Globalization: Yesterday, Today, and Tomorrow is distinguished by its focus on the systemic aspects of globalization processes. Political, economic, geographic, ecological, social, cultural, ethnic, religious and historical processes are analyzed and their single and joint impacts on globalization are discussed. The purpose is to complement more objective or 'technical' globalization narratives with more direct accounts of social and emotional issues.

There are a number of publications dealing with particular aspects of globalization. However, the growing complexity has increased the interrelatedness among all countries. Recurrent economic and political crises that have global repercussions demand new approaches. This book provides a wider range of views on globalization than some other globalization journals and books. In particular, we believe that seeking perspectives that cross organizational, geographic and cultural boundaries may aid in reducing misunderstandings and diminish the negative aspects of globalization. The global financial crisis has only emphasized the need to develop local solutions in a global environment and at the same time to search for global solutions to common problems. New approaches are required that demonstrate an appreciation of the 'local' in particular political, economic, social, cultural and geographic contexts, while simultaneously promoting effective change in response to pressing global issues.

Globalization: Yesterday, Today, and Tomorrow provides a multi-faceted analysis of globalization that is based on the understandings of authors working in both Western and non-Western traditions. We believe that current events such as the global financial crisis illustrate that discussion should not be limited to particular geographic regions or narrowly-defined methods of analysis. The perspectives in the book are the result of cooperation with scholars from different countries. They provide visions of global processes from both the developed and developing countries, including those in Africa, Asia, America, Australia and Oceania, West, East, Central and South Europe, Russia etc.

Globalization is a very broad concept not only with respect to the diversity of regions, cultures, and actors, but also with respect to the diversity of analytical approaches that can be employed to study it. The articles in this book embrace the need to cover a variety of aspects and dimensions of globalization, and to see both its local and its global manifestations. From our perspective, globalization studies imply research that is not just limited to the most popular spheres of economic and political globalization, but also includes the study of global problems such as climatic change, cultural globalization, and so on.

In summary, the special character of the Globalization: Yesterday, Today, and Tomorrow is that it delivers a broad international and multicultural spectrum of issues associated with globalization, including the impact of globalization on particular cultural-geographic regions.

Organization

The 18 articles are grouped into three sections. As suggested by the title, these address aspects of the past, present and future of globalization. Part I (4 articles) addresses globalization in history. Part II (10 articles) addresses contemporary globalization. Part III (4 articles) addresses globalization in the future.

PART I—GLOBALIZATION IN HISTORY

1. Leonid Grinin and Andrey Korotayev contribute to the history of globalization an analysis of the nature of global processes and causes of increasing integration. They propose a history of globalization that draws on a special methodology and a world-system approach based on the development of spatial links over seven periods of time starting with the Agrarian Revolution (four before and three after the great geographic discoveries). The time periods range from

before the 4th millennium BCE to the 21st century. The types of spatial links described range from local and regional links to global and planetary links through continental and intercontinental ones. Evidence is presented for each period. This includes for instance the existence of large-scale trade in metals as early as 4th millennium BCE, and the social impact of intercontinental trade in the late 1st millennium BCE. The evidence also includes a density and diversity of transcontinental links sufficient to transmit disease (bubonic plague) from the Far East to the Atlantic in two decades (1330s-1340s), and the comparability of some aspects of global integration prior to the great geographic discoveries with more recent periods. The authors note that globalization began at least as early as 4th thousand BCE. The proposed system of spatial links addresses shortcomings in previous systems that tended to underestimate the scale of spatial links in the pre-industrial era.

2. William Thompson contributes to the history of political and economic globalization an analysis of the significance of global events. He argues that the way we make sense of world politics and episodes of accelerated globalization depends on our historical scripts, and that these vary considerably. It is not so much a matter of disagreeing about what happened in the past as it is the one of disagreeing about which past events were most significant to an understanding of international relations processes. Validating one person's historical script versus someone else's is a highly problematic exercise. Counterfactuals, however, can be utilized to at least suggest or reinforce the asserted significance of different versions of political-economic history. A series of eight counterfactuals encompassing the past 1000 years are harnessed to buttress the utility of framing the development of the modern world economy around a chain of lead economies and system leaders extending back to Sung China and forward to the United States. These potential turning points matter in part because they did not go down the counterfactual path but might have. They matter even more because of the path that was pursued at each point. They matter because they created a political-economic structure for world politics that has first emerged, then evolved and, so far, endured. The implications of what did happen (not what did not happen) are still with us today.

3. Christopher Chase-Dunn contributes to the globalization in history section a discussion on the continuities and transformations of systemic logic. Modes of accumulation in the world historical evolutionary perspective are described and the prospects for systemic transformation in the next several decades evaluated. The article also considers the meaning of the recent global financial meltdown by comparing it with earlier debt crises and periods of collapse. Has this been

just another debt crisis like the ones that have periodically occurred over the past 200 years, or is it part of the end of capitalism and the transformation to a new and different logic of social reproduction? The author considers how the contemporary network of global counter-movements and progressive national regimes are seeking to transform the capitalist world-system into a more humane, sustainable and egalitarian civilization and how the current crisis is affecting the network of counter-movements and regimes, including the Pink Tide populist regimes in Latin America, and the anti-austerity movements. The ways in which the New Global Left is similar to, and different from, earlier global counter-movements are also described. The discussion contributes to the development of a comparative and evolutionary framework that examines what is really new about the current global situation and what constitutes therefore collectively rational responses.

4. Randall Collins completes Part I with a geopolitical analysis of key globalization events in the past, present, and future. As historical sociologists in the tradition of Weber have documented, the state's existence has depended on its military power, which varies in degree of monopolization, of legitimacy, and of extent of territory controlled. Geopolitical principles (comparative resource advantage, positional or marchland advantage, logistical overextension) have determined both the Chinese dynastic cycles, and the balance of power in European history. In 1980 the author was successful in using these geopolitical principles to predict the strains which brought about the collapse of the Soviet empire, which was itself a continuation of the older Russian empire. The same geopolitical principles continue to apply to recent wars in Afghanistan, Iraq, and Pakistan. Guerrilla wars differ from conventional wars by relying especially on geopolitical principles of promoting enemy overextension. Geopolitics encompasses both war and diplomacy, the means by which coalitions among states are organized. The rule of international law depends on a dominant coalition upheld by favorable geopolitical conditions; and on the extension of bureaucracy via state penetration, but now on a world-wide scale. Randall Collins answers two key questions regarding historical globalization processes: "Is the world of the early 21st century moving towards a new era of international rule of law to support universal human rights?" and "Where does the opposition to universal human rights come from?" His answers to both of these questions demonstrate that international rule of law is not an alternative to geopolitics, but is successful only under specific geopolitical conditions.

PART II—CONTEMPORARY GLOBALIZATION

5. Shmuel Eisenstadt examines some specific aspects of contemporary globalization as they bear on the crystallization of new distinct civilizational formations. He describes new and very intensive processes of contemporary globalization characterized by growing interconnectedness between economic, cultural and political processes. The full impact of these intertwined globalization processes can only be understood within the new historical context. The author notes that different religions are now acting in a common civilizational setting. In this context competition and struggles between religions often became vicious—yet at the same time there are tendencies toward the development of interfaith meetings and encounters rooted in the original program of modernity. For instance, many of the criticisms of the Enlightenment project made by Sayyid Qutb, possibly the most eminent fundamentalist Islamic theologian, are similar to the major religious and 'secular' critics of Enlightenment from de Maistre, the romantics, and those who, in Charles Taylor's words emphasized the 'expressivist dimension of human experience'. These premises implied the possibility of cooperation between different faiths. Movements to reform globalization philosophies and civilizational premises have taken place in a variety of local and regional contexts, including 'centers' such as the European Union and various 'peripheries'.

6. Roland Robertson investigates the return of religion to the study of world politics and globalization. He argues that religion has been neglected in international relations ever since the Peace of Westphalia in 1648. This neglect has largely occurred because of the primacy given to changes and events in the West, particularly since the formal separation of church and state and its imposition on or emulation by Eastern societies. The study of international relations was insulated from the study of religion and vice versa. The apparent eruption of Islam onto the world scene as symbolized and expressed by the events of 9/11 was greeted with surprise in many academic disciplines. However, the degree to which the global conflict between the two major actors—namely al-Qaeda and the regime in the USA—has assumed heavily religious terms cannot responsibly be questioned. The recent heightened concern with religion in globalization, and the globalization of religion, provides the opportunity to undertake historical discussion from new perspectives which overcome the normal Western view that religion is not important in Realpolitik. Moreover, it is argued that much of the neglect of religion in work on world affairs has largely been the product of the inaccurate perception of ongoing secularization. The overall discussion is framed by some objections to the limiting consequences of disciplinarity.

7. Ervin Laszlo argues that the values and associated behaviors of the dominant culture of the contemporary world gave rise to a globally extended system that is not sustainable in its present form. If a cataclysmic breakdown caused by unchecked global warming is to be averted, the influential culture that shapes today's world must change. Humanity can no longer afford to be dominated by a narrowly materialist and manipulative culture focused on ego-centered, company-centered, or nation-centered short-term benefit, with no regard to the wider system that frames existence on this planet. Consciously moving toward a harmonious system of cooperative societies focused on the shared objective of sustaining the systems of life on the planet is an urgent necessity. To this end a mutation is needed in the cultures of the contemporary world, so as to create the values and aspirations that would bring together today's individually diverse and largely self-centered societies in the shared mission of ensuring the sustainability of the global system of humanity in the framework of the biosphere. The global system is highly diverse today, but it is insufficiently coordinated. Creating a higher level of unity within its diversity is intrinsically feasible: it calls for system-maintaining cooperation among the diverse societies that make up the system.

8. Axel Dreher, Noel Gaston, Pim Martens, and Lotte Van Boxem discuss the measurement of globalization with a view to advancing the understanding of globalization indices. Can globalization be better understood by measuring it? What are the intellectual and political implications of the existing globalization indices? What are the attributes and limitations of globalization indices? In what fields can they be used? Is the objective assessment of both the causes and consequences of globalization an essential agenda for contemporary societies. Do positive economic, social and political analyses require data and are globalization indices a most promising means for providing it? A central theme is the size of the gap between the quantitative and the qualitative analysis of globalization. They argue that if globalization indices are to make a substantive contribution, they ought to bridge some existing gaps in our understanding of globalization. In confronting new questions on the essential nature of globalization, interdisciplinary cooperation is required. It would be fruitful for academics from the quantitative side (modeling, conclusive statements, certainty and proofs) and qualitative side (analysis, discussion, conceptual revision, background and textual form) to sit together and work on the challenges. Despite the different methodologies, choice of variables and weights, and so on, new cooperative frameworks are needed.

9. Rafael Reuveny critically examines the proposition that a free global market benefits the environment. The aim is to introduce the insights to be gained from a proposed research program focusing on the WTO and the environment in the context of climate change. This research programe explores the link between climate change, which has recently emerged as the greatest environmental threat, and world trade, which has grown continuously since WWII. The growth of world trade, facilitated by the GATT-WTO regime, evokes an important question. Is this regime good for the environment, or has it contributed to the increase of greenhouse gases, the primary driver of climate change? While this question cannot be fully answered in this paper alone, it is important to consider it now because many of the expected damages caused by climate change may be considerable and nonreversible. After discussing the state of knowledge on the effects of trade on the environment, the author evaluates whether the biosphere can accommodate perpetual economic growth. The possibility is considered that the global community could decide to create a new World Environmental Protection Agency that would give priority to environmental considerations of trade policy.

10. Scott Frey argues that globalization and sustainability are contradictory tendencies in the current world-system. He examines this dichotomy in the context of a particular case—the transfer by transnational corporations of electronic waste from the more developed countries to the less developed zones of the world-system. He argues that such exports (which total more than 30 million tons per year) reduce sustainability and put humans and the environment in recipient countries at substantial risk. The discussion proceeds in several steps. Environmental justice and sustainability in the world-system are first examined. This is followed by a discussion of the e-waste trade in the world-system. The extent to which this trade has negative health, safety, and environmental consequences in Guiyu, China (home to an estimated 150,000 e-waste workers) is outlined. The neo-liberal contention that such exports are economically beneficial to the core and periphery is critically examined. Policies proposed as solutions to the problem of e-waste traffic in Guiyu and the world-system are critically reviewed. The paper concludes with an assessment of the likelihood that existing 'counter-hegemonic' globalization forces will overcome the tensions between globalization and environmental justice and sustainability.

11. Zinsê Mawunou and Chunmei Zhao discuss the great power politics around Africa's development. The policies of the European Union (EU) and China are compared and contrasted. After the African countries got independence, the European Union (formerly the EEC) and China demonstrated a willingness to

contribute to the improvement of socio-economic development of Africa. Both partnered a longstanding series of measures to enhance Africa's socio-economic prosperity. More recently the longstanding relationships with EU countries have been challenged by the unprecedented scale of China's involvement on the continent. This article analyzes these trends by providing an introduction to diverse EU – Africa agreements and conventions, on the one hand, and the present Chinese strategy of cooperating with Africa, on the other. It is concluded that African countries have been able to take advantage of the Chinese strategy to improve their economic situation. Secondly, the article describes the economic outcomes associated with the involvement of these actors in Africa. The article concludes with some suggestions to those actors as well as to African countries emphasizing that even though the partnerships with the EU and China move Africa in the right direction there are gaps that should be addressed.

12. Michael Linke discusses the importance to globalization of cross-border logistics flows supported by advanced communications networks such as the UN Single Window concept. Customs procedures (including complex country-specific fees, tariffs, and taxes) typically involve 15 or more agencies and multiple copies of paper documents. The result is what many businesses view as an inhibitive and stifling system that is overly complicated and slows the process of trade. The UN Single Window concept is a facility that allows parties involved in trade and transport to lodge standardized information and documents at a single entry point to fulfil all import, export, and transit-related regulatory requirements. Because the information is electronic, data elements need only be submitted once. Overall, the effect of a single window on a government is far reaching. The changes will propagate through the economy and allow business to engage in international trade more easily. Several approaches and implementations are described, although all of them need proper planning to achieve worldwide penetration. Enterprise Architecture Management (EAM) as a specialized IT strategy discipline can help to manage this complex challenge of integrating application landscapes into different existing UN integration frameworks.

13. Tessaleno Devezas reviews the recent global crisis under the light of the Kondratieff long wave theory. The variation in four economics-related agents is analysed with an eye to their combined long term effects. The author states that the holistic analysis of long term effects provides insight into what happened in the past in the global economy and sheds some light about possible future trajectories. The four agents considered are: world population, its global output (GDP), gold price and the Dow Jones index. Although many other measures exist, these four agents function as indicators that represent significant aspects

of the world economic realm. The application of analytical tools such as spectral analysis, moving averages and logistic curves to time series data about the historical unfolding of these four agents suggests that the recent global crisis may be a mix of two tendencies. The first tendency is a self-correction mechanism that brought the global output back to its original learning natural growth pattern. The second tendency is a new pattern in the world economic order. The evidence for Kondratieff long waves suggests that the present decade (2010-2020) will probably be one of worldwide economic expansion, corresponding to the second half of the expansion phase of the fifth K-wave.

14. Jim Sheffield seeks to answer two broad questions: "How should New Zealand respond to the multiple, intertwined and fast-changing impacts of globalization?" and "What strategies are available to this small South Pacific country and how may these be facilitated?" The answers are based on findings from empirical research on the facilitation of aspects of national policies in the domains of science funding, economic development and regional growth. A well-funded science sector encourages entrepreneurial and innovative activity to be located in New Zealand and facilitates international knowledge transfer. Economic development improves competitiveness in global markets, including those in the Asia-Pacific region. Regional planning in Auckland, New Zealand's major growth area, attracts skilled migrants and reduces the loss of New Zealand-born citizens to Australia and other countries. The facilitation of these local solutions in a global environment is framed within the theoretical perspective of pluralism and communicative action. Facilitating national policies required extensive consultation among a large number of stakeholders in different organizations. The context was pluralistic—the objectives of social actors were divergent and power was diffused. Electronic meeting technology was employed. The focus question is: 'Does electronic discourse increase the success of local solutions in a global environment?'

PART III—GLOBALIZATION IN THE FUTURE

15. Ervin Laszlo writes with moral urgency about global bifurcation and the decision window. He opens with the saying that our generation is the first in history that can decide whether it is the last in history. His response is to remind us that our generation is also the first in history that can decide whether it will be the first generation of a new phase in history. We have reached a watershed in our social and cultural evolution. The sciences of systems tell us that when complex open systems, such as living organisms, and also ecologies and societies of organisms, approach a condition of critical instability, they face a moment of truth: they

either transform, or break down. Two scenarios are developed. The first scenario, business as usual, describes the impact of unchecked global warming, waves of destitute migrants fleeing from areas of ecological disaster, and the destabilization that follows failed military solutions. The second scenario, timely transformation, describes the emergence and growth of popular movements for sustainability and peace, increased action by non-governmental organizations to revitalize regions ravaged by ecological disaster, reduced military budgets, and a quest for social and ecological responsibility. The choice between these scenarios is not yet made. The question is, how much time is there to make a timely transformation?

16. Ian Yeoman explores two possible identities—fluid and simple—of the future tourist and the scenarios that favour the emergence of each identity. The scenario favouring the fluid identity enfolds like this. The globalisation of tourism and increases in real wealth have meant tourists can take a holiday anywhere in the world, whether it is the North Pole or the South Pole and everywhere in between including a day trip into outer space with Virgin Galactic. Increases in disposable income allow a real change in social order, living standards and the desire for quality of life with tourism at the heart of this change. This scenario is about the concept of self which is fluid and malleable. The self cannot be defined by boundaries within which the choice and the desire for self and new experiences drive tourist consumption. The scenario favouring the simple identity enfolds like this. A collapse in pension funds means people work longer and are less wealthy in retirement. As wealth decreases a new thriftiness and desire for simplicity emerge. Web technologies are employed to search out bargains, advice on the use of scarce leisure time, and personal recommendations. This paper examines the values, behaviours, trends and thinking of the future tourist, either with a fluid or simple identity.

17. Vladimir Klimenko and Alexey Tereshin analyze world energy and climate in the 21st century in the context of historical trends. Clear constraints to future growth are imposed based on the theory of institutional change. The paper deals with global energy perspectives and forthcoming changes in the atmosphere and climate under the influence of anthropogenic and natural factors. In the framework of the historical approach to energy development the forecast of the future global energy consumption for the present century is elaborated, and its resource base and the global impact of the power sector on the atmosphere and climate against the background of natural factors' influence are studied. It is shown that, following the historical path of global energy evolution, the global

energy consumption will remain within 28–29 billion tons of coal equivalent (tce) by the end of the century, with CO_2 emissions peaking in the middle of this century. In this scenario, the CO_2 concentrations will not exceed 500 ppm, and the global temperature should rise by 1.5 °C by 2100, with the growth rate not exceeding the adaptation limits of the biosphere.

18. Leonid Grinin and Andrey Korotayev analyze the global causes of the contemporary global financial-economic crisis and the possibility of eliminating the most acute problems that generated the crisis. In the first part of paper they consider both the negative role of the world financial flows and their important positive functions, including the 'insurance' of social guaranties at the global scale. On one hand, anarchic and extremely rapid development of new financial centers and financial flows contributed to the outbreak of the crisis. The latter was amplified by the non-transparency of many financial instruments, which led to the concealment of risks and their global underestimation. On the other hand, new financial technologies decrease risks in a rather effective way, and they expand possibilities to attract and accumulate enormous capitals, actors, and markets. The modern financial sector also contributes to the provision of insurance for social funds at the global scale. The participation of pension and insurance funds in financial operations leads to the globalization of the social sphere. Countries poor in capital, but with a large young population, are increasingly involved in a very important (though not readily apparent) process of supporting the elderly portion of the population in the West through the vigorous unification of the world's financial flows. These flows are becoming more standardized, more mobile, and more anonymous. They represent perhaps the greatest threat and the greatest promise of globalization.

The second part of the article considers some global scenarios of the World System's near future and describes a few characteristics of the forthcoming 'Epoch of New Coalitions'. The article attempts to answer the following questions: What are the implications of the economic weakening of the USA as the World System center? Will the future World System have a leader? Will it experience a global governance deficit? Will the world fragmentation increase? The authors suppose that in 15–25 years our world will be both similar to the present and already substantially different from it. Global changes are forthcoming, but not all of them will take a distinct shape. Contrary to that, new contents may be covered by old outdated surfaces (as in the Late Middle Ages the emerging centralized state was not quite distinctly seen behind the traditional system of relationships between the crown, major seniors, and cities). One may say that these will be such changes that could prepare the world to the transition to a new phase of globalization (it will be very fortunate if there are grounds to call it the phase of sustainable globalization) whose contours are not yet clear.

Chapter 1

THE ORIGINS OF GLOBALIZATION

Leonid Grinin & Andrey Korotayev

The main aim of this article is to analyze the processes and scales of global integration in an historical perspective, starting with the Agrarian Revolution. There have already been numerous studies on this subject, but there are still many points that need further research, clarification, and new interpretation. Most researchers into globalization are convinced that its origins are to be traced back to a point deep in history, although there are diverse views as regards the exact starting point. The article analyzes different approaches to this problem.

The subject of this article relates to the integration that began a few thousand years BCE in the framework of the Afroeurasian world-system and whose links became so well-developed long before the Great Geographic Discoveries that they could be described as global (albeit in a limited sense). However, among some researchers there is still a tendency to underestimate the scale of those links in the pre-Industrial era. Thus, it appears necessary to provide additional empirical support for our thesis.

A special methodology is also required, *i.e.*, the use of the world-system approach. We analyze some versions of periodization of globalization history. We also propose our own periodization using as its basis the growth of the scale of intersocietal links as an indicator of the level of globalization development.

Keywords: globalization, world-system, Afroeurasian world-system, World System, global communication, cycles of political hegemony, agrarian revolution, industrial revolution, technologies.

INTRODUCTION: THE AIMS OF THE ARTICLE

The present article has been prepared within an emerging field that can be termed the "History of Globalization". This aspect of Globalization Studies deals with the historical dimension of globalization. Its main goal is to analyze processes and scales of global integration in an historical perspective, starting with the Agrarian Revolution. Those integration processes (depending on the position of a particular researcher) may be regarded as preparatory stages of the globalization, or as its initial phases. There have already been several studies carried out on this subject (see Foreman-Peck, 1998; Held *et al.*, 1999; O'Rourke & Williamson, 1999; Hopkins, 2002, 2004; Sharp, 2008; Lewis & Moore, 2009). However, there are still many points that need further research, clarification, and new interpretation.

Most researchers into globalization recognize that its origins are to be traced more or less deeply in history, though there are diverse views as regards the exact starting point.[1] Yet, it is clear that it would be very productive to search for the origins of globalization in the depths of history. We contend that to a certain extent World History can be regarded as a movement toward increasingly large social systems, their integration, and, thus, as a history of globalization. Therefore, in history and sociology the investigation is broadening with respect to the historical development of globalization processes (see Grinin, 2011b, 2011c, 2012a; Grinin & Korotayev, 2009a, 2009b, 2012). It is no coincidence that growing interest in globalization has increased awareness of the trend often described as the "historical dimension of globalization". Among such movements *Global History* is most worth noting; and its core, according to Mazlish and Iriye (2005), is just the history of globalization.

According to various authors, globalization has been going on either since the first movement of people out of Africa into other parts of the world; or since the 3rd millennium BC (when, according to Frank, the World System emerged [Frank 1990, 1993; Frank, Gills 1993]); or since the so-called Axial Age (Jaspers, 1953) in the 1st millennium BC; or only from the Great Geographical Discoveries; or in the 19th century; or after the year 1945; or only since the late 1980s (see also Footnote 1). Each of these dates has its own justification.

---

1. Some scholars say that it started already in the Stone Age, some others maintain that it began in the 3rd millennium BCE; there also such datings as the Axial Age of the 1st millennium BCE, the Great Geographic Discoveries period, the 19th century, 1945, or even the late 1980s. Each of these datings has certain merits. For their review see, *e.g.*, Tracy 1990; Menard 1991; Bentley 1999; O'Rourke and Williamson 1999, 2000; Lewis, Moore 2009; Conversi 2010; Held *et al.* 1999; Чумаков 2011; Келбесса 2006: 176; Пантин 2003, *etc.*

Some researchers discuss the problem in the context of whether one should speak about globalization before the start of the Great Geographical Discoveries, as a result of which the idea of the Earth as a globe was no longer just the opinion of a group of scientists and became general or public knowledge (Chumakov, 2011). In spite of this point of view, there is no doubt that the historical dimension of globalization is quite challenging (see Grinin, 2011f).

The subject of this article is related to the integration that began a few thousand years BCE in the framework of the Afroeurasian world-system and whose links *became so well developed long before the Great Geographic Discoveries that they could be described as global (albeit in a limited sense).* However, among some researchers there is still a tendency to underestimate the scale of those links in the pre-Industrial era. Thus, it appears necessary to provide additional empirical support for above mentioned point. It is also necessary to apply a special type of methodology *i.e.,* the world-system approach.

There have been several periodizations proposed for the history of globalization. The most wide-spread type is trinomial, which is sometimes considered to be the most logical. Gellner (1988), for example, believes that three periods are the optimum number for a periodization. Such an example is as follows: (1) Archaic globalization; (2) Early modern globalization;[2] (3) Modern globalization (Hopkins 2003; Bayly, 2004).

Trinomial periodizations are also used by those who suggest that globalization begins with the period of the Great Geographic Discoveries. For example, Friedman (2005) divides the history of globalization into three periods: "Globalization 1" (1492–1800), "Globalization 2" (1800-2000) and "Globalization 3" (2000—present). He states that Globalization 1 involves the globalization of countries, Globalization 2 involves the globalization of companies and Globalization 3 involves the globalization of individuals.

However, the apparent convenience of trinomial periodizations does not necessarily mean that they are more relevant. We believe that the number of periods within the given periodization should rather be determined by the nature of the process in question.

---

2. This phase is also denoted as "proto-globalization"; but this notion does not appear quite appropriate, because, logically, a stage with a designation starting with "proto-" should not be placed in the middle; it should rather be the first stage. Hence, it appears much more logical to denote the stage of archaic globalization as protoglobalization phase.

There are periodizations constructed on other grounds—for example, the one developed by Chumakov (2011) who determined the periodization of the evolution of global links on the basis of their scale (which reflects rather logically the general trend toward the growth of this scale): 1) the "Period of Fragmentary Events" (till 5000 BP); 2) the "Period of Regional Events" (till the 15th century CE); 3) the "Period of Global Events" (till the mid-20th century). The 4th period (the "Period of Cosmic Expansion") in this periodization starts in 1957. This periodization appears valid, but a few points need considerable clarification and re-interpretation. First of all, as will be demonstrated below, beginning with the second half of the 1st millennium BCE, many events did not only expand beyond regional levels, but could be measured on continental and transcontinental scales. Even before the 1st millennium BCE some events had regional and continental impact (see below). Evidence in support of this approach is presented below and a brief exposition can be found in Tables 1 and 2.

In the framework of this article we are attempting to complete the following tasks:

1. To demonstrate that several thousand years ago (at least since the formation of the system of long-distance large-scale trade in metals in the 4th millennium BCE) the scale of systematic trade relationships had already significantly outgrown the local level and had become regional (and even transcontinental to a certain extent);

2. To show that by the late 1st millennium BCE the scale of processes and links within the Afroeurasian world-system did not only exceed the regional level and reach the continental level, but also went beyond continental limits. Thus we contend that within this system marginal systemic contacts between agents of various levels (from societies to individuals) may be defined as transcontinental;[3]

3. To demonstrate that even prior to the Great Geographic Discoveries the scale of global integration to a certain extent could be comparable with that in more recent periods. In particular, demographically even 2000 years ago the effectively integrated part of humankind encompassed 90% of all the world population.

In the present article we are not attempting to describe the whole history of globalization in detail. Therefore, a summary of our interpretation of its main phases may be found in Table 1. In particular, we are basing our thesis on the following observation: though the Great Geographic Discoveries made it possible to transform intersocietal links into global ones in a full sense of this notion, the period between 1500 and

---

3. Note that here we are dealing not only with overland contacts, as since the late 1st millennium BCE in some cases we encounter oceanic contacts—the most notable case is represented here by the Indian Ocean communication network.

1800 CE was not yet fully global for a number of reasons. Firstly, not all the territories of the Earth had been discovered (Antarctica being the most important of these); secondly, many societies in Australia, Oceania, and some parts of Inner Africa had not experienced global contacts in any significant way; thirdly, some of the larger countries of East Asia had quite consciously isolated themselves from the rest of the world; fourthly, the extent of trade at this time could hardly be called global (see O'Rourke & Williamson, 1999, 2000). In this regard, we consider the period from the late 15th century to the early 19th century as a special era when oceanic (intercontinental) links were being established. Chronologically this period is almost identical with the one that was identified by Hopkins (2003) and Bayly (2004) as a period of proto-globalization or early modern globalization. However, we believe our designation of this period reflects the scale and character of links in this period in a more accurate way. Indeed the period starting in the early 19th century may well be described as "a very big globalization bang" (O'Rourke & Williamson, 2000). That is why we use the term "global" for links in this period which continued till the 1970s, after which the level of intersocietal interconnectedness began to grow very rapidly (especially from the early 1990s). It was during this period that it was recognized that we had entered a new period of interconnectedness that was termed "globalization" (*mondialisation* in French). In order to distinguish this period from the previous we have described it as "planetary", which reflects, firstly, the implications of space exploration (these are the space/satellite communication technologies which now secure unprecedented opportunities for communication with respect to their speed, density, and diversity). Secondly, we detail the involvement in the globalization process of those societies in Asia, Africa, and other regions which were weakly connected with the rest of the world, because these links were rather limited and often imposed in a coercive manner. Thirdly, we discuss the notion that modern globalization has not realized all its potential, that this process continues, and when this period ends in the 21st century, the level of interrelatedness will be truly planetary, and almost any place in the world can and will be connected with almost any other place.

Of the 7 periods outlined above (and below in Table 1), all except the first and second refer to historical globalization.

Note that this table does not take into account the information networks of the technological diffusion that acquired a transcontinental scale from the first emergence of the Afroeurasian world-system (see Korotayev, 2005, 2006, 2007, 2012; Korotayev, Malkov, & Khaltourina, 2006a, 2006b; Гринин, & Коротаев, 2009; Grinin, & Korotayev, 2012). See some other qualifications below.

Type of spatial links (globalization level)	Period
Local links	1. Till the 7th –6th millennium BCE
Regional links	2. From the 7th –6th millennium till the second half of the 4th millennium BCE
Regional-continental links	3. From the second half of the 4th millennium BCE to the first half of the 1st millennium BCE
Transcontinental links	4. From the second half of the 1st millennium BCE to the late 15th century CE
Oceanic (intercontinental) links	5. From the late 15th century to the early 19th century
Global links	6. From the early 19th century to the 1960s and 1970s
Planetary links	7. From the last third of the 20th century to the mid-21st century

Table 1. *The growth of globalization level in the historical process*

In Table 2 we present a description of the correlations between the globalization periods and such characteristics as spatial links, political organization and level of technology which are very important for the history of globalization.

Our analysis suggests that the abovementioned marginal level of integration within the framework of the Afroeurasian world-system was not something insignificant or virtual. In fact, it substantially influenced the general direction of development and significantly accelerated the advancement of many social systems whose rate of development would have been otherwise much slower. It is clear that it took signals a relatively long time to get from one end of the world-system to another—actually, many orders of magnitude longer than now—but still such signals were transmitted through the pre-Modern Afroeurasian world-system, and they caused very significant transformations. However, the speed was not always so slow. For example, the bubonic plague pandemic (which killed millions) spread from the Far East to the Atlantic Ocean within two decades in the 1330s and 1340s (see McNeill, 1976; Dols, 1977; Borsch, 2005). Such rapid and vigorous movements were directly related to the growing number of contacts between societies and their diversification, which in turn opened the way to a rapid diffusion of pathogens.[4]

---

4. Note that the Mongol warriors went from the Pacific zone to the Atlantic zone of Eurasia with a rather similar speed.

Type of socio-spatial links	Period	Forms of political organization	Level of technology (production principles and production revolutions)
Local links	Up to the second half of the 4th millennium BCE (≈ 3500 BCE)	Pre-state (simple and medium complexity) political forms, the first complex polities	Hunter-gatherer production principle, beginning of the agrarian production principle
Regional links	The second half of the 4th millennium BCE – the first half of the 1st millennium BCE (≈ 3500– 490 BCE)	Early states and their analogues; the first empires	The second phase of the agrarian revolution; agrarian production principle reaches its maturity
Continental links	The second half of the 1st millennium BCE – the late 15th century CE (≈ 490 BCE – 1492 CE)	Rise of empires and first developed states	Final phase of the agrarian production principle
Intercontinental (oceanic) links	The late 15th century – the early 19th century (≈ 1492– 1821)	Rise of developed states, first mature states	The first phase of the industrial production principle and industrial revolution
Global links	The early 19th century – the 1960s and 1970s	Mature states and early forms of supranational entities	The second phase of the industrial revolution and the final phase of the industrial production principle
Planetary links	Starting from the last third of the 20th century	Formation of supranational entities, washing out of state sovereignty, search for new types of political unions and entities, planetary governance forms	The start and development of scientific-information revolution whose second phase is forecasted for the 2030s and 2040s

Table 2. *The correlation between spatial links, political organization and level of technology*

I—THE AFROEURASIAN WORLD-SYSTEM: A GENERAL OVERVIEW

For an analysis of the origins of globalization the traditions of various schools of thought should be examined. However, we believe that the world-system approach is one of the most promising in this respect, as it was originally constructed to solve this kind of task. This approach may be used much more widely in this area due to its undoubted merits. First of all, this approach is systemic and capable of analyzing processes on very wide temporal and spatial scales. As Chase-Dunn and Hall (1997) emphasize, within this approach the main unit of analysis is not a particular society, or a particular state (as is usual in ordinary historical studies), but a world-system. Secondly, the object of world-system analysis is in many respects identical with the one of Global Studies. Thirdly, taking into consideration the interdisciplinary character of Global Studies, they can only be enriched by the integration of new approaches. As regards the aims of the present article, the world-system approach and its issues and terminology appear to be quite appropriate for the achievement of its goals.

The world-system approach originated in the late 1960s and 1970s in the works of Braudel (1973), Frank (1990, 1993), Frank and Gills (1993), Wallerstein (1987), Chase-Dunn and Hall (1994), Amin *et al.* (2006), and Arrighi and Silver (1999), and was substantially developed afterwards. The formation of this approach was connected to a considerable degree to the search for actual socially evolving units which are larger than particular societies, states, and even civilizations, but which, on the other hand, have the qualities of real systems.

The most widely known version of the world-system approach was developed by Wallerstein (1974, 1987, 2004), who believes that the modern world-system was formed in the "long 16th century" (c.1450–1650). In his opinion, before that period there had been a very large number of other world-systems. Those world-systems are classified by Wallerstein into three types: 1) minisystems; 2) world-economies; 3) world-empires. Minisystems were typical for foragers. The two other types (world-economies and world-empires) are typical for agrarian (and especially complex and supercomplex agrarian) societies.

World-economies are politically decentralized systems of societies interconnected by real economic ties. Wallerstein also uses the so-called "bulk goods criterion" to identify the "reality" of economic ties, that is those ties which are manifested in massive flows of such basic goods as wheat, ore, cotton, tools, and mass consumption commodities etc. If the trade between two regions is limited to an exchange of "preciosities", then, according to Wallerstein, we have no grounds to consider them parts of one world-system in general, and one world-economy in particular.

If a world-economy becomes centralized politically within one empire, Wallerstein (1974) maintains that we should speak about a world-empire, not world-economy. In general, world-economies were characterized by a higher socioeconomic dynamism than world-empires, but almost all the pre-capitalist world-economies were sooner or later transformed into world-empires.[5]

Wallerstein (1974) also contends that there was just one significant exception from this rule, which was analyzed in considerable detail in his first "world-system" monograph. In "the long 16th century" the Western European world-economy blocked the tendency toward its transformation into a world-empire and experienced a capitalist transformation that led to the creation of a world-economy of a new, capitalist type. This new world-system had already experienced a rapid expansion in "the long 16th century" and, after a phase of relative stabilization (in the second half of the 17th century and 18th century), it encompassed the whole world by the 19th century.

Though the version of the world-system approach developed by Frank (1990, 1993; Frank & Gills 1993) is much less well known than Wallerstein's, we believe it might have even more scientific value. Frank brings our attention to the point that within Wallerstein's approach the very notion of a "world-system" loses much of its sense. Indeed, if the pre-capitalist world consisted of hundreds of "world-systems", it is not quite clear why each of them should be denoted as a "WORLD-system".

Frank's approach is somewhat more logical. He contends that we should speak only about one World System (and he prefers to denote it using capital initial letters). According to Frank, the World System originated many millennia before the "long 16th century" in the Near East. This idea is clearly expressed in the title of the volume he edited with Gills—*The World System: Five Hundred Years or Five Thousand?* (Frank & Gills, 1993). This World System had gone through a long series of expansion and contraction phases until in the 19th century it encompassed the whole world.

We believe a synthesis of the both the main versions of the world-system approach is quite possible, and in the present article we will analyze the processes that contributed to the emergence and growth of the Afroeurasian world-system, which

---

5. World-empires also frequently disintegrated and could be replaced with world-economies, but this was just a beginning of a new cycle ending with the formation of a new world-empire in place of the world-economy.

may be considered the direct predecessor of the modern planetary World System. It was already more than 2000 years ago when the Afroeurasian world-system became connected from one end to the other by trade links. By the late 13th century it had reached its culmination point (for the pre-capitalist epoch), and in the late 15th century it began its explosive expansion with the result that between the 16th and 19th centuries it became a truly planetary World System[6].

In addition to the Afroeurasian world-system, there were several other world-systems on the Earth (such as the New World, Oceania, and Australia) prior to the transformation of the Afroeurasian world-system into the modern planetary World System (*e.g.*, Grinin & Korotayev, 2012a). However, from the time of its formation and in the course of the subsequent millennia, the Afroeurasian world-system was constantly leading on a global scale, because it had the most noticeable tendency toward expansion, growth of complexity, and the highest growth rates. It is also important to note that already by the early 1st millennium CE it comprised more than 90% of the world's population (Durand, 1977).

The notion of "world-system" (as it is used in the present article) can be defined as *a maximum set of human societies that has systemic characteristics, and a maximum set of societies that are significantly connected among themselves in direct and indirect ways. It is important that there are no significant contacts and interactions beyond borders of this set, and there are no significant contacts and interactions between societies belonging to the given world-system and societies belonging to the other world-systems.* If there are still some contacts beyond those borders, then those contacts are insignificant, which means that even after a long period of time they do not lead to any significant changes within the world-system.[7] However, this definition appears to be the most appropriate for a period when there were only a few world-systems on our planet. For the modern unique World System the definition appears to be closer to such notions as "planetary system", "global system", or "humankind as a system".

The important peculiarities of the Afroeurasian world-system stemmed from its scale and very ancient age: (1) the special complexity (supercomplexity) of its structure; (2) the primary/autochthonous character of the majority of its social and technological innovations; (3) the succession of qualitative transformations as a result of (4) the especially high speed of these changes. One should also take into account

---

6. Correspondingly, when we speak about one out of a few world-systems, we use the term "world-system", whereas we use Frank's notion of "the World System" when we speak about the unique global system covering the whole of our planet.

7. For example, the early Scandinavians' travels to the New World and even their settlement there did not result in any significant change either in the New World, or in Europe (see Слезкин, 1983: 16).

some specific geographic conditions, in particular, the enormous Eurasian Steppe Belt, which resulted in the especially large role played by the barbarian (and especially nomadic) periphery; the especially important role of water communications, due to the number of communication networks with particular high levels of contact density which emerged (the Mediterranean network, the Baltic Sea network, the trade route from the Varangians to the Greeks, the Indian Ocean network etc.).

A Brief Overview of the Main Phases of the Afroeurasian World-System's Evolution

The processes of intersocietal interaction began many thousands of years ago. That is why it appears impossible to use such a term such as perfect isolation even with respect to Paleolithic cultures. Already for the Upper Paleolithic there are numerous archeological, paleolinguistic and other data on information-cultural and trade-material contacts covering hundreds and even thousands of kilometers (see Korotayev & Kazankov 2000; Korotayev *et al.*,2006). For example, Mediterranean sea shells are found at the Paleolithic sites of Germany, Black Sea shells are being discovered at the в Mezine site on a bank of the Desna River 600 kilometers far from that sea (see Кларк, 1953; Румянцев, 1987). However, we observe a new phase of intersocietal integration after the start of the Agrarian Revolution (see Childe, 1952; Reed, 1977; Harris & Hillman, 1989; Cohen, 1977; Rindos, 1984; Cowan & Watson, 1992; Ingold, 1980; Cauvin, 2000; Mellaart, 1975, 1982; Smith 1976; Grinin 2007b).

In the 10ᵗʰ—8ᵗʰ millennia BCE the transition from foraging to food production took place in the West Asia (the Fertile Crescent) area, as a result of which a significant growth in the complexity of its social systems can be observed, which marked the beginning of the formation of the Afroeurasian world system. **In the 8ᵗʰ—5ᵗʰ millennia BCE** one can note the Afroeurasian world-system' expansion and the formation of effective informational, cultural, and even trade links between its parts.

In the 4ᵗʰ and 3ʳᵈ millennia, first in Southern Mesopotamia, and then in most other parts of Afroeurasian world-system the formation of a large number of cities can be seen. Writing systems, large-scale irrigation agriculture, and new technologies of tillage developed. The first early states and civilizations were formed on this basis. A large number of very important technological innovations were introduced to most parts of Afroeurasian world-system: the wheel, the plow, the pottery wheel, and harness etc. The emergence and diffusion of copper and bronze metallurgy increased military capabilities and contributed to the intensification of regional hegemony struggles. New civilization centers emerged outside the Middle Eastern core (*e.g.,* the Minoan and Harappan civilization).

In the late 3rd and the 2nd millennia BCE in Mesopotamia one can observe the succession of such large-scale political entities as the Kingdom of Akkad, the 3rd Dynasty of Ur, and the Old Babylonian and Assyrian Kingdoms. The hegemony struggle at the core of the Afroeurasian world-system moved to a new level with the clash between the New Kingdom of Egypt and the Hittite Empire. These political macroprocesses were exacerbated by invasions from tribal peripheries (the Gutians, Amorites, and Hyksos etc.) with a gradual increase in the role of nomadic herders in such invasions. In the 2nd millennium BCE a new Afroeurasian world-system center emerged in the Far East with the formation of the first Chinese state of Shang/Yin. In general, these processes led to the enormous expansion of Afroeurasian world-system. **In the late 2nd and 1st millennia BCE** the knowledge related to iron metallurgy spread throughout the Afroeurasian world-system, which led to a significant growth in agricultural production in the areas of non-irrigation agriculture of Europe, North Africa, the Middle East, South Asia, and the Far East. This also led to the rise of crafts, trade, urbanization, and military capabilities. In the 1st millennium BCE the hegemony struggles moved far beyond the Near East. The fall of the New Assyrian Empire in the 7th century BCE paved the way for the formation of enormous new empires (Median, and later Persian). The Greek-Persian wars marked the first clash between European and Asian powers. In the second half of the 4th century BCE Alexander the Great's campaign created (albeit for a short period of time) a truly Afroeurasian empire encompassing vast territories in all three parts of the Old World—Asia, Africa, and Europe.

In the 2nd millennium BCE the Harappan civilization disappeared in a rather enigmatic way. However, in the 1st millennium BCE the Indoarians who had migrated to this region from Central Asia created a new and more powerful civilization there.

In the late 1st millennium BCE the formation of new empires can be seen: the Roman Republic and the Chinese Empire (Qin, and later Han). Then there developed an unusually long network of trade routes (the so-called Silk Route) between the western and eastern centers of the Afroeurasian world-system.

In the 1st millennium BCE and the early 1st millennium CE in connection with climatic change and some important technological innovations (the saddle, and the stirrup etc.) a new type of nomadic society emerged. These new nomads were able to cover enormous distances and to transform themselves very quickly into a type of mobile army. As a result, the whole vast landmass of the Eurasian steppe belt became the nomadic periphery of the Afroeurasian world-system. The Scythian "kingdom" in Europe and the more recent "empire" of Hsiung-nu that emerged to the north from China were two of the first powerful nomadic polities of such a type.

In the first centuries CE, as a result of mass migrations and the military invasions of peoples from the barbarian periphery the ethnic and cultural landscape of the Afroeurasian world-system experienced very significant changes. The Western Roman Empire disappeared as a result of the barbarians' onslaught. The Han Empire in China had collapsed even earlier. As a result of such stormy events within the Afroeurasian world-system a considerable number of new states (including states of the imperial type) emerged such as the Frankish, the Byzantine, and the Sassanid empires, the Gupta Empire in India, and the Tang Empire in China etc.).

During the first millennium CE new world religions emerged and a wide diffusion of old and new world and super-ethnic religions took place. Buddhism spread very widely into many regions of Central, South-East, and East Asia (including China, Korea, Japan, and Tibet). Confucianism prevailed in East Asia, while Christianity was embraced by the whole of Western and Eastern Europe and even proliferated in some areas of Africa and Asia. Finally, starting with the 7th century one can observe the explosive spread of Islam which encompassed the whole of the Near and Middle East. As a result, the very extensive Islamic Khalifate emerged (however, it quite soon disintegrated).

The first half of the 2nd millennium CE. The Crusades in the 11th – 13th centuries CE were one of the most important world-system events. Among other things they opened a channel for the spice trade with Europe. A very important role was played by the Mongolian conquests of the 13th century which brought unprecedented destruction and political disturbance to the region. However, the consequent emergence of an unprecedentedly large Mongolian empire secured the spread of a number of extremely important types of technologies throughout the Afroeurasian world-system (including its European part). This empire also established a network of trade roots connecting East Asia with Europe that was unpaparlleled as regards its scale and efficiency. The barbarian semiperiphery became incorporated in the civilized environment of Islam, Buddhism, and Confucianism, which contributed to the vigorous penetration of its world-system links far into the Eurasian North and deep into Africa. On the other hand, the expansion of trade contacts between the East and the West contributed to the already mentioned above spread of the Black Death pandemic in the 14th century.

Another important event was the establishment of close contacts between South India and the other parts of Afroeurasian world-system as a result of a gradual penetration of Islamic polities and a partial Islamization of its population. In the 15th century a new political and military force emerged in West Asia, *i.e.*, the Ottoman Empire. The Turks obstructed the Levantine spice trade and, thus, accelerated the search for a sea route to India.

New qualitative changes within the Afroeurasian world-system were connected with the beginning of the Great Geographic Discoveries and the Afroeurasian world-systems' transformation into the planetary capitalist World System. These events marked the inception of a qualitatively new phase in globalization history which will be discussed below.

II—WORLD-SYSTEM LINKS AND PROCESSES

The Systemic Character of the World-System Processes

World-system processes and transformations can be understood much better if their systemic properties are taken into account. Such systemic properties account for the synchronicity or asynchronicity of certain processes, and the presence of positive and negative feedback which can be traced over very long periods of time, for example, in demographic indicators. We believe that special attention should be paid to the idea of Chase-Dunn and Hall (1997) that a world-system is constituted not only of intersocietal interactions, but of the whole set of such interactions, whereas the level of analysis which is most important for our understanding of social development is not that of societies and states, but of the world-system as a whole. In this way, a fundamental property of the system (*i.e.*, the whole is more than just a sum of its parts) is realized in these world-systems. Changes and transformations in certain parts of a world-system can produce changes in its other parts through what may be called *impulse transformation*. It may be manifested in various forms and produces sometimes rather unexpected consequences. Thus, the obstacles to the delivery of spices to Europe created by the Turkish conquests of the 15th century stimulated the search for a sea route to India, which finally changed the whole set of relationships within the Afroeurasian world-system. Due to its systemic properties, the processes that started in a certain part of the Afroeurasian world-system, could be rapidly transmitted to most other areas (the already mentioned above rapid diffusion of the Black Death pandemic in the 14th century could serve here as a example).

A very interesting manifestation of the Afroeurasian world-system's systemic properties is represented by the synchronized processes that took place in various parts of the Afroeurasian world-system. One can mention as an example an East/West synchrony in the growth and decline of the population sizes of the largest cities from 500 BCE to 1500 CE in West Eurasia and those in East Eurasia (Chase-Dunn & Manning, 2002). There is a similar synchrony in the territorial sizes of the largest empires (Hall, Chase-Dunn & Niemeyer, 2009). Barfield (1989) also argues that large steppe confederacies usually cycle synchronously with the rise and fall of the large sedentary agrarian states that they raid. These cycles are one hypothesized mechanism of the systemic linkages between East and West Asia (Hall, Chase-Dunn, & Niemeyer 2009). Such synchronized processes within the Afroeurasian world-system have also been detected by researchers of the Bronze Age and earlier periods (Chernykh 1992; Frank 1993; Frank & Thompson, 2005). Other salient examples of such synchronized processes are the Axial Age transformations of the 1st millennium BCE (Jaspers, 1953) or the military revolution and formation of a new type of statehood in Europe and Asia in the late 15th and 16th centuries CE which provided an enormous influence on the formation of the modern World-System (see Grinin, 2012a). However, transformations were similar across different regions only in a broad sense and that development has always been spatially uneven (Chase-Dunn & Hall, 1997).

When considering the general trends of Afroeurasian world-system development, it is necessary to note the following points:

1. The Afroeurasian world-system (phase) transition to a new phase produced an effect of dispersion (through borrowing, modernization, coercive transformation, and incorporation etc.) of relevant innovations throughout territories which, however, were unprepared for such an independent transformation. This can be seen in many of those processes that supported the Afroeurasian world-system development, such as the dissemination of statehood, or world religions.

2. The Afroeurasian world-system development was frequently accompanied (and even supported) by the decline/underdevelopment of some of its parts. On the other hand, the flourishing of some societies could lead to a temporary decrease in the overall level of development/complexity of the Afroeurasian world-system, as was observed some time after the Mongolian conquests.

The Most Important Types of World-System Links

Diffusion of Innovations

The Afroeurasian world-system movement at every new level of development was inevitably connected with the expansion and strengthening of communication links and networks. Chase-Dunn and Hall (1997) single out the following main types of world-system spatial links: bulk-goods exchange, prestige-goods exchange, political-military interaction, and information exchange. They also state that world religions furnished major innovations in the information networks and technologies of ideological power (Chase-Dunn & Hall 1997). Therefore, civilization-cultural (ideological) interactions could be identified as a special type of the world-system link, as they differ substantially from the usual flow of information. Cultural-ideological interactions played a very important role within the Afroeurasian world-system, especially during the period of its maturity. In particular, from the 8th century CE all the civilized sectors of the Afroeurasian world-system (with a partial exception of South Asia) consisted of actively interacting world religion areas.[8] Initially, world-system analysis paid attention mainly to the bulk goods trade (Wallerstein, 1974). However, during the period of the Afroeurasian world-system formation the most important role was played by information links, especially the dissemination of innovations (Korotayev, 2005, 2007, 2012; Korotayev, Malkov, & Khaltourina 2006a; Grinin 2007b, 2012a; Grinin & Korotayev, 2009).

Development of Trade Links

A relatively large scale of trade in strategic economically important items could be observed in the framework of the emerging Afroeurasian world-system in West Asia. In particular, obsidian (which was in high demand for the manufacture of stone tools) was already being transported from the Anatolian Plateau throughout the Afroeurasian world-system by the 7th millennium BCE. This is likely to have been accompanied by a trade in foodstuffs, leather, and textiles (Lamberg-Karlovsky & Sabloff, 1979). Inthe 5th and 4th millennia BCE we have evidence for a large-scale trade in metals (Chernykh, 1992; Frank, 1993). There is even more evidence for large-scale trade in the 3rd and 2nd millennia BCE (Wilkinson, 1987; Frank, 1993). In the 1st millennium BCE long distance trade (including sea trade) became even more highly developed (Chase-Dunn & Hall, 1997).

---

8. For more detail on the influence of the world religions on the evolution of Afroeurasian world-system (see Korotayev, 2004).

In the second half of the 1st millennium CE in the Indian Ocean Basin (in the area stretching from the East African Coast to South-East Asia, including Indonesia and China) the formation of a prototype of the oceanically-connected World-System can be observed. In this enormous network of international trade an important role was played by Persian, Arab, and Indian etc. merchants (see Bentley, 1996). It is important to note that the trade in this region was not restricted to luxury items, but included a considerable number of bulk goods, such as dates, timber, and construction materials etc.(ibid.)

In the 13th and 14th centuries there was an emergence of a vigorous transcontinental trade network through the territories of the Mongolian states which directly connected all the Afroeurasian world-system's main zones. As is noted by Abu-Lughod (1989), the organization of this world-system trade network was more complex and had a larger volume than any previously existing network.

III—THE WORLD SYSTEM GENESIS AND TRANSFORMATIONS: A DETAILED ANALYSIS

Origins of the Afroeurasian World-System

There are a considerable number of points of view regarding the initial date of the possible formation of the Afroeurasian world-system. For example, Frank (1993) and Frank and Thompson(2005) date its origins to the 4th and 3rd millennia BCE; whereas Wilkinson (1987) and Berezkin (2007) consider the 2nd millennium to be its beginning. The authors of the present article date the emergence of Afroeurasian world-system to a considerably earlier period, *i.e.,* the 10th—8th millennia BCE (Korotayev & Grinin, 2007, 2012; Grinin & Korotayev, 2009, 2012). Other world-system researchers believe that it only came into existence in the late 1st millennium BCE (Chase-Dunn & Hall, 1997, 2008; Hall, Chase-Dunn, & Niemeyer, 2009).

The approaches to this issue differ considerably depending on the world-system criteria employed: *i.e.,* bulk goods (a more rigid criterion), or prestige goods and information networks (softer criteria). The more rigid the approach, the more recent the dating that it produces. However, the datings also depend on the general approaches taken towards the emergence of the Afroeurasian world-system. For example, if together with Chase-Dunn and Hall (1997) we believe that by the operational beginnings of the Silk Route there were three main independent world-systems (West Asian,

Chinese, and South Asian) which merged later into a single world-system (the Afroeurasian world-system), then it appears logical to date the emergence of the single Afroeurasian world-system to the late 1st millennium BCE. However, if we are basing our suppositions on the fact that the West Asian world-system was in the lead from the very beginning technologically, socially, and economically, and that it was much more innovative than other world-systems, and that the West Asian world-system enormously influenced the development of South Asia and the Far East, whereas the influence in the opposite direction by the late 1st millennium BCE was negligible (and hence we should speak about the incorporation of South and East Asia into Afroeurasian world-system, rather than a merger of three equally important world-systems), then the origins of the Afroeurasian world-system must have datings which are much more ancient by several millennia.[9]

Hence, whatever dating we provide for the beginning of the Afroeurasian world-system, it is clear that the roots of its formation can be traced several millennia back in time to the beginnings of the agrarian (Neolithic) revolution in West Asia in the 10th—8th millennia BCE. Within this prolonged process of the Afroeurasian world-system genesis and transformation a few major phases can be identified as below.

1) The 8th-4th millennia—the formation of the contours and structure of the Middle Eastern core of Afroeurasian world-system (the first phase). This is the period which covers the finalization of the first stage of the agrarian revolution in the Near East. The second phase of the Agrarian Revolution was connected with the formation of large-scale irrigation and later intensive plow agriculture in the 4th—1st millennia BC (Korotayev & Grinin, 2007). This initial period included the beginning of the formation of long-distance and permanent information/exchange contacts. These processes were accompanied by the formation of medium-complexity early agrarian societies, relatively complex polities, and settlements which, in regard to their sizes and structure, were remotely similar to cities.

In the 5th millennium BCE the Ubaid culture emerged in Southern Mesopotamia. It was in this culture that the material and social basis of the Sumerian civilization was developed to a considerable degree. The Uruk culture that succeded the Ubaid was characterized by the presence of a considerable number of moderately large settlements. Thus, by the end of the period in question the Urban Revolution had taken place within the Afroeurasian world-system. This revolution can be regarded as a tran-

---

9. One may also take into account the point that it was the Near East where one could observe for the first time in human history the transition to the cultivation of cereals, large-scale intensive agriculture, urban settlements, metallurgy, regular armies, writing, states, empires, and so on.

sition phase in the Afroeurasian world-system towards a qualitatively new level of social, political, cultural, demographic, and technological complexity (Березкин, 2007). By the end of the period in question the emergence of urbanized societies could be seen (Bernbeck & Pollock, 2005), as well as the first early states, their analogues (Grinin & Korotayev, 2006; Grinin, 2003, 2008a), and civilizations. Thus, by the end of the period in question the Urban Revolution took place within Afroeurasian world-system; this revolution can be regarded as a phase transition of Afroeurasian world-system to a qualitatively new level of social, political, cultural, demographic, and technological complexity (Березкин 2007). By the end of the period in question one could observe the emergence of urbanized societies (Bernbeck, Pollock 2005: 17), as well as the first early states, their analogues (Grinin, Korotayev 2006; Grinin 2003, 2008a), and civilizations.

At the beginning of this period the scale of links within the Afroeurasian world-system may be regarded as regional because this world-system was only the size of a region. With the expansion of the Afroeurasian world-system, the scale of its world-system links also expanded. So some years later (after the 7–6 millennia BCE) they transformed into regional-continental ones. However, during this period the Afroeurasian world-system still covered a minor part of the Globe. Hhence, on the global scale local links still prevailed during this period.

2) The 3rd and 2nd millennia BCE—the development of the Afroeurasian world-system centers in the Bronze Age (the second phase). This is a period of relatively rapid increase in the growth of intensive agriculture and in the population of the Afroeurasian world-system. An equally swift process of emergence and growth in the cities in the Afroeurasian world-system was observed in the second half of the 4th millennium and the first half of the 3rd millennium BCE. Later the Afroeurasian world-system urbanization process slowed down significantly until the 1st millennium BCE (Korotayev, 2006; Korotayev & Grinin, 2006, 2012). One of the most important outcomes of this period was the growth in the political integration of Afroeurasian world-system core societies, which was a consequence of rather complex military-political and other interactions. First of all, in the Afroeurasian world-system core one could observe the growth of political complexity from cities and small polities to large early and developed states (Grinin & Korotayev, 2007; Grinin, 2008a). Secondly, the first empires emerged. Thirdly, from the 3rd millennium BCE cycles of political hegemony consisting of upswings and downswings occurred (Frank & Gills, 1993; Chase-Dunn *et al.*, 2010).

In the late 3rd millennium and the 2nd millennium BCE in Mesopotamia the Akkadian Empire was succeeded in turn by the 3rd Dynasty of Ur Kingdom, the Old Babylonian

Kingdom, and the Assyrian Kingdom. In the second half of the 2nd millennium BCE a vigorous hegemonic struggle took place between Assyria, Egypt, and the Hittite Kingdom.

Within the West Asian region the prestige goods trade network achieved a high level of development and was often supported by states. Some parts of Europe were now included in the Afroeurasian world-system communication network. Trade links with South Asia were established throughout the Persian Gulf.

Key West Asian technologies such as the cultivation of West Asian cereals, the breeding of cattle and sheep, some important types of metallurgy, transportation, and military technologies, penetrated East Asia (possibly through the Andronovo intermediaries). This is marked archaeologically by the transition from the Yangshao to the Longshan culture (see Березкин, 2007). In this way the formation of the main Afroeurasian world-system centers took place and these centers continued to develop throughout the subsequent history of Afroeurasian world-system. However, during this period this development was notable for the technological (and other) leadership of its West Asian center and the strengthening of (still rather weak) communication links between various centers.

Thus, within the Afroeurasian world-system the links became not only interregional, but also the contours of transcontinental links became quite visible. Nevertheless, on a global scale regional links still prevailed.

3) The 1st millennium BCE till 200 BCE—the Afroeurasian world-system as a belt of expanding empires and new civilizations (the third period). This is the time of the early Iron Age. Already in the first part of this period the agrarian revolution within Afroeurasian world-system had been completed as a result of the spread of the technology for plow/non-irrigation agriculture based on the use of cultivation tools with iron working parts (see Korotayev & Grinin, 2006, 2012b). Over this production base enormous changes in trade and military-political spheres occurred, accompanied by a new urbanization and state development upswing in such a way that a group of developed states emerged (see Grinin & Korotayev, 2006; Grinin, 2008a]). Within the Afroeurasian world-system a constant growth in the belt of empires could be remarked on: the New Babylonian, Median, Achaemenid, Macedonian Empire (and its descendants) in the world-system center; the Maurya Empire in South Asia; and the Carthaginian Empire in the West. By the end of the period the formation of empires both in the Far West (Rome) and the Far East (China) of the Afroeurasian world-system had taken place. This was the Axial Age period, the period of the emergence of second generation civilizations. From this point on, development of all the Afroeurasian world-system centers proceeded at a vigorous pace. The West Asian center was finally integrated with the Mediterranean world, whereas the European areas of the barbarian periphery were linked more and more actively to the Afroeurasian world-system centers in military, trade, and cultural aspects. In South Asia a new civilization formed, and the first world religion – Buddhism – emerged. Trade links were established in the space stretching from Egypt to Afghanistan and the Indus Valley (Bentley, 1996; 1999), and this entire territory became connected militarily and politically. The East Asian center of the Afroeurasian world-system also developed very rapidly. In this period Confucianism emerged as its own super-ethnic quasi-religion. Thus all the world-system centers were developing at a rapid pace, *and the complexity, and density of links within this world-system continued to increase on continental and intercontinental scales.*

4) 200 BCE–the early 7ᵗʰ century CE. The Afroeurasian world-system is integrated into the steppe periphery (the fourth phase). In this period within this world-system links became transcontinental and could even be regarded as global.

Around the 2ⁿᵈ century BCE relatively stable trade links (albeit involving preciosities rather than bulk goods) were established between the "marcher empires" of Afroeurasian world-system through the so-called Silk Route, a significant part of which went through the territories of nomadic periphery and semiperiphery[10]. Thus, in this period the periphery completed the circle of Afroeurasian world-system trade links. The Afroeurasian world-system expansion proceeded for a long period of time largely due to the expanding interaction between civilizations and their barbarian peripheries. The larger and more organized civilizations grew, the more active and organized their peripheries became. In the given period this process was sharply amplified, and in the Great Migration epoch the barbarian periphery itself acquired a world-system scale and synchronicity of influence. The disintegration of the Western Roman Empire, the weakening of the Eastern Roman Empire, the rapid dissemination of Christianity in the western part of Afroeurasian world-system, and the new rise of the Chinese Empire in its eastern part all prepared the Afroeurasian world-system for major geopolitical changes and its movement to a new level of complexity. On the other hand, the growth of the Afroeurasian world-system population by the end of the 1ˢᵗ millennium BCE into the hundreds of thousands led to an increased level of pathogen threat. Therefore, the Antonine and Justinian pandemics caused catastrophic depopulation throughout the Afroeurasian world-system in the 2ⁿᵈ and 6ᵗʰ centuries, contributing (in

---

10. In particular, many note the important roles of steppe nomads in these linkages (Barfield 1989; Chase-Dunn and Hall 1997, Ch. 8; Christian 1994, 2000; Frank 1992; Hall 2005; Kradin 2002; Kradin *et al.* 2003; Lattimore 1940; Liu & Shaffer 2007; Mair 2006; Sherratt 2006; Teggart 1939).

addition to an onslaught by the barbarian peripheries) in a very substantial way to the significant slowdown of the Afroeurasian world-system demographic and economic growth in the 1st millennium CE (Korotayev, Malkov, Khaltourina 2005*b*).

5) The 7th-14th centuries—the Afroeurasian world-system apogee: world religions and world trade (the fifth phase). On one hand, in this period the level of development of the world-system links reached the upper limits of what could be achieved on an agrarian basis. On the other hand, there arose the formation of important preconditions for the transformation of the Afroeurasian world-system into the planetary capitalist World System.

First, one should remark on the formation and development of all the world religions at this time. At certain periods within this phase the Afroeurasian world-system developed into a supersystem of contacting and competing third generation civilizations, which created firm cultural-information links between all the Afroeurasian world-system centers, including South Asia, which had remained in relative isolation during the preceding period. Note also the unprecedented sweep of military-political contacts and the growth in development of state structures.

Second, the important aspects are: a) the formation of particularly complex oceanic trade links in the second half of the 1st millennium in the Indian Ocean Basin (see above); b) the creation of vigorous major transcontinental land routes through the territory of the Mongol states which directly connected with the main Afroeurasian world-system centers (see above); c) the initial formation (by the end of this period) of an urbanized zone stretching from Northern Italy through Southern Germany to the Netherlands, where commodity production became the dominant area of the economy (Bernal, 1965; Wallerstein, 1974; Blockmans, 1989).

In fact, by 1500 there were more than 150 cities with a population of more than 10 000 in Europe (Blockmans ,1989). A very high level of urbanization had been reached in Holland, where by 1514 more than half of the population lived in cities (Hart, 1989),and a similar level of urbanization could also be found in the Southern Netherlands (Brugge, Ghent, and Antwerp). However, in Northern Italy in the Po River valley this level might have been even higher (Blockmans, 1989). From the 14th century the growth of cities could have increased as a result of the emergence of developed statehood, the concomitant process of the formation of developed state capitals (*e.g.,* Grinin, 2008a, 2012a; Grinin & Korotayev 2012; Гринин & Коротаев, 2009), and the growth of cities of all types and sizes..

6) The 15th-18th centuries—the transformation of the Afroeurasian world-system into the planetary World System (the sixth phase). This phase is connected with the beginning (the first phase) of the industrial revolution which determined the transformation of the Afroeurasian world-system simultaneously into the planetary (on the one hand) and capitalist (on the other hand) World-System (see Knowles, 1937; Dietz, 1927; Henderson, 1961; Phyllys, 1965; Cipolla, 1976; Stearns, 1993, 1998; Lieberman, 1972; Mokyr, 1985, 1993; More, 2000; Grinin, 2007b, 2012a; Гринин & Коротаев, 2009). This now corresponds closely to Wallerstein's (1974) notion of the world-system, as its development now involved mass movements of bulk goods throughout, whereas some territories (especially in the New World) had become entirely specialized in their particular type of production. A really high level of intensity of the emerged planetary world-system links could be evidenced, for example, by a really high effect produced by the price revolution that resulted from the mass importation into the Old World of the New World gold and silver.

However, as the agrarian production principle still prevailed, an extreme development could be seen in previous trends, especially in the non-European centers of the world-system. In particular, East Asia still continued its development along its own trajectory, demonstrating a high level of achievements in the development of state and cultural structures, and outstanding demographic growth etc.

In the 16th and 17th centuries the so called "military revolution" took place in Europe (see Гринин & Коротаев, 2009; Grinin, 2012a). It involved the formation of modern regular armies armed with sophisticated firearms and artillery and required the reorganization of its entire financial and administration system. In turn the growth of the military might of European powers contributed to the initial modernization of some non-European states□ (the Ottoman Empire, Iran, and the Mughal Empire in India), on the one hand, and to an artificial self-isolation from Europe of some other Asian states (China, Japan, Korea, and Vietnam), on the other.

7) From the beginning of 19th century to the 20th century—the industrial World System and mature globalization (subsequent phases). The Great Geographic Discoveries greatly extended the Afroeurasian world-system's contact zone. As a result of this (as well as Europe's technological breakthroughs) a new structure for this world-system began to be formed. The trade-capitalist core emerged in Europe, whereas previous world-system centers (in particular, the one in South Asia) were transformed into exploited periphery. This process became even more active in the subsequent phase of the World-System evolution. Thus the phenomenon which was the world-system periphery experienced a significant transformation.

The subsequent World System development is connected directly to the second phase of the industrial revolution in the last third of the 18th century and the first half of the 19th century (see Grinin, 2007b; Гринин, 2007). Changes in transportation and communication produced an especially revolutionizing effect on the development of world-system links. They contributed to the transformation of a World System which was still based primarily on information links being regularly exchanged from the Atlantic to the Pacific along with various commodities and services, into a new type of World System which had powerful and regular information flows instead of the previously fragmentary and irregular system. This new World System was now based on a truly international and global division of labor.

In the 20th century the development of the World System (after world wars and decolonization) was closely related to the scientific-information revolution of the second half of the 20th century (see Grinin, 2012a), which in conjunction with many other processes finally led to a rapid growth in globalization processes, especially those involving powerful financial flows, and their qualitative transformation (see Grinin & Korotayev, 2010a, 2010b; Korotayev *et al.*, 2011). As a result the world became tightly interconnected as has been recently demonstrated in a convincing way by the recent global financial-economic crisis. By the late 20th century the view that our world is experiencing globalization (whatever meaning is assigned to this word) became a generally accepted. However, the analysis of contemporary globalization processes is beyond the scope of the present article, although it is discussed in another contribution to this volume (see Гринин, 2009; Гринин & Коротаев 2009; Grinin, 2007a, 2008b, 2012b).

REFERENCES

Березкин Ю. Е. 2007. О структуре истории: временные и пространственные составляющие. История и Математика: Концептуальное пространство и направления поиска / Ред. П. В. Турчин, Л. Е. Гринин, С. Ю. Малков, А. В. Коротаев, с. 88–98. М.: ЛКИ/УРСС.

Бернал Дж. 1956. Наука в истории общества / пер. с англ. М.: Ин. лит-ра.

Гринин Л. Е. 2007. Производственные революции и периодизация истории. Вестник Российской Академии наук 77(4): 309–315.

Гринин Л. Е. 2009. Государство и исторический процесс. Политический срез исторического процесса. 2-е изд. М.: ЛИБРОКОМ.

Гринин Л. Е. 2011. Государство и исторический процесс. Эпоха формирования государства. 2-е изд. М.: ЛКИ.

Гринин Л. Е., Коротаев А. В. 2009. Социальная макроэволюция: Генезис и трансформации Мир-Системы. М.: ЛИБРОКОМ.

Келбесса В. 2006. Глобализация и локализация. Глобалистика. Международный междисциплинарный энциклопедический словарь / Ред. И. И. Мазур, А. Н. Чумаков. М.; СПб.; Н.-Й.: ИЦ «Элима»; ИД «Питер».

Кларк Дж. Г. Д. 1953. Доисторическая Европа. М.: Ин. лит-ра.

Ламберг-Карловски К., Саблов Дж. 1992. Древние цивилизации. Ближний Восток и Мезоамерика. М.: Наука.

Новейший словарь иностранных слов и выражений. 2001. Минск: АСТ; Харвест.

Пантин В. И. 2003. Циклы и волны глобальной истории. Глобализация в историческом измерении. М.: Новый век.

Румянцев А. М. 1987. Первобытный способ производства (политико-экономические очерки). М.: Наука.

Уткин А. И. 2003. Глобализация. Глобалистика: энциклопедия / Ред. И. И. Мазур, А. Н. Чумаков. М.: Радуга.

Хелд Д., Гольдблатт Д., Макгрю Э., Перратон Дж. 2004. Глобальные трансформации: Политика, экономика, культура / пер. с англ. М.: Праксис.

Чумаков А. Н. 2011. Глобализация. Контуры целостного мира. 2-е изд. М.: Проспект.

Ясперс К. 1994. Смысл и назначение истории. М.: Республика.

Abu-Lughod J. 1989. Before European Hegemony: The World System A.D. 1250–1350. New York, NY: Oxford University Press.

Amin S., Arrighi G., Frank A. G., Wallerstein I. 2006. Transforming the Revolution: Social Movements and the World-System. Delhi: Aakar.

Bayly C. A. 2004. The Birth of the Modern World, 1780–1914: Global Connections and Comparisons. Maiden, MA and Oxford: Blackwell, 2004.

Bentley J. H. (1999), Asia in World History. Education About Asia 4: 5–9.

Bernbeck R., Pollock S. 2005. A Cultural-Historical Framework. Archaeologies of the Middle East: Critical Perspectives / Ed. by S. Pollock, R. Bernbeck, pp. 11–40. Oxford: Blackwell.

Blockmans W. T. 1989. Preindustrial Europe. Theory and Society 18(5): 733–755.

Borsch S. J. 2005. The Black Death in Egypt and England. Cairo: The American University of Cairo Press.

Cauvin, J. 2000. The Birth of the Gods and the Origins of Agriculture. Cambridge, UK: Cambridge University Press.

Chase-Dunn C., Hall T. D. 1997. Rise and Demise: Comparing World-Systems. Boulder, CO: Westview.

Chase-Dunn C., Manning S. 2002. City Systems and World-Systems: Four Millennia of City Growth and Decline. Cross-Cultural Research 36(4): 379–398.

Chase-Dunn C., Niemeyer R., Alvarez A., Inoue H., Love J. 2011. Cycles of Rise and Fall, Upsweeps and Collapses: Changes in the Scale of Settlements and Polities since the Bronze Age. Evolution: Cosmic, Biological, and Social / Ed. by L. E. Grinin *et al.* Volgograd: Uchitel.

Chernykh E. N. 1992. Ancient Metallurgy in the USSR: The Early Metal Age. Cambridge: Cambridge University Press.

Childe G. 1952. New Light on the Most Ancient East. 4th ed. London: Routledge & Kegan Paul.

Childe, V. G. 1952. New Light on the Most Ancient East. 4th ed. London: Routledge & Paul.

Cipolla, C. M. 1976 (ed.). The Industrial Revolution. 1700–1914. London: Harvester.

Cohen, M. N. 1977. The Food Crisis in Prehistory. Overpopulation and the Origins of Agriculture. New Haven and London: Yale University Press.

Conversi, D. 2010. The Limits of Cultural Globalisation? Journal of Critical Globalisation Studies 3: 36–59.

Cowan, S. W., and Watson, P. J. (eds.) 1992. The Origins of Agriculture. Washington & London: Smithsonian Institution Press.

Dietz, F. 1927. The Industrial Revolution. New York, NY: Holt.

Dols M. W. 1977. The Black Death in the Middle East. Princeton, NJ: Princeton University Press.

Durand J. D. 1977. Historical Estimates of World Population. An Evaluation. Population and Development Review 3(3): 253–296.

Foreman-Peck, J. 1998. Historical Foundations of Globalization. Cheltenham: UK: Edward Elgar.

Frank A. G. 1993. Bronze Age World System Cycles. Current Anthropology 34(4): 383–419.

Frank A. G., Gills B. K. (Eds.) 1993. The World System: Five Hundred Years or Five Thousand? London: Routledge.

Friedman T. L. 2005. "It's a Flat World, After All". New York Times Magazine 3-4-2005.

Gellner E. 1988. Plough, Sword and Book. The Structure of Human History. Chicago, IL: The University of Chicago Press.

Grinin L. E. 2007a. Globalization and the Transformation of National Sovereignty. In Sheffield, J., and Fielden, K. (eds.), Systemic Development: Local Solutions in a Global Environment. Goodyear: ISCE Publishing.

Grinin L. E. 2007b. Production Revolutions and Periodization of History: A Comparative and Theoretic-mathematical Approach. Social Evolution & History, Vol. 6 No. 2, September 2007 75–120

Grinin L. E. 2008a. Early State, Developed State, Mature State: The Statehood Evolutionary Sequence. Social Evolution & History 7(1): 67–81.

Grinin L. E. 2008b. Globalization and Sovereignty: Why do States Abandon their Sovereign Prerogatives? Age of Globalization 1: 22–32.

Grinin L. E. 2012a. Macrohistory and Globalization. Volgograd: Uchitel.

Grinin L. E. 2012b. New Foundations of International System or Why do States Lose Their Sovereignty in the Age of Globalization? Journal of Globalization Studies. Volume 3, Number 1 / May 2012.

Grinin L. E., Korotayev A. V. 2010a. Will the Global Crisis Lead to Global Transformations. 1. The Global Financial System: Pros and Cons. Journal of Globalization Studies 1/1: 70–89.

Grinin L. E., Korotayev A. V. 2010b. Will the Global Crisis Lead to Global Transfor-mations? 2. The Coming Epoch of New Coalitions. Journal of Globalization Studies 1/2: 166–183.

Grinin L. E., Korotayev A. V. 2012. Afroeurasian World-System: Genesis, Transformations, Characteristics. Routledge Handbook of World-Systems Analysis. Edited by Salvatore Babones and Christopher Chase-Dunn. London: Routledge, 2012. P. 30–39.

Harris, D., and Hillman, G. 1989. An Evolutionary Continuum of People-plant Interaction. Foraging and Farming. The Evolution of Plant Exploitation (pp. 11–27). London: Unwin Hyman.

Hart M. T. 1989. Cities and Statemaking in the Dutch Republic, 1580–1680. Theory and Society 18(5): 663–687.

Henderson, W. O. 1961. The Industrial Revolution on the Continent: Germany, France, Russia, 1800–1914. [London]: F. Cass.

Hopkins, A. G. 2002. Globalization in World History. New York: Norton.

Hopkins, A.G., ed., 2003. Globalization in World History. New York City, NY: Norton.

Ingold, T. 1980. Hunters, Pastoralists, and Ranchers: Reindeer Economies and Their Transformations. Cambridge, UK: Cambridge University Press.

Knowles, L. C. A. 1937. The Industrial and Commercial Revolutions in Great Britain during the Nineteenth Century. London: Routledge.

Korotayev A. 2005. A Compact Macromodel of World System Evolution. Journal of World-Systems Research 11/1: 79–93.

Korotayev A. 2006. The World System Urbanization Dynamics: A Quantitative Analysis. History & Mathematics: Historical Dynamics and Development of Complex Societies / Ed. by P. Turchin, L. Grinin, A. Korotayev, V. C. de Munck, p. 44–62. Moscow: KomKniga/ URSS.

Korotayev A. 2007. Compact Mathematical Models of World System Development, and How they can Help us to Clarify our Understanding of Globalization Processes). Globalization as Evolutionary Process: Modeling Global Change / Ed. by G. Modelski, T. Devezas, W. R. Thompson, p. 133–160. London: Routledge.

Korotayev A. 2012. Globalization and Mathematical Modeling of Global Development. Globalistics and Globalization Studies / Ed. by L. Grinin, I. Ilyin, and A. Korotayev. Moscow—Volgograd: Moscow University—Uchitel, 2012. P. 148–158.

Korotayev A., Berezkin Yu., Kozmin A., Arkhipova A. 2006. Return of the White Raven: Postdiluvial Reconnaissance Motif A2234.1.1 Reconsidered. Journal of American Folklore 119: 472–520.

Korotayev A., Grinin L. 2006. Urbanization and Political Development of the World System: A Comparative Quantitative Analysis. History and Mathematics. Historical Dynamics and Development of Complex Societies / Ed. by P. Turchin *et al.* Moscow: URSS.

Korotayev A., Grinin L. 2012. Global Urbanization and Political Development of the World System. Globalistics and Globalization Studies / Ed. by L. Grinin, I. Ilyin, and A. Korotayev. Moscow—Volgograd: Moscow University—Uchitel, 2012. P. 28–78.

Korotayev A., Kazankov A. 2000. Regions Based on Social Structure: A Reconsideration. Current Anthropology 41/5: 668–690.

Korotayev A., Malkov A., Khaltourina D. 2006a. Introduction to Social Macrodynamics: Compact Macromodels of the World System Growth. Moscow: KomKniga/URSS.

Korotayev A., Malkov A., Khaltourina D. 2006b. Introduction to Social Macrodynamics: Secular Cycles and Millennial Trends. Moscow: KomKniga/URSS.

Korotayev A., Zinkina J., Bogevolnov J., Malkov A. 2011. Global Unconditional Convergence among Larger Economies after 1998? Journal of Globalization Studies 2/2 (2011): 25–62.

Kremer M. 1993. Population Growth and Technological Change: One Million B.C. to 1990. The Quarterly Journal of Economics 108: 681–716.

Lewis D., Moore K. 2009. The Origins of Globalization. London: Routledge.

Lieberman, S. (ed.) 1972. Europe and the Industrial Revolution. Cambridge, MA: Schenkman.

McNeill W. H. 1976. Plagues and Peoples. New York, NY: Monticello.

Mellaart, J. 1975. The Neolithic of the Near East. London: Thames and Hudson.

Mellaart, J. 1982. Drevneishie tsivilizatsii Blizhnego Vostoka [The Most Ancient Civilizations of the Near East]. Moscow: Nauka.

Menard R. 1991. Transport Costs and Long-Range Trade, 1300-1800: Was There a European 'Transport Revolution' in the Early Modern Era? in J. D. Tracy (ed.), Political Economy of Merchant Empires, (Cambridge: Cambridge University Press): 228-75.

Mokyr, J. 1985. The Economics of the Industrial Revolution. London: George Allen & Unwin.

Mokyr, J. (ed.). 1993. The British Industrial Revolution: an Economic Perspective. Boulder, CO: Westview.

More, C. 2000. Understanding the Industrial Revolution. London: Routledge.

O'Rourke K. H., Williamson J. G. 2000. When did globalization begin? Cambridge, MA: NBER (NBER Working Paper 7632).

O'Rourke, K. H., Williamson, J. G. 1999. Globalization and History: The Evolution of a Nineteenth-Century Atlantic Economy. Cambridge, MA: MIT Press.

Phyllys, D. 1965. The First Industrial Revolution. Cambridge, UK: University Press.

Reed, Ch. A. (ed.). 1977. Origins of Agriculture. The Hague: Mouton.

Rindos, D. 1984. The Origins of Agriculture: an Evolutionary Perspective. Orlando, CA: Academic Press.

Sharp, P. 2008. Why Globalization Might Have Started in the Eighteenth Century. VoxEU May 16, 2008.

Smith, Ph. E. L. 1976. Food Production and Its Consequences. Menlo Park, CA: Cumming Publishing Company.

Stearns, P. N. 1993. Interpreting the Industrial Revolution. In Adams, M. (ed.), Islamic and European Expansion. The Forging of a Global Order (pp. 199–242). Philadelphia, PA: Temple University Press.

Stearns, P. N. 1998 (ed.). The Industrial Revolution in the World History. 2nd ed. Boulder, CO: Westview.

Tracy J. D. 1990. Introduction. in J. D. Tracy (ed.), The Rise of Merchant Empires, Cambridge: Cambridge University Press): 1-13.

Wallerstein I. 1974. The Modern World-System. Vol. I. Capitalist Agriculture and the Origins of the European World-Economy in the Sixteenth Century (Studies in Social Discontinuity). New York, NY: Academic Press.

Chapter 2

THE LEAD ECONOMY SEQUENCE IN WORLD POLITICS (FROM SUNG CHINA TO THE UNITED STATES): SELECTED COUNTERFACTUALS

William R. Thompson

How we make sense of world politics and episodes of accelerated globalization depends on our historical scripts. Validating one person's historical script versus someone else's is a highly problematic exercise. Counterfactuals, however, can be utilized to at least suggest or reinforce the asserted significance of different versions of political-economic history. A series of eight counterfactuals encompassing the past 1000 years are harnessed to buttress the utility of framing the development of the modern world economy around a chain of lead economies and system leaders extending back to Sung China and forward to the United States.

Keywords: counterfactual, lead economy, alternative history, transition.

Counterfactual analysis is credited with various types of utility (Chamberlain, 1986; Ferguson, 1997a; Tetlock & Belkin, 1996; Weber, 1996; Parker & Tetlock, 2006; Tetlock & Parker, 2006; Levy, 2008; Lebow, 2010). For some, alternative history is entertaining. For others, it represents a challenge to conventional notions about causality. Some users believe that they can test theories with counterfactuals. Still others find their utility in probing future possibilities. I wish to employ a sequence of counterfactuals for another purpose altogether. Historical scripts in international politics that provide political-economic infrastructures for charting political and economic globalization vary considerably. It is not so much a matter of disagreeing about what happened in the past as it is the one of disagreeing about which past events were most significant to an understanding of international relations processes. Ultimately, there may be no way to convert analysts from one historical script to another. Appreciation of what is most significant in history tends to be a highly subjective undertaking. Quite often, it seems to hinge on what sort of history we were taught in grade school. Declaring that one historical script is superior to another, then, can resemble attempting to communicate with hearing-impaired individuals. There are simply too many cognitive roadblocks to overcome.

It would be highly desirable if we could put historical scripts to empirical test just as we do with rival theories. But we cannot. However, there may be at least one approach to indirect testing. If a historical script has a definite starting point and important possible turning points along the way, one way to assess the value of such a story is to impose counterfactuals on the important milestones in the chronology. If the counterfactuals stay within the rules of minimal revisions and they suggest that vastly different realities could have emerged with small twists, it does not confirm the significance of the historical script. But it should be regarded as at least reinforcing the script. If counterfactuals lead to alternative realities that do not differ all that much, one would have to be a bit suspicious that the chosen turning points were all that significant in the first place.

Accordingly, I develop or harness other people's alternative scenarios for eight significant points in a sequence of systemic leadership and lead economies that have driven globalization processes for almost a thousand years. Beginning in Sung China of the 11th-12th century and traversing Genoa, Venice, Portugal, the Netherlands, Britain, and the United States, the claim is that each actor (or at least most of the actors) in succession played an unusually critical role in creating a structure of leadership that became increasingly global in scope across time. Along the way, a number of wars also performed roles as catalytic opportunities for the emergence of renewed leadership. Who won and lost these wars provides the basic fulcrum for developing counterfactual understandings of what was at stake. If things had worked out differently, markedly different structures of world politics and globalization possibilities would have been developed. In that sense, it can be claimed that the significance of what did occur, the armature of the economic leadership historical script, has been reinforced, albeit indirectly.

COUNTERFACTUALS AND HISTORICAL SCRIPTS

Counterfactuals are said to possess a bad flavor in history circles.[1] They are often dismissed as without value or worse. But historians have their own problems and we need not dwell on their intra-disciplinary disputes. Social scientists have not quite fully embraced counterfactuals either. The two main reasons for this recalcitrance appear to be their implications for causality presumptions and their ultimate utility. Causally-speaking counterfactuals have some potential to be upsetting. We proceed on the basis of X 'causing' Y. When someone comes along and suggests that the Y outcome may have hinged on some minor flap of 'butterfly wings' or that, at best, X might have led to a half dozen different and equally plausible Y outcomes, the foundation of positivist social science is seemingly threatened.

An extreme case is Williamson Murray's (2000) very brief Churchill counterfactual. In 1931 a New York City cab driver collided with Winston Churchill on a street corner and injured him. Murray goes on to suggest that if Churchill had been killed in the accident that a strategically beleaguered Britain would have surrendered in 1940, turned over their fleet to the Germans who, in turn, would have conquered Europe by 1947 and gone on to fight the U.S. forces in South America. Just how these events would have come about are not explicated in the Murray scenario. But the overarching assumption is that one man stood in the way of a European victory by the Germans. Remove the one man and all is lost—or won, depending on one's perspective.[2]

There is a simple theory of the Great Man lurking in this tale. We do not usually base our social science theories on singular individuals. The 1945 outcome is most

---

1. Judging by the number of historians who have written counterfactuals, this complaint may be exaggerated.

2. A similar effort by Large (2000) has Annie Oakley shooting a cigar held by an impetuous Kaiser Wilhelm II in 1889. If her aim had been less accurate and she had killed the Kaiser, the author suggests that Germany might not have pursued an aggressive Weltpolitik policy in World War I. This particular counterfactual is saved by the author's last line in which he notes that Oakley wrote the Kaiser after the war asking for a second try. Fiefer (2002) advances the thesis that if Lenin had been unable to get to Russia in 1917, the Bolsheviks would have failed to take over the Russian government and there would have been no Russian Civil War, no Stalin, and no Cold War.

usually explained, most briefly, by the observation that the winning side had access to a great deal more material resources than the losing side. In retrospect, if not inevitable, the Allied victory was highly probable based on this asymmetry of power. To be told that much of that asymmetry made little difference and that it all hinged on a taxi driver's error a decade or so earlier is downright irritating, if not disturbing. So, not only do counterfactuals complicate our ability to test theories by requiring potentially the construction of many possible rival hypotheses (what if Roosevelt, Stalin, or Eisenhower had died, Rommel been triumphant in the North African desert, or Hitler had been more successful as an artist?) that would be exceedingly difficult to test, they also undermine the possibility of reasonably parsimonious theory construction. World War II engaged many millions of people quite directly. The presence or absence of just how many different individuals might have made some difference? Since most of our theories exclude specific personalities, how are we to proceed? If counterfactuals such as Murray's were the rule, we could literally paralyze ourselves attempting to cope with their analytical implications. Not surprisingly, the easiest solution is to simply evade counterfactuals altogether.

There is, however, at least one way in which counterfactuals might play a useful role in the study of world politics. Analysts of world politics (and globalization) share no common understanding of the history of their subject matter. I do not mean to suggest that there is disagreement about whether a World War I occurred. Rather, there is an extensive disagreement about what time periods matter for developing a theoretical understanding of international relations. For the hardest-core realist, historical time periods are not all that critical. Any should do equally well because nothing much has changed. Liberals focus on integrating tendencies toward greater interdependence and thus are apt to start with the late 19th century globalization upsurge, even though earlier globalization upsurges are readily discernible. Others dispute the value of 1494, 1648, 1815, or 1945 starting points for 'modernity' in international relations.

A late 15th century starting point keys on the French drive into Italy as an act ushering in a period of increasing Western European system-ness thanks, in part to the Spanish resistance and the long Habsburg-Valois feud that became a regional armature of conflict for the next century and a half. A mid-16th century starting point emphasizes a legalistic transition from empires to states as the central actor of international politics. The post-Napoleonic 1815 is usually meant to capture the significance of emergent industrialization for altering the fundamental nature of international relations—or, if not its nature at least its form. The dropping of two atomic bombs on

Japan in 1945 is a salient turning point for some who stress the distinctions between nuclear and pre-nuclear international politics.[3]

The adherence to multiple starting points need not matter much. Yet, it seems to do so. Analysts who start at different points in time tend to adopt vastly different perspectives on what world politics is about. No doubt, there is more to these disagreements than simply different preferences for starting points. But the fact that analysts have much different historical scripts underlying their analyses seems less than coincidental.

The Lead Economy Sequence (from Sung China to the United States)

There are, to be sure, non-trivial reasons for initiating one's international relations historical script at one point or another. Nuclear weapons, industrial revolutions, and system-ness are not to be treated lightly. But another way of looking at these more recent points is that they are simply that—more recent transition points—in a longer term process that changed fundamentally a millennium ago. Weapon innovations, industrial productivity, and system-ness are also related to the earlier transition point. The argument is not that the earlier transition point is necessarily more significant than more recent ones. Rather, the point is that the nature of world politics underwent a fundamental change millennium that turned out to have rather major structural implications for world politics. None of the more recent transition points have eliminated the significance of the earlier point. They are, on the contrary, under-recognized by-products of the earlier fundamental transition in systemic processes.

What happened a thousand years ago to transform the basic nature of world politics? The Chinese, ruled by the Sung Dynasty, created the first 'modern' economy, characterized by monetarization and paper money, extensive commercial transactions on land, via canals/rivers, and on sea, maritime technology that involved multi-masted junks guided by advanced navigation skills unlike anything known elsewhere, unprecedented iron production fueled by military demand, and the development of gunpowder weaponry. Without going into the details of economic innovation, the Sung appear to have been the first land-based state to transcend the limitations of agrarian economies via radical innovations in a host of economic activities ranging from agriculture through manufacturing to energy and transportation. In this respect, China,

---

3. No doubt, some might include 1989/91 for ushering in a post-Cold War era and for the genuinely American-centric analyst, September 11, 2001 might be seen as a critical turning point in perceived U.S. vulnerability at least.

roughly a thousand years ago, deserves the appellation of the first modern economy.[4]

While this breakthrough has major implications for economic development, what does it matter for world politics? The answer is that it is the origin of a sequential process in which a lead economy emerges as the primary source for radical economic innovations that drive productivity, transportation, and commerce. Earlier states had managed to monopolize various types of innovation before but there was no continuity to the process. Innovations were both less radical in general and more isolated in time and space. What took place in Sung China initiated a process that can be traced through the next millennium and is still very much with us in even more developed and complex form.

Given its considerable economic lead in about the 11th-12th century, Sung China might have been expected to inaugurate movement toward an increasingly Sinocentric world system. It did not. In contrast to the image that we now possess of continuity in Chinese imperial predominance in East Asia, the Sung accomplished many of their breakthroughs in a competitive and threatening East Asian multipolar system. That East Asia contained multiple powerful actors a millennium ago may have contributed to the Sung economic breakthrough in transcending agrarian constraints. Military threat certainly encouraged iron production for armor and weapons and gunpowder applications. The inability to trade overland due to the hostility of neighbors may well have encouraged maritime developments. Yet this same threatening environment proved to be overwhelming. The Sung first lost North China with its ore and saltpeter deposits that were critical to iron and gunpowder production to the Manchurian Jurchens. South China was eventually overrun by the Mongols in the 13th century.

The East Asian threat environment and outcomes in combat between the Chinese and their rivals set back the early Chinese lead in economic productivity and military innovation. It did not extinguish the innovations altogether but it did accelerate their diffusion in the western direction. Mongol armies co-opted gunpowder and Chinese engineers and spread the military innovations throughout Eurasia. The success of Mongol imperial domination created an opportunity for some Europeans (Venice and

4. See, among others, Hartwell (1966), Gernet (1982), McNeill (1982), Jones (1988), Modelski and Thompson (1996), Maddison (1998), and Hobson (2004) on the Sung economic revolution. De Vries and van der Woude (1997) make a good case for the 17th century Dutch deserving the first modern economy appellation. They certainly have a point in the sense in contrasting what the Dutch accomplished vis-à-vis the subsequent British industrial revolution. Menzies (2008: 214) briefly argues for 15th century northern Italy as the first European industrial 'nation', based on borrowed Chinese technology. Certainly, the case for an Italian-Netherlands-Britain European sequence of increasingly revolutionary industrialization deserves consideration.

Genoa for the most part) to control the western ends of increased Eurasian east-west trade. Accompanying this increased trade were a number of ideas about technological innovation in maritime commerce and manufacturing that helped stimulate subsequent navigational and industrial revolutions in the Mediterranean and in Western Europe. The technical ability to escape the Mediterranean and sail around the world was further encouraged in various ways by the indirectly Mongol-induced Black Death, the demise of the Mongol empire, and increasing problems in engaging in trade on land in Eurasia in the absence of a singular imperial regime. Portugal was encouraged ultimately to stumble into the Indian Ocean as a means of breaking the Venetian-Mamluk maritime monopoly on Asian spices coming into European markets.[5]

Venetian, Genoese, and Portuguese innovations in developing maritime commercial networks and infrastructure (boats, bases, and governmental regulation) were impressive but were based on limited resource bases. The political implications of a sequence of lead economies took on a more overt appearance as the sequential lead moved on to the 17th century Dutch, the 18th—19th century British, and the 20th century United States. Perhaps the most overt consequences were in the outcomes of repeated attempts to take over the European region. The lead economies by no means stopped single-handedly the ambitions of the Spanish, the French, and the Germans through 1945. But they were certainly significant as coalition organizers/subsidizers/strategic leaders, concentrations of economic wealth, conduits for extra-European resources, and developers of tactical and weaponry innovations in the military sphere. Without the lead economies, markedly different outcomes in the warfare of the later 16th—early 17th, later 17th—early 18th, later 18th—early 19th and the first half of the 20th centuries are not difficult to imagine. It does not seem an exaggeration to state that our most basic understanding of the 'reality' of world politics owes a great deal to the lead economy sequence that began to emerge in Sung China a millennium ago.

A corollary of this generalization is that the 1494, 1815, and 1945 transition points were dependent to varying degrees on the Sung breakthrough. The movement of the French into Italy in the 1490s reflected the general deterioration of the late-medieval Italian lead over the rest of Europe thanks in part to Italian city-state control of the western distribution of Eurasian east-west trade. That is, the French moved into a decaying Italian city-state subsystem and not when it was still thriving earlier in the 15th

5. On the post-Sung, Mediterranean transitional period, see Modelski and Thompson (1996: 177–208). Different views, sometimes in agreement and sometimes not, may be found in Lane (1973), McNeill (1974), Scammell (1981), Lewis (1988), Abu-Lughod (1989), Tracy (1990), and Findlay and O'Rourke (2007). Angus Maddison's (2001: ch. 2) interpretation of this period increasingly resembles the leadership long cycle view expressed in Modelski and Thompson (1996).

century. The British-led Industrial Revolution, culminating in a number of production breakthroughs in iron and textiles in the late 18th century and on, was dependent on information developed earlier on the other end of the Eurasian continent. Such a statement does not imply that the European industrial revolution could not have occurred in the absence of earlier Chinese developments—only that it did not have to do so. The 1945 revolution in military technology embodied in nuclear weapons, of course, was also a resultant of the interaction of the earlier gunpowder revolution and the later industrial revolution.

A case can therefore be made for strong linkages among contemporary (read 'modern') world politics, economic development, and military weaponry that can be traced back to Sung China in the 11th and 12th centuries. Where do counterfactuals fit into this bigger picture? Basically, they reinforce the importance of this interpretation of the history of world political economy while, at the same time, emphasizing the fragility of historical contingencies. But even the fragility underscores the significance of a historical understanding of the continuing evolution of world politics. Contemplating what might have been gives us all the more reason to pay attention to what did transpire. A third value of counterfactuals is that they help to defeat the deterministic complaint so often levied against systemic interpretations. Things did not have to work out the way they did. A variety of other, alternative trajectories are conceivable.[6] Yet the plausibility of alternative realities does not detract from the fundamental fact that a historical trajectory or path was traveled that was critical to both the development of world political system-ness and some of its most important structural features.

EIGHT COUNTERFACTUALS

Eight counterfactuals follow. Others are imaginable. Indeed, the potential number of alternative turns are rather numerous, if not infinite. But the eight that have been developed place maximum attention on the Sung to United States historical script and its possible twists at most of the major potential turning points. Note that each successive counterfactual is rendered less likely if preceding counterfactuals

---

6. I feel personally compelled to make this point because I have engaged in an academic debate with Ned Lebow over the implications of Archduke Ferdinand not dying in Sarajevo in 1914 (Lebow, 2000-2001, 2003; Thompson, 2003; and continued in Goertz & Levy, 2007). Lebow argues that it is possible that World War I would never have occurred if Ferdinand had escaped assassination. I argue that World War I was probable due to certain systemic processes, including a number of 'ripe' rivalries, leader-challenger transitional dynamics, and increasing polarization. None of this means that World War I could not have taken a different form. For a completely different perspective, see the argument made by Schroeder (2004). But see also Taylor (1972 [1932]).

had actually materialized to alter the future.

Counterfactual no. 1: The Sung did not need to have lost North China to the Jurchen steppe warriors (see, e.g., Yates, 2006). They had allied with the Jurchen initially to defeat a mutual enemy, the Kitan empire, later called Liao. In the process, the Jurchen realized how vulnerable Sung areas were to attack and, after Liao was defeated, turned to raiding their former allies. The initial goal was the customary hit-and-run extortion but the Jurchen forces managed to capture the Sung capital and emperor after a string of disastrous battles. The Sung forces retreated to South China abandoning North China to the Jurchen conquerors.[7] If, however, the Sung had defeated the Jurchen and maintained control of the North—a possibility that was not inconceivable with better political and military managers, they would have been in a good or at least much better position to have defeated the Mongols in the next steppe-sedentary iteration a century or more later.[8] A decisive defeat of the Mongols would have had a considerable impact on subsequent history. In East Asia, Sung economic and military progress could have continued unabated with less pressure from northern and western threats. Subordinated Mongols would mean that some two-thirds of Eurasia from Korea to Hungary would not have come under Mongol control. An accelerated diffusion of industrial and military technology throughout Eurasia would have been less probable. A Chinese set-back would have been avoided and the opportunity for a European catch-up might have disappeared altogether. No Black Death might, paradoxically, have led to overpopulation problems in Europe.[9] Western Europe might still have developed economically but surely at a much slower rate, especially if the introduction of gunpowder and cannons had come much later. The need for competitive states in Western Europe to pay for increasing levels of military expenditures would also have developed much more slowly. It is conceivable that the Protestant revolt against Catholic hegemony would have failed eventually, depending on whether the

7. See Lorge (2005: 51-56) for an account of the initial Sung-Jurchen combat. Haeger (1975) frames the policy debate within Sung circles as one of non-accommodation versus appeasement with policy-makers preferring negotiation and concessions prevailing.

8. Despite an unimpressive response to Mongol attacks in the early 13th century, it still took two decades for the Mongols to defeat the Jurchen (Lorge, 2005: 70) before moving on to the Sung in the mid-13th century who, in turn, were not finally defeated until 1276. Peterson (1975) argues that if the Sung had realized that the Mongols would prove to be an even greater threat than the Jurchen, they might have pursued much different and less passive policies that could have altered the outcome substantially, even without controlling North China. Most pertinent to counterfactual considerations, the appropriate response was debated at the time, with advocates of a harder line strategy losing to moderates who preferred not acting at all.

9. One interpretation of the Black Death is that eliminating roughly a third of the European population meant that the survivors had more income per capita to spend on long-distance trade goods than might otherwise have been the case.

Netherlands gained its independence and England still joined the Protestant ranks. Without the American silver that the Spanish distributed throughout Europe in military expenditures, fewer resources would have been available in Northern Europe for economic development.

Farther east Muscovy would not have been favored by Mongol rulers. Kiev might have become the Russian center or an enlarged Polish-Lithuania and/or an expanded Sweden might have eventually absorbed eastern territory all the way to Siberia. Even the Ottoman Empire might have been able to expand to the northeast and continued to be an expansive empire past its late 17th century peak. It is hard to say what might have become of European forays down the coast of Africa or to the Americas. They might not have occurred at all or if they did, they might have come about at a slower pace and centuries later. In general, though, we would have much less reason to expect a European ascendancy to have taken place. Even if for some reason China had not become the most salient region in the world (as opposed to Western Europe), we should expect greater symmetry in the world's power distribution to have evolved after 1800 than in fact did emerge.[10]

Counterfactual no. 2: The Mongol attack on Eurasia was neither premeditated nor inevitable. Temujin or Genghis Khan acknowledged that he had little idea how vulnerable his opponents were at the outset. Only gradually did he realize that there was little to stop his attacks and that he could dream about conquering the 'world'.[11] Removing a single individual from history is a favorite ploy of alternative history. Whether everything would have been different if one individual was removed from the scene 'prematurely' is often a dubious proposition. But in the case of the Mongols, a great deal did rest on Temujin.[12] Quite a few attempts to murder him very early on could easily have worked out differently.[13] In his absence, it seems unlikely that the coali-

---

10. Pomeranz (2006), for one, is skeptical that China would have duplicated the British industrial revolution.

11. Jackson (2005: 46) suggests that the earliest evidence that Mongols believed that they were engaged in world domination dates only from the 1240s, a generation after the initiation of the Mongol expansion.

12. Lorge (2005: 67) offers an antidote to an overly enthusiastic 'great man' interpretation of Temujin when he describes him as 'not a particularly brilliant general or accomplished warrior, nor was he physically very brave. His abilities in all three areas were respectable, he could not have become a steppe leader otherwise, but he most distinguished himself as a politician, both strategically and charismatically. Chinggis's armies overran most of Asia because he had managed to unite separate and often warring steppe tribes and turn their preexisting military capabilities outward. His tactics were not innovative, and it seems the only substantive change he imposed upon the steppe armies was to spread a decimal organization system throughout his entire forces'.

13. Weatherford (2004: 3-77) retells a number of stories from the Secret History of the Mongols that indicate that Temujin was exceedingly lucky to have survived attempts to eliminate him beginning with

---

tions and military organizations that he created would have been very likely, particularly since they required an abrupt departure from standard operating practices that presumably was motivated by Temujin's inability to successfully manipulate or rely on traditional organizational forms.

Any developments that might have been associated with a Sung victory over the Jurchen and Mongols would also have been equally likely with an aborted Mongol takeover of Eurasia. In the absence of a Genghis Khan, the most likely nomad-sedentary pattern would have resembled the traditional trade and raid alternation that existed prior to the rise of Temujin to unprecedented power as the leader of Central Eurasian nomads. China would not have been occupied by the Mongols. Chinese decision-makers would have been far less likely to develop their Mongol phobia which led to greater official insularity from the outside world and a preoccupation with the northwestern frontier after the first third of the 15th century and into the 18th century. The Ming decision to withdraw from the outside world would have been less likely. But then so, too, would the probability of the existence of a Ming dynasty.

While it is likely that Chinese vulnerability to northern invasions would have continued, there still would have been a much greater probability that any Europeans venturing into Asian waters in the 16th century would have encountered a stronger Chinese naval presence than was actually the case. As it was, Chinese naval technology in the early 16th century was still adequate to the task of beating back the initial Portuguese intrusion into Chinese waters. An alternative future might have seen all European coercive maritime intrusions in the general Asian area repelled early on.

Chinese technology would have diffused more slowly to the West. It is certainly conceivable that eastern Eurasia would have improved its technological edge over western Eurasia. If so, any maritime European ventures to the East might well have been restricted to the small enclaves they initially occupied in the 16th through 18th centuries. The European dominance of Asia in the 19th and 20th century would have been far less likely without an asymmetrical, European industrial edge. Alternatively, technological changes at both ends of Eurasia might have proceeded along parallel tracks and timing. The end result would, of course, have been a vastly different history everywhere in Eurasia encompassing the last half-millennia, if not longer.

---

being abandoned by his own family at a very early age, through his capture for slaying his half-brother, and escapes from various clashes with rival clans and tribes—all before his emergence as leader of the Mongols. Alternatively, Peterson (1975) discusses how the Sung might have reacted more proactively than they did to the initial appearance of the Mongols.

Counterfactual no. 3: The European push into the Atlantic was stimulated by a variety of factors. It required larger ships with more masts and sail, rudders, and better navigational capabilities. To some extent these hinged on Chinese naval technology diffusing westward and major improvements in Mediterranean and southern European maritime technology. Information about Chinese naval technology would probably have diffused in any event but perhaps at a slower rate. Alternatively, there is the possibility that Chinese fleets might have circumnavigated Africa as opposed to proceeding no further than eastern Africa in the 14th century. If Chinese movement into the Mediterranean had, had a parallel impact to the Portuguese movement into the Indian Ocean, a much different version of the gradual Western ascendancy in the East is quite likely.[14] For the first three centuries or so of western expansion in Eurasia, the Portuguese, Dutch, and English were just able to hang onto precarious bases along the coast until technological developments involving steam engines and improved weapons gave them a decisive edge.

The motivation to seek profits in the east-west trade had a great deal to do with greed which we can assume is pretty much a constant in world history. The western European push in the late 15th century, nevertheless, was motivated in part by a desire to circumvent the Venetian-Mamluk monopoly which, in turn, was an outcome traceable to Genoese-Venetian conflict over how best to monopolize the Black Sea position on the overland Silk Routes. The Black Sea position was initially advantaged by the Pax Mongolica and then disadvantaged when the Mongols lost their control over a respectable proportion of Eurasia. The resulting higher costs on overland trade made the maritime routes connecting east and west via the Persian Gulf and Red Sea in the west more attractive—hence, the Venetian-Mamluk lock became more probable after the Genoese position in the Black Sea (wrested earlier from the Venetians) became less attractive.[15] Genoese investment in Portuguese and Spanish explorations into the near Atlantic was also a concomitant of Genoa losing in the Eastern Mediterranean (to the Venetians) and moving west looking for new profitable opportunities (e.g., slaves and sugar production) in the Western Mediterranean and beyond.

Where does that leave the Portuguese circumnavigation of Africa? Portugal broke the Venetian-Mamluk lock on Asian spices coming into the Mediterranean for a few decades at least. The push into the Indian Ocean required considerable technological

14. Menzies (2008) argues for what will seem to many others to sound very counterfactual. He claims that a Chinese fleet visited Italy in the 1430s and stimulated the Italian Renaissance. However, one could argue that the European push into the Atlantic predated the 1430s by several hundred years.

15. The story is complicated further by the Genoese practice of supplying new slaves for the Mamluk military organization from the Black Sea area becoming less viable as Mamluk military competition with Mongols waned.

innovation in ship construction and navigation skills (Devezas & Modelski, 2008) and took several generations to accomplish. It might have been forestalled by an earlier Castilian conquest of Portugal and the Spanish focus on eliminating Moorish control in the Iberian Peninsula (not accomplished until 1492). If the Portuguese had been more successful in seizing Moroccan territory—their first objective in 1415—they might have been less likely to have kept moving down the African coastline looking for vulnerabilities to exploit. They would have been less likely to have found gold and spices in West Africa which allowed them to keep going farther south.

If the Portuguese had not entered the Indian Ocean in force in the early 16th century, it is quite likely that no other Europeans would have in that century—at least before 1595 and the Dutch effort to do so. But would the Dutch have chosen to go around the Cape of Good Hope if the Portuguese had not already done so? The Dutch effort was stimulated by a Spanish edict forcing them to look for alternatives to Mediterranean markets that were being denied to them. Why not circumvent the Mediterranean markets and go to the source? But the 'why not' might have come a little slower if it had not already been accomplished by the Portuguese in the 1490s.

It is also possible to argue that southwestern Europeans were most likely to 'discover' the Americas in the late 15th century because they were situated closer to the Americas than anybody else. That may well be true but it is possible that the discoveries could have been delayed considerably if many of the encouraging factors in the late 15th century had been relatively absent or inoperable. Without American silver, European trade with Asia could not have proceeded as it did. The Europeans initially lacked sufficient coercive advantages and had few commodities, other than silver, that were desired in the east. If they could neither buy nor fight their way in, European participation in Asian markets would have been quite marginal at best. That suggests quite strongly that the European occupation and subordination of India, the Philippines, Indonesia, and, indirectly, China, once again, would probably not have taken place. The current world would be much less unequal in terms of income distribution between states.

Counterfactual no. 4: The 1588 Spanish attempt to land troops in England was not well executed but could have succeeded. The decision to conquer England stemmed from frustrations encountered in suppressing the Dutch Revolt. The logic was that if English support could be neutralized, the revolt would fail. The 1588 Armada was intended to provide cover for troopships that would ferry some 27,000 Spanish veterans across the Channel. The soldiers were not quite ready to embark when the Armada fleet arrived. English attacks managed to drive the Spanish fleet north thereby interrupting the invasion plan. If the English attacks had been less disruptive or if the soldiers had, had another day or two, the invasion could have been initiated. Defending England on land were only a few thousand soldiers with any experience but not necessarily very reliable and some highly dubious militia units.

A Spanish conquest of England in 1588 could have been even more revolutionary than the Norman one in 1066. Spain was already predominant in Europe. Assuming the assumptions about the loss of English support would have doomed the Dutch Revolt, Spain and/or its allies would have controlled all of Western Europe within a few years. Protestantism would have been on the defensive in England and throughout northern Europe. A Thirty Years War would have been far less likely. North and South America would have been under Spanish rule.[16] The combination of the Portuguese and Spanish empires, following Philip II's acquisition of the Portuguese throne in the early 1580s would probably not have broken apart in 1640.

The Spanish might also have been able to suppress or delay the 17th century challenge for regional leadership and Spanish relative decline in the second half of the 17th century.[17] Even if the Spanish had failed to stop the French ascent, the probability of English-Dutch opposition to Louis XIV's territorial expansion would have been substantially reduced. In sum, Spanish hegemony in Europe and elsewhere would have been considerably reinforced. When or if Spain's predominance had run its course, it would most likely have been simply replaced by France—meaning that Western Europe's fabled competitiveness could easily have disappeared, with major repercussions for consequent economic and military developments that drove Europe to the center of the world system by the 19th century. In this respect, the 'Rise of the West' might have been derailed altogether or at least postponed considerably.

Counterfactual no. 5: Goldstone (2006) has William of Orange successfully invading England in 1688 and capturing the English crown but then has him die in 1690 from a wound sustained in Irish fighting in 1690. The wounding actually occurred but in reality was less than fatal. William proceeded to eliminate resistance to his rule in England and Ireland. More importantly, the larger motivation for this conquest of England was realized. In 1688 France was preparing to attack Austria before resuming its intention of absorbing the Netherlands. England under the Catholic ruler James could be expected to again follow the French lead, as in the early 1670s, with a mari-

---

16. Somerset's (2004) counterfactual has the American colonies revolting eventually from a Catholic England not controlled by Spain.

17. Parker (2000) thinks Spanish hegemony was doomed in any event thanks to Habsburg in-breeding and successively weaker rulers. See Martin and Parker (1999) for some equivocation about the likelihood of Spanish success had they landed in England.

time attack on the Netherlands. As Dutch stadtholder, William's invasion of England with Dutch troops not only neutralized the English threat, it also brought England solidly into the coalition to thwart Louis XIV. By 1713, a financially exhausted Netherlands had become Britain's junior partner in managing the international relations of Western Europe and, increasingly, long-distance commerce as Britain emerged into its first global system leader iteration.

Actually, Goldstone acknowledges that his scenario works whether the 1690 wound had been fatal or if William's invasion had failed due to an English naval interception at sea (thwarted by prevailing winds) or greater resistance on land than had occurred. Of the two possibilities, the latter seems more promising for counterfactual construction purposes.[18] In any event, a French and English attack on the Netherlands in the late 1680s from land and sea could have been too much for the Dutch to withstand. Goldstone suggests that at best the Netherlands would have been subordinated to French regional predominance that would have included a French king on the Spanish throne (without a War of Spanish Succession) and French access to the Spanish empire. France might well have maintained its hold on Canada and, should there still have been a revolutionary war in the British colonies in North America, French intervention could easily have been on behalf of Britain rather than the American revolutionaries.

To the extent that the French Revolution was predicated on French state bankruptcy due to the escalating military costs of the 18th century, the Revolution might have been avoided if France had sustained fewer costs and more successes in places such as North America, the Caribbean and India. Presumably, antagonism with Germans and Austrians would have persisted but the ultimate outcome would have been a gradual shift eastward of the French boundaries due to French military successes along and beyond the Rhine. Latin America and the Caribbean would have remained within a French-Spanish colonial empire. India, at best, might have been partitioned with Britain. As late as 1900, Western Europe would have remained subject to French predominance with possible Austrian expansion into the Balkans without a strong German protector.

---

18. Pestana (2006) notes that if William had died in 1690, Mary would still have assumed the English throne which might not have changed history all that much.

Goldstone adds in a strong technological component as well.[19] Catholic hegemony in England does not stifle scientific research but the socio-political environment becomes less encouraging. Hugenots fleeing French persecution no longer view Britain as a welcome haven. The British navy's growth, no longer fueled by Anglo-French antagonism, does not become a major catalyst for industrial experimentation and organization. A number of direct and indirect advances in iron manufacture, steam engine construction, and textile spinning machines are precluded as a consequence. The expansion of coal as a source of energy is restricted. The potential and implications of Newtonian science are never realized or fully developed. Europe would have been powerful in some parts of the world (the Americas) but not necessarily in Asia. Moreover, the combination of the lack of changes in political and economic structures implies that British democratization might not have progressed much either—with major ramifications for democratization elsewhere as well.[20]

Counterfactual no. 6: The first counterfactual published as a book (Geoffroy-Chateau 1836) focused on Napoleon passing on a Russian attack and instead going on to conquer the world.[21] Zamoyski (2004) envisions a successful second French attack into Russia after an earlier 1812 withdrawal from Moscow. Russia acknowledges defeat and surrenders its Baltic and Polish territory. Finland is returned to Sweden. Russian troops are dispatched to Spain to fight in the guerrilla warfare there. Prussia is demoted to a Brandenburg dukedom. Britain, losing in the Baltic and Eastern Mediterranean to combined French-Russian forces, accepts a negotiated peace. Most of Europe, outside of the Austrian empire, becomes first the Confederation of Europe and then the Empire of Europe, with Napoleon as emperor. Interstate rivalries within Europe are gradually extinguished and replaced by a regional bureaucratic framework focusing increasingly on regulatory functions.[22] In part because Russian decision-makers

19. The Goldstone scenario is predicated on the assumption that only England and to a lesser extent the Netherlands were pulling free from a continental propensity toward monarchical absolutism and conformity. Eliminate the 'pulling fee' element and you unravel the probable development of western science and technology. At the same time, England was not all that much different from the rest of Europe so that slight alterations in political and military fortunes would have led to a less exceptional development trajectory.

20. Another interesting Goldstone assumption is that industrialization and representative democracy are not general processes but, essentially, rare events based on 'a unique combination of factors that came together by chance in one location and generally not elsewhere' (Goldstone 2006: 193).

21. See Shapiro (1998). A now dated but annotated bibliography of alternative histories can be found in Hacker and Chamberlain (1986).

22. Trevelyan (1972 [1932]) also has Napoleon's imperial system surviving in much of Western Europe after Napoleon wins the Battle of Waterloo. Carr (2000), on the other hand, suggests that if Napoleon had won at Waterloo, interstate warfare would simply have continued throughout the 19th century. Horne (2000) thinks that even if Napoleon had won at Waterloo, it would not have ended the Napoleonic Wars until Napoleon was defeated decisively—but this would not have taken too long to accomplish given the number of troops available to the continental opponents of the French.

proved incapable of returning their country to its 18th century form, industrialization sets in successfully and earlier than it might have in an alternative universe. Nevertheless, by the end of the 19th century, economic growth was proceeding most quickly outside of Europe and Russia with dominant economic centers emerging in North America, Brazil, southern Africa and some parts of Asia.

Counterfactual no. 7: Imagine what is called World War I being waged without Britain or the United States. We would not call it World War I but regard it presumably as a wider-scale version of the Franco-Prussian War in 1870–1871 in which German predominance in Europe was introduced, if not established. A German-Austro-Hungarian war versus France and Russia presumably would have led to a similar collapse in the East and a less familiar defeat of France. It is even conceivable that the Central Powers could have won the day with Britain in but without the infusion of U.S. resources from 1917 on. Neither British nor U.S. involvement in World War I was ever inevitable. Britain might have remained aloof in 1914, as the Germans hoped.[23] The United States presumably entered late in the war to get a seat at the victors' negotiation table but would it still have intervened if it was clear that the Central Powers were winning?

One of the main implications of this scenario is that to the extent World War II was a continuation of unresolved issues in World War I, World War II might not have come about at all.[24] The process is similar to the story of a time traveler that accidentally eliminates one of her ancestors only to find that she has eliminated herself in the process. That clearly does not mean that the 20th century would have been pacific. It might still have managed to kill as many or perhaps even more people as a function of the industrialization of warfare but the format and maybe even the alignments might have been considerably different. If so, it might have been very difficult to reach the kind of world that sprang from the defeat of Germany and Japan in 1945. To be sure, the pace of relative decline (Britain's for instance) would have been slower and the pace of ascent (the United States and Russia/Soviet Union) might have been much slower. The twentieth century (and after) could conceivably have remained multipolar and characterized by many smaller or more localized wars through its entirety. The total wars of the twentieth century required the full participation of the great powers in two major exercises in blood-letting. In the absence of the total wars, we might not

23. Ferguson (1997b) offers a detailed scenario for such an outcome and goes on to suggest that early German hegemony in Europe would have been better for Britain, possibly for Russia, and would have excluded the first U.S. intervention into European affairs. It might have simply led to an early version of the European Union.

24. However, Blumetti (2003) offers a scenario in which the war ends in 1916 without U.S. participation but in which a second world war is still waged.

recognize a world of weaker states, less advanced technology, and more complex, cross-cutting interactions among the more powerful states in this version of reality.[25]

Counterfactual no. 8: The last counterfactual has a different outcome for World War II. One way in which this alternative outcome might have come about is if the German attack on the Soviet Union in 1941 had been successful relatively quickly, thereby allowing the Germans to turn on Britain and take it as well.[26] Downing (2001 [1979]) has an extensive scenario that focuses on an early German defeat of the Soviet Union but leaves the implications fairly open-ended with Britain and the United States continuing to prepare for an assault on German positions at some vulnerable point, perhaps in Egypt. Lucas (1995) also has the Germans capture Moscow before the 1941 winter set in which leads to an incorporation of the Soviet Union into the Third Reich. Burleigh (1997) argues that if the Lucas scenario had played out, the Germans might have installed puppet regimes in separatist parts of the Baltic, Byelorussia, the Caucasus, and the Ukraine. On the other hand, these were some of the same territories, the Germans sought as part of the Lebensraum program and could anticipate some degree of German colonization and SS genocide. Burleigh (1997) also contends that we should expect the German war aims to have focused on other parts of the globe once their share of Eurasia was in hand. Africa, the Dutch colonial empire in Southeast Asia, and the United States would have become the next targets of an expanding German empire. In contrast, Montefiore (2004) has Stalin executed by his lieutenants (Molotov and Beria) but then Molotov leads a nationalistic resistance and counter-attack against the Germans in a way that the Georgian Stalin could not have. The subsequent scenario plays out in typical Cold War fashion except that Molotov survives to rule continuously after the war up to the Soviet invasion of Afghanistan in 1979. He is replaced by Gorbachev in 1986. Herwig (2006) has the Germans defeating the Soviet Union but a similar post-1945 future is salvaged by the U.S. deploying atomic weapons against the Germans. The subsequent Pax America is then due to U.S. actions alone—as opposed to a Soviet-Anglo/American war effort. Blumetti (2003) also has a German victory in 1942 that does not prevent a Soviet resurgence in 1944-1945.

Some scenarios have Germany occupying Britain before taking on the Soviet Union (Macksey, 1980, 1995) but if Germany had managed to defeat the Soviet Union decisively and quickly, there might have been little to interfere with a renewed focus

---

25. Without the exhaustion of British resources in two world wars and the pressure of a new American system leader, decolonization, presumably, would at least have been delayed.

26. In addition to having the British surrender early (Roberts 2002), another way is to have the Germans skip the Soviet attack altogether. Keegan (2000) pushes a scenario that has Germany move into the Middle East for the oil that it hoped to acquire in the Soviet Union. Fromkin (2000) echoes this gambit in a sketchy way. An inventory of alternative options is found in Alexander (2000).

on Britain.[27] If both the Soviet Union and Britain had been taken out of the World War II equation, it is hard to imagine a 1945 scenario in which the United States emerged as the most prosperous and powerful leader of an anti-Axis coalition. At best, much of the world would be divided between Germany, the United States and Japan in an extremely uneasy cold war. At worse, the three might have continued fighting indefinitely until or unless one party came up with atomic weapons before the others. But keep in mind the American lead in the nuclear race presumes that the German effort was hard-pressed while Germany was under a multiple-front attack. A different outcome might have occurred if Germany had been less hard-pressed. Roberts (1997: 320) also notes that many of the scientists who later worked on the U.S. atomic bomb were in Britain in 1940 and most would have been captured if the Germans had occupied Britain early on.[28]

A different approach to World War II is to have the Pacific theater work much differently along the lines of Japan not attacking the United States in 1941. John Lukacs' (2003) counterfactual scenario is premised on the assumption that Japanese and U.S. decision-making circles were both divided on the wisdom of going to war in late 1941. We know that the Japanese attacked Pearl Harbor in December precipitating an unsurprising U.S. movement into a Pacific War, quickly globalized by a German declaration of war on the United States. But what if ongoing Japanese-U.S. negotiations had achieved some level agreement that caused the Japanese not to attack? In Lukacs' story, German successes in the Soviet Union and North Africa encourage the Japanese to attack Britain in Southeast Asia. A bombing of Hong Kong harbor leads to the sinking of two U.S. ships and a declaration of war on Japan by the United States in 1942. The rest of the scenario proceeds along lines similar to what actually transpired with the U.S. ultimately defeating the Japanese and gradually becoming more active in the European theater as well.[29] Black (2004), alternatively, simply gives the United States more time to prepare for a concentrated effort to enter the European theater.

---

27. Roberts (1997: 300) notes that there was precious little left to defend Britain, aside from some surplus mustard gas left over from World War I, in May of 1940 when the invasion was first proposed to Hitler.

28. A reader of an earlier version of this paper, Joachim Rennstich, notes that to the extent that post-1945 Soviet and U.S. nuclear and space capabilities benefited from scientists and information captured at the end of World War II, a German victory would have led to less or slower diffusion of technology in this sphere as well.

29. Rose (2000) has the Japanese attack on Pearl Harbor eliminating all three of the U.S. carriers that formed the core of the post-Pearl Harbor U.S. Navy in the real world, without really elaborating the consequences. Cook (2000) has the Japanese win at Midway but the U.S. still prevails eventually in the Pacific War. Some of the scenarios in Tsouras's (2001) edited work are similar but with different outcomes. Black (2004) uses a premise similar to Lukacs' which keeps the Japanese from attacking and gives the United States two more years to build up its military forces to fight in Europe.

Tsouras's scenario (Tsouras, 2001; see as well Tsouras, 2002) is more interesting. He has Japan, following up clashes in the 1920s and 1930s, attacking the Soviet Union in 1941 in coordination with the German Barbarossa attack.[30] By March 1942, the Soviet Union is forced to withdraw from this version of World War II with the Germans occupying Moscow and the Japanese in control of Vladivostok and its surrounding province. Tsouras halts his scenario at this point but it is clear that the nature of the geopolitical landscape has changed dramatically. Germany controls most of Europe and North Africa. Japan is occupying much of East Asia. An isolated United States and a Britain that might not have survived long in the circumstances are confronted with a tripolar structure in which the German and Japanese poles are vastly stronger than they were in reality. One can easily imagine the advent of a new type of cold war until or unless somebody was prepared to strike across the Atlantic and/or Pacific.[31]

CONCLUSION

We have now looked at a number of alternative scenarios relating to events occurring in the last one thousand years. The initial claim is that a sequence of lead economies beginning with Sung China created a critical structure for world politics that was intermittently punctuated by bouts of intensive warfare. These combat episodes were important in facilitating the rise of some key actors, the decline of others, and thwarting outcomes that would have led to vastly different worlds. Although little attention was paid to some of the intermediate parts of the sequence (specifically, the Genoa-Venice-Portugal string), the other parts of the sequence lived up to expectations. Each one, with some slight twists of chance, could have led to markedly different world political realities.

So what? After all, is that not what counterfactuals are almost guaranteed to deliver—some discernible change in reality that demonstrates how fragile reality really is? Yes and no. It is not clear that all possible turning points are equally linked to multiple alternative realities that matter. How much did it matter whether the Genoese initially out-maneuvered the Venetians for control of the Black Sea in the 13th century (thereby establishing a better position to take advantage of the Mongol Pax) or the Venetians later surpassed the Genoese in control of Mediterranean trade (thereby es-

---

30. A Blumetti (2003) variation has Japan concentrating on the British Empire in a 'southern' strategy scenario and a postwar tripolar world in 1945 with Germany, Japan, and the United States as the leading powers.

31. For alternative scenarios to the Cold War that did actually emerge, see Almond (1997), Haslam (1997), and O'Connell (2003).

tablishing a better position to take advantage of the Red Sea route for Asian spices)? The answer is not that the two Italian city-state were entirely interchangeable but it is possible that outcomes would have been similar if they had reversed their order in the sequence. It is even possible to imagine another Italian city state, such as Pisa, taking their place. What was important was that some Italian city states took the initiative to organize European/Mediterranean markets for receiving and demanding Asian goods.

What if the French had not intervened in Italy in 1494? The Ottomans had flirted with the idea of landing troops in Italy a few years earlier. It is conceivable that the European reaction to such a move might have led to something similar to what did transpire in European international relations of the first half of the 16th century. Imagine if the Thirty Years War had been the Sixty Years War. How would international relations have changed?[32] If atomic bombs had not been dropped on Hiroshima and Nagasaki, do we know that the Cold War would have been nastier than it was? Maybe yes, maybe no. But no Sung intensive economic growth spurt and possibly no European industrialization. No Mongol Pax and possibly continuing Chinese ascendance as the world's lead economy and, again, less diffusion of Chinese technological gains to a wider world. A Spanish victory in 1588, a defeat for William III in 1688, a Napoleonic victory sometime in the early 19th century, a less-than-World War I, or a German victory in the 1940s and we should expect rather major consequences for the world politics of each respective era.

These potential turning points matter in part because they did not go down the counterfactual path but might have. They matter even more because of the path that was pursued at each point. They matter because they created a political-economic structure for world politics that has first emerged, then evolved and, so far, endured. The implications of what did happen (not what did not happen) are still with us today. As a consequence, they are a fundamental part of the history of world politics and accelerations of globalization that deserve greater recognition as a sequence of possible forks in the road that might have turned out differently but instead contributed mightily to constructing our past and present reality. If so, the lead economy sequence deserves much greater recognition than it has received to date. The various fragilities associated with the sequence also remind us that future contingencies are apt to be equally chancy. Humility in projecting our interpretations very far into the

---

32. My hunch is not all that much but I start from the premise that the Thirty Years War's overall significance has always been exaggerated. It was important to central Europe but less so as one moves away from this not-always-so-critical sub-region.

future is well advised. Moreover, little seems inevitable about the next iteration in the lead economy sequence.[33]

REFERENCES

Abu-Lughod, J. L. 1989. Before European Hegemony: The World System, A.D. 1250–1350. New York: Oxford University Press.

Alexander, B. 2000. How Hitler Could Have Won World War II: The Fatal Errors That Led to Nazi Defeat. New York: Three Rivers Press.

Almond, M. 1997. 1989 Without Gorbachev: What if Communism Had Not Collapsed? In Ferguson, N. (ed.), Virtual History: Alternatives and Counterfactuals (pp. 392–415). London: Picador.

Black, C. 2004. The Japanese Do Not Attack Pearl Harbor. In Roberts, A. (ed.), What Might Have Been: Leading Historians on Twelve 'What Ifs' of History (pp. 153–165). London: Weidenfeld and Nicolson.

Blumetti, R. 2003. What If? Alternative Historical Time Lines. New York: iUniverse.

Burleigh, M. 1997. Nazi Europe: What if Nazi Germany Had Defeated the Soviet Union? In Ferguson, N. (ed.), Virtual History: Alternatives and Counterfactuals (pp. 321–347). London: Picador.

Carr, C. 2000. Napoleon Wins at Waterloo. In Cowley 2000: 220–221.

Chamberlain, G. B. 1986. Afterword: Allohistory in Science Fiction. In Waugh, Ch. G., and Greenburg, M. H. (eds.), Alternative Histories: Eleven Stories of the World as it Might Have Been (pp. 281–300). New York: Garland.

Cook, Th. F., Jr. 2000. Our Midway Disaster: Japan Springs a Trap, June 4, 1942. In Cowley 2000: 313–339.

Cowley, R. 2000. (Ed.). What If?: The World's Foremost Historians Imagine What Might Have Been. New York: Berkley Books.

Cowley, R. 2002. (Ed.). What If? 2: Eminent Historians Imagine What Might Have Been. New York: Berkley Books.

Devezas, T., and Modelski, G. 2008. The Portuguese as System-Builders in the 15th–16th Centuries: A Case Study on the Role of Technology in the Evolution of the World System. In Modelski, G., Devezas, T., and Thompson, W. R. (eds.), Globalization as Evolutionary Process: Modeling Global Change (pp. 30–57). London: Routledge.

de Vries, J., and van der Woude, A. 1997. The First Modern Economy: Success, Failure and Perseverance of the Dutch Economy, 1500–1815. Cambridge: Cambridge University Press.

---

33. This observation implies that there will be a next iteration in the sequence and that, too, needs to remain open-ended.

Downing, D. 2001 [1979]. The Moscow Option: An Alternative Second World War. London: Greenhill.

Fiefer, G. 2002. No Finland Station: A Russian Revolution without Lenin? In Cowley 2002: 210–235.

Ferguson, N. 1997a. Virtual History; Towards a 'Chaotic' Theory of the Past. In Ferguson, N. (ed.), Virtual History: Alternatives and Counterfactuals (pp. 1–90). London: Picador.

Ferguson, N. 1997b. The Kaiser's European Union: What if Britain Had 'Stood Aside' in August 1914? In Ferguson, N. (ed.), Virtual History: Alternatives and Counterfactuals (pp. 228–280). London: Picador.

Findlay, R., and O'Rourke, K. H. 2007. Power and Plenty: Trade, War and the World Economy in the Second Millennium. Princeton, NJ: Princeton University Press.

Fromkin, D. 2000. Triumph of the Dictators. In Cowley 2000: 308–309.

Geoffroy-Chateau, L.-N. 1836. Napoleon et la Conquete du Monde. Paris: Dellaye.

Gernet, J. 1982. A History of Chinese Civilization. Cambridge: Cambridge University Press.

Goertz, G., and Levy, J. S. 2007. (Eds.). Explaining War and Peace. New York: Routledge.

Goldstone, J. A. 2006. Europe's Peculiar Path: Would the World Be 'Modern' if William III's Invasion of England in 1688 Had Failed? In Tetlock, Ph. E., Lebow, R. N., and Parker, G. (eds.), Unmaking the West: 'What-Ifs?' Scenarios That Rewrite World History (pp. 168–196). Ann Arbor: University of Michigan Press.

Hacker, B. C., and Chamberlain, G. B. 1986. Pasts That Might Have Been, II: A Revised Bibiography of Alternative History. In Waugh, Ch. G., and Greenburg, M. H. (eds.), Alternative Histories: Eleven Stories of the World as It Might Have Been (pp. 301–363). New York: Garland.

Haeger, J. W. 1975. 1126–27: Political Crisis and the Integrity of Culture. In Haeger, J. W. (ed.), Crisis and Prosperity in Sung China. Tucson, AZ: University of Arizona Press.

Hartwell, R. 1966. Markets, Technology and the Structure of Enterprise in the Development of the Eleventh Century Chinese Iron and Steel Industries. Journal of Economic History 26: 29–58.

Haslam, J. 1997. Stalin's War or Peace: What if the Cold War Had Been Avoided? In Ferguson, N. (ed.), Virtual History: Alternatives and Counterfactuals (pp. 348–367). London: Picador.

Herwig, H. H. 2006. Hitler Wins in the East but Germany Still Loses World War II. In Tetlock, Ph. E., Lebow, R. N., and Parker, G. (eds.), Unmaking the West: 'What-ifs?' Scenarios That Rewrite World History (pp. 323–360). Ann Arbor, MI: University of Michigan Press.

Hobson, J. M. 2004. The Eastern Origins of Western Civilisation. Cambridge: Cambridge University Press.

Horne, A. 2000. Ruler of the World: Napoleon's Missed Opportunities. In Cowley 2000: 203–219.

Jackson, P. 2005. The Mongols and the West. New York: Pearson Longman.

Jones, E. L. 1988. Growth Recurring. Oxford: Clarendon Press.

Keegan, J. 2000. How Hitler could have Won the War: The Drive for the Middle East, 1941. In Cowley 2000: 295–305.

Lane, F. C. 1973. Venice: The Maritime Republic. Baltimore, MD: Johns Hopkins University Press.

Large, D. C. 2000. Thanks, But No Cigar. In Cowley 2000: 290–291.

Lebow, R. N. 2000–2001. Contingency, Catalysts and International System Change. Political Science Quarterly 115: 591–616.

Lebow, R. N. 2003. A Data Set Named Desire: A Reply to William R. Thompson. International Studies Quarterly 47: 475–478.

Lebow, R. N. 2010. Forbidden Fruit: Counterfactuals and International Relations. Princeton, NJ: Princeton University Press.

Levy, J. S. 2008. Counterfactuals and Case Studies. In Box-Steffensmeier, J., Brady, H., and Collier, D. (eds.), Oxford Handbook of Political Methodology (pp. 627–644). New York: Oxford University Press.

Lewis, A. R. 1988. Nomads and Crusaders, A.D. 1000–1368. Bloomington, IN: Indiana University Press.

Lorge, P. 2005. War, Politics and Society in Modern China, 900–1795. London: Routledge.

Lucas, J. 1995. Operation WOTAN: The Panzer Thrust to Capture Moscow, October-November 1941. In Macksey, K. (ed.), The Hitler Options: Alternate Decisions of World War II (pp. 54–81). London: Greenhill Books.

Lukacs, J. 2003. No Pearl Harbor? FDR Delays the War. In Cowley, R. (ed.), What Ifs? of American History: Eminent Historian Imagine What Might Have Been (pp. 179–188). New York: Berkley Books.

Macksey, K. 1980. Invasion: The German Invasion of England. July 1940. London: MacMillan Publishing Co.

Macksey, K. 1995. Operation Sea Lion: Germany Invades Britain, 1940. In Macksey, K. (ed.), The Hitler Options: Alternate Decisions of World War II (pp. 13–34). London: Greenhill Books.

Maddison, A. 1998. Chinese Economic Performance in the Long Run. Paris: OECD.

Maddison, A. 2001. The World Economy: A Millennial Perspective. Paris: OECD.

Martin, C., and Parker, G. 1999. The Spanish Armada. Manchester: University of Manchester Press.

McNeill, W. H. 1974. Venice: The Hinge of Europe, 1081–1797. Chicago, IL: University of Chicago Press.

McNeill, W. H. 1982. The Pursuit of Power. Chicago, IL: University of Chicago Press.

Menzies, G. 2008. 1434: The Year A Magnificent Chinese Fleet Sailed to Italy and Ignited the Renaissance. New York: William Morrow.

Modelski, G., and Thompson, W. R. 1996. Leading Sectors and World Powers: The Coevolution of Global Politics and Economics. Columbia, SC: University of South Carolina Press.

Montefiore, S. S. 2004. Stalin Flees Moscow in 1941. In Robert, A. (ed.), What Might Have Been: Leading Historians on Twelve 'What Ifs' of History (pp. 134–152). London: Weidenfeld and Nicolson.

Murray, W. 2000. What a Taxi Driver Wrought. In Cowley 2000: 306–307.

O'Connell, R. L. 2003. The Cuban Missile Crisis: Second Holocaust. In Cowley, R. (ed.), What Ifs of American History: Eminent Historians Imagine What Might Have Been (pp. 253–274). New York: Berkley Books.

Parker, G. 2000. The Repulse of the English Fireships: The Spanish Armada Triumphs, August 8, 1588. In Cowley 2000: 141–154.

Parker, G., and Tetlock, Ph. E. 2006. Counterfactual History: Its Advocates, Its Critics, and Its Uses. In Tetlock, Ph. E., Lebow, R. N., and Parker, G. (eds.), Unmaking the West: 'What-If' Scenarios That Rewrite World History (pp. 363–392). Ann Arbor, MI: University of Michigan Press.

Pestana, C. G. 2006. Nineteenth Century British Imperialism Undone with a Single Shell Fragment: A Response to Jack Goldstone's 'Europe's Peculiar Path'. In Tetlock, Ph. E., Lebow, R. N., and Parker, G. (eds.), Unmaking the West: 'What-ifs?' Scenarios That Rewrite World History (pp. 197–202). Ann Arbor, MI: University of Michigan Press.

Peterson, Ch. A. 1975. First Sung Reactions to the Mongol Invasion of the North, 1211–17. In Haeger, J. W. (ed.), Crisis and Prosperity in Sung China (pp. 215–252). Tucson, AZ: University of Arizona Press.

Pomeranz, K. 2006. Without Coal? Colonies? Calculus?: Counterfactuals and Industrialization in Europe and China. In Tetlock, Ph. E., and Parker, G. (eds.), Unmaking the West: 'What-ifs?' Scenarios That Rewrite World History (pp. 241–276). Ann Arbor, MI: University of Michigan Press.

Roberts, A. 1997. Hitler's England: What if Germany Had Invaded Britain in May 1940? In Ferguson, N. (ed.), Virtual History: Alternatives and Counterfactuals (pp. 281–320). London: Picador.

Roberts, A. 2002. Prime Minister Halifax. In Cowley 2002: 279–290.

Rose, E. 2000. The Case of the Missing Carriers. In Cowley 2000: 340.

Scammell, G. V. 1981. The World Encompassed: The First European Maritime Empires, ca. 800–1650. Berkeley, CA: University of California Press.

Schroeder, P. W. 2004. Embedded Counterfactuals and World War I as an Unavoidable War. In Wetzel, D., Jervis, R., and Levy, J. S. (eds.), Schroeder, P. W. Systems, Stability and Statecraft: Essays on the International History of Modern Europe (pp. 158–191). New York: Palgrave.

Shapiro, S. 1998. What is Alternate History? In Dozois, G., and Schmidt, S. (eds.), Roads Not Taken: Tales of Alternate History (pp. xi–xiv). New York: Del Rey.

Somerset, A. 2004. The Spanish Armada Lands in England. In Roberts, A. (ed.), What Might Have Been: Leading Historians on Twelve 'What Ifs' of History (pp. 15–26). London: Weidenfeld and Nicolson.

Taylor, A. J. P. 1972 [1932]. If Archduke Ferdinand Had Not Loved His Wife. In Squire, J. C. (ed.), If It Had Happened Otherwise (pp. 313–320). London: Sidgwick and Jackson.

Tetlock, Ph. E., and Belkin, A. 1996. Counterfactual Thought Experiments in World Politics: Logical, Methodological and Psychological Perspectives. In Tetlock, Ph. E., and Belkin, A. (eds.), Counterfactual Thought Experiments in World Politics: Logical, Methodological, and Psychological Perspectives (pp. 3–38). Princeton, NJ: Princeton University Press.

Tetlock, Ph. E., and Parker, G. 2006. Counterfactual Thought Experiments: Why We Can't Live Without Them and How We Must Learn to Live With Them. In Tetlock, Ph. E., Lebow, R. N., and Parker, G. (eds.), Unmaking the West: 'What-If?' Scenarios That Rewrite World History (pp. 14–44). Ann Arbor, MI: University of Michigan Press.

Thompson, W. R. 2003. A Streetcar Named Sarajevo: Catalysts, Multiple Causation Chains, and Rivalry Structures. International Studies Quarterly 47: 453–474.

Tracy, J. D. 1990. (Ed.). The Rise of Merchant Empires: Long-Distance Trade in Early Modern World, 1350–1750. Cambridge: Cambridge University Press.

Trevelyan, G. 1972 [1932]. If Napoleon Had Won the Battle of Waterloo. In Squire, J. C. (ed.), If It Had Happened Otherwise (pp. 299–312). London: Sidgwick and Jackson.

Tsouras, P. 2001. (Ed.). Rising Sun Victorious: The Alternate History of How the Japanese Won the Pacific War. London: Greenhill.

Tsouras, P. 2002. (Ed.). Third Reich Victorious: Alternate Decisions of World War II. London: Greenhill.

Weatherford, J. 2004. Genghis Khan and the Making of the Modern World. New York: Three Rivers Press.

Weber, S. 1996. Counterfactuals, Past and Future. In Tetlock, Ph. E., and Belkin, A. (eds.), Counterfactual Thought Experiments in World Politics (pp. 268–288). Princeton, NJ: Princeton University Press.

Yates, R. D. S. 2006. The Song Empire: The World's First Superpower. In Tetlock, Ph. E., Lebow, R. N., and Parker, G. (eds.), Unmaking the West: 'What-Ifs?' Scenarios That Rewrite World History (pp. 205–240). Ann Arbor, MI: University of Michigan Press.

Zamoyski, A. 2004. Napoleon Triumphs in Russia. In Roberts, A. (ed.), What Might Have Been: Leading Historians on Twelve 'What Ifs' of History (pp. 79–91). London: Weidenfeld and Nicolson.

Chapter 3

CONTINUITIES AND TRANSFORMATIONS IN THE EVOLUTION OF WORLD-SYSTEMS

Christopher Chase-Dunn

This paper discusses continuities and transformations of systemic logic and modes of accumulation in world historical evolutionary perspective and the prospects for systemic transformation in the next several decades. It also considers the meaning of the recent global financial meltdown by comparing it with earlier debt crises and periods of collapse. Has this been just another debt crisis like the ones that have periodically occurred over the past 200 years, or is it part of the end of capitalism and the transformation to a new and different logic of social reproduction? I consider the contemporary network of global counter-movements and progressive national regimes that are seeking to transform the capitalist world-system into a more humane, sustainable and egalitarian civilization and how the current crisis is affecting the network of counter-movements and regimes, including the Pink Tide populist regimes in Latin America, and the anti-austerity movements. I describe how the New Global Left is similar to, and different from, earlier global counter-movements. The point is to develop a comparative and evolutionary framework that can discern what is really new about the current global situation and that can inform collectively rational responses.

THE COMPARATIVE EVOLUTIONARY
WORLD-SYSTEMS PERSPECTIVE

This paper will employ three different time horizons in the discussion of continuities and transformations.

1. 50,000 years;
2. 5,000 years;
3. 500 years.

Hall and Chase-Dunn (2006; see also Chase-Dunn and Hall 1997)) have modified the concepts developed by the scholars of the modern world-system to construct a theoretical perspective for comparing the modern system with earlier regional world-systems. The main idea is that sociocultural evolution can only be explained if polities are seen to have been in important interaction with each other since the Paleolithic Age. Hall and Chase-Dunn propose a general model of the continuing causes of the evolution of technology and hierarchy within polities and in linked systems of polities (world-systems). This is called the iteration model and it is driven by population pressures interacting with environmental degradation and interpolity conflict. This iteration model depicts basic causal forces that were operating in the Stone Age and that continue to operate in the contemporary global system (see also Chase-Dunn and Hall 1997: Chapter 6; Fletcher *et al.*, 2011). These are the continuities.

The most important idea that comes out of this theoretical perspective is that transformational changes in institutions, social structures and developmental logics are brought about mainly by the actions of individuals and organizations within polities that are semiperipheral relative to the other polities in the same system. This is known as the hypothesis of semiperipheral development.

As regional world-systems became spatially larger and the polities within them grew and became more internally hierarchical, interpolity relations also became more hierarchical because new means of extracting resources from distant peoples were invented. Thus did core/periphery hierarchies emerge. Semiperipherality is the position of some of the polities in a core/periphery hierarchy. Some of the polities that are located in semiperipheral positions became the agents that formed larger chiefdoms, states and empires by means of conquest (semiperipheral marcher polities), and some specialized trading states in between the tributary empires promoted production for exchange in the regions in which they operated. So both the spatial and demographic scale of political organization and the spatial scale of trade networks were expanded by semiperipheral polities, eventually leading to the global system in which we now live.

The modern world-system came into being when a formerly peripheral and then semiperipheral region (Europe) developed an internal core of capitalist states that were eventually able to dominate the polities of all the other regions of the Earth. This Europe-centered system was the first one in which capitalism became the predominant mode of accumulation, though semiperipheral capitalist city-states had existed since the Bronze Age in the spaces between the tributary empires. The Europe-centered system expanded in a series of waves of colonization and incorporation (See Figure 1). Commodification in Europe expanded, evolved and deepened in waves since the thirteenth century, which is why historians disagree about when capitalism became the predominant mode. Since the fifteenth century the modern system has seen four periods of hegemony in which leadership in the development of capitalism was taken to new levels. The first such period was led by a coalition between Genoese finance capitalists and the Portuguese crown (Wallerstein 2011[1974]; Arrighi, 1994). After that the hegemons have been single nation-states: the Dutch in the seventeenth century, the British in the nineteenth century and the United States in the twentieth century (Wallerstein, 1984a). Europe itself, and all four of the modern hegemons, were former semiperipheries that first rose to core status and then to hegemony.

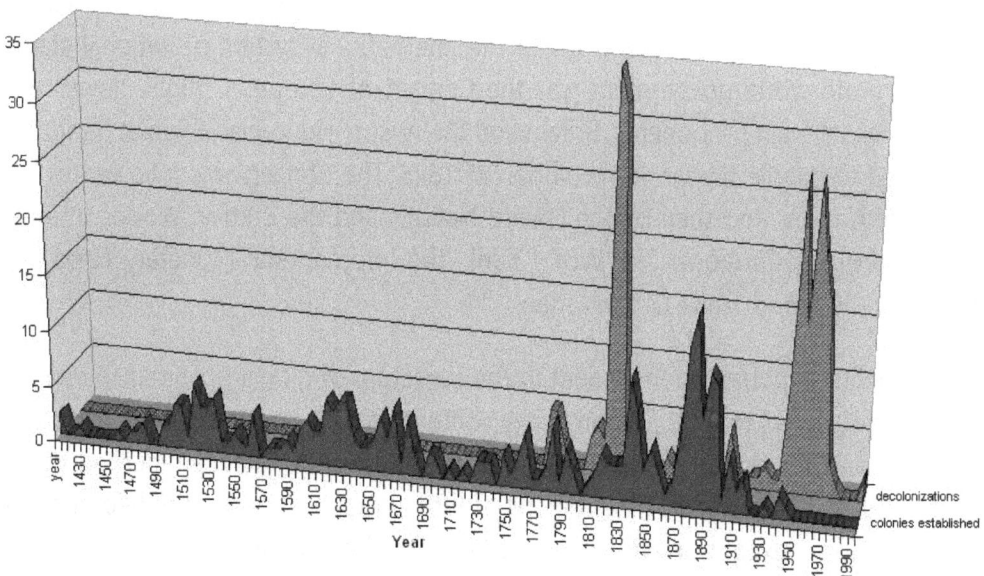

Figure 1 *Waves of Colonization and Decolonization Since 1400—Number of colonies established and number of decolonizations (Source: Henige (1970))*

In between these periods of hegemony were periods of hegemonic rivalry in which several contenders strove for global power. The core of the modern world-system has remained multicentric, meaning that a number of sovereign states ally and compete with one another. Earlier regional world-systems sometimes experienced a period of core-wide empire in which a single empire became so large that there were no serious contenders for predominance. This did not happen in the modern world-system until the United States became the single super-power following the demise of the Soviet Union in 1989.

The sequence of hegemonies can be understood as the evolution of global governance in the modern system. The interstate system as institutionalized at the Treaty of Westphalia in 1644 is still a fundamental institutional structure of the polity of the modern system. The system of theoretically sovereign states was expanded to include the peripheral regions in two large waves of decolonization (see Figure 1), eventually resulting in a situation in which the whole modern system became composed of sovereign national states. East Asia was incorporated into this system in the nineteenth century, though aspects of the earlier East Asian tribute-trade state system were not completely obliterated by that incorporation (Hamashita, 2003).

Each of the hegemonies was larger as a proportion of the whole system than the earlier one had been. And each developed the institutions of economic and political-military control by which it led the larger system such that capitalism increasingly deepened its penetration of all the areas of the Earth. And after the Napoleonic Wars in which Britain finally defeated its main competitor, France, global political institutions began to emerge over the tops of the international system of national states. The first proto-world-government was the Concert of Europe, a fragile flower that wilted when its main proponents, Britain and the Austro-Hungarian Empire, disagreed about how to handle the world revolution of 1848. The Concert was followed by the League of Nations and then by the United Nations and the Bretton Woods international financial institutions (The World Bank, the International Monetary Fund and eventually the World Trade Organization).

The political globalization evident in the trajectory of global governance evolved because the powers that be were in heavy contention with one another for geopolitical power and for economic resources, but also because resistance emerged within the polities of the core and in the regions of the non-core. The series of hegemonies, waves of colonial expansion and decolonization and the emergence of a proto-world-state occurred as the global elites tried to compete with one another and to contain resistance from below. We have already mentioned the waves of decolonization. Other important forces of resistance were slave revolts, the labor movement, the extension of citizenship to men of no property, the women's movement, and other associated rebellions and social movements.

These movements affected the evolution of global governance in part because the rebellions often clustered together in time, forming what have been called "world revolutions" (Arrighi *et al.*, 1989). The Protestant Reformation in Europe was an early instance that played a huge role in the rise of the Dutch hegemony. The French Revolution of 1789 was linked in time with the American and Haitian revolts. The 1848 rebellion in Europe was both synchronous with the Taiping Rebellion in China and was linked with it by the diffusion of ideas, as it was also linked with the emergent Christian Sects in the United States. 1917 was the year of the Bolsheviks in Russia, but also the same decade saw the Chinese Nationalist revolt, the Mexican Revolution, the Arab Revolt and the General Strike in Seattle led by the Industrial Workers of the World in the United States. 1968 was a revolt of students in the U.S., Europe, Latin America and Red Guards in China. 1989 was mainly in the Soviet Union and Eastern Europe, but important lessons about the value of civil rights beyond justification for capitalist democracy were learned by an emergent global civil society.

The current world revolution of 20xx (Chase-Dunn & Niemeyer, 2009) will be discussed as the global countermovement in this paper. The big idea here is that the evolution of capitalism and of global governance is importantly a response to resistance and rebellions from below. This has been true in the past and is likely to continue to be true in the future. Boswell and Chase-Dunn (2000) contend that capitalism and socialism have dialectically interacted with one another in a positive feedback loop similar to a spiral. Labor and socialist movements were obviously a reaction to capitalist industrialization, but also the U.S. hegemony and the post-World War II global institutions were importantly spurred on by the World Revolution of 1917 and the waves of decolonization.

TIME HORIZONS

So what does the comparative and evolutionary world-systems perspective tell us about continuities and transformations of system logic? And what can be said about the most recent financial meltdown and the contemporary global countermovement from the long-run perspectives? Are recent developments just another bout of financial expansion and collapse and hegemonic decline? Or do they constitute or portend a deep structural crisis in the capitalist mode of accumulation. What do recent events signify about the evolution of capitalism and its possible transformation into a different mode of accumulation?

50,000 YEARS

From the perspective of the last 50,000 years the big news is demographic and ecological. After slowly expanding, with cyclical ups and downs in particular regions, for millennia the human population went into a steep upward surge in the last two centuries. Humans have been degrading the environment locally and regionally since they began the intensive use of natural resources. But in the last 200 years of industrial production ecological degradation by means of resource depletion and pollution has become global in scope, with global warming as the biggest consequence. A demographic transition to an equilibrium population size began in the industrialized core countries in the nineteenth century and has spread unevenly to the non-core in the twentieth century. Public health measures have lowered the mortality rate and the education and employment of women outside of the home is lowering the fertility rate. But the total number of humans is likely to keep increasing for several more decades. In the year 2000 there were about six billion humans on Earth. But the time the population stops climbing it will be 8, 10 or 12 billion.

This population big bang was made possible by industrialization and the vastly expanded use of non-renewable fossil fuels. Fossil fuels are captured ancient sunlight that took millions of years to accrete as plants and forests grew, died and were compressed into oil and coal. The arrival of peak oil production is near and energy prices will almost surely rise again after a long fall. The recent financial meltdown is related to these long-run changes in the sense that it was brought on partly by sectors of the global elite trying to protect their privileges and wealth by seeking greater control over natural resources and by over-expanding the financial sector. But non-elites are also implicated. The housing expansion, suburbanization, and larger houses with fewer people in them have been important mechanisms, especially in the United States, for incorporating some of the non-elites into the hegemonic globalization project of corporate capitalism. The culture of consumerism has become strongly ensconced both for those who actually have expanded consumption and as a strong aspiration for those who hope to increase their consumption to the levels of the core.

5,000 YEARS

The main significance of the 5,000-year time horizon is to point us to the rise and decline of modes of accumulation. The story here is that small-scale human polities were integrated primarily by normative structures institutionalized as kinship relations—the so-called kinship-based modes of accumulation. The family was the economy and the polity, and the family was organized as a moral order of obligations that allowed social labor to be mobilized and coordinated, and that regulated distribution. Kin-based accumulation was based on shared languages and meaning systems, consensus-building through oral communication, and institutionalized reciprocity in sharing and exchange. As kin-based polities got larger they increasingly fought with one another and polities that developed institutionalized inequalities had selection advantages over those that did not. Kinship itself became hierarchical within chiefdoms, taking the form of ranked lineages or conical clans. Social movements using religious discourses have been important forces of social change for millennia. Kin-based societies often responded to population pressures on resources by "hiving-off"—a subgroup would emigrate, usually after formulating grievances in terms of violations of the moral order. Migrations were mainly responses to local resource stress caused by population growth and competition for resources. When new unoccupied or only lightly occupied but resource-rich lands were reachable the humans moved on, eventually populating all the continents except Antarctica. Once the land was filled up a situation of "circumscription" raised the level of conflict within and between polities, producing a demographic regulator (Fletcher *et al.*, 2011). In these circumstances technological and organization innovations were stimulated and successful new strategies were strongly selected by interpolity competition, leading to the emergence of complexity, hierarchy and new logics of social reproduction.

Around five thousand years ago the first early states and cities emerged in Mesopotamia over the tops of the kin-based institutions. This was the beginning of the tributary modes of accumulation in which state power (legitimate coercion) became the main organizer of the economy, the mobilizer of labor and the accumulator of wealth and power. Similar innovations occurred largely independently in Egypt, the Yellow (Huang-Ho) river valley, the Indus river valley, and later in Mesoamerica and the Andes. The tributary modes of production evolved as states and empires became larger and as the techniques of imperialism, allowing the exploitation of distant resources, were improved. This was mainly the work of semiperipheral marcher states (Alvarez *et al.*, 2011) Aspects of the tributary modes (taxation, tribute-gathering, accumulation by dispossession) are still with us, but they have been largely subsumed and made subservient to the logic of capitalist accumulation. Crises and countermovements were often involved in the wars and conquests that brought about social change and evolution of the tributary modes.

A tributary mode became predominant in the Mesopotamian world-system in the early Bronze Age (around 3000 BCE). The East Asian regional world-system was still predominantly tributary in the nineteenth century CE. That is nearly a 5,000-year run. The kin-based mode lasted even longer. All human groups were organized around different versions of the kin-based modes in the Paleolithic, and indeed since human culture first emerged with language. If we date the beginning of the end of the kin-based modes at the coming to predominance of the tributary mode in Mesopotamia (3000 BCE) this first qualitative change in the basic logic of social reproduction took more than 100,000 years.

500 YEARS

This brings us to the capitalist mode, here defined as based on the accumulation of profits returning to commodity production rather than taxation or tribute. As we have already said, early forms of capitalism emerged in the Bronze Age in the form of small semiperipheral states that specialized in trade and the production of commodities. But it was not until the fifteenth century that this form of accumulation became predominant in a regional world-system (Europe and its colonies). Capitalism was born in the semiperiphery but in Europe it moved to the core, and the forereachers that further evolved capitalism were former semiperipheral polities that rose to hegemony. Economic crises and world revolutions have been important elements in the evolution of capitalism and global governance institutions for centuries.

Thus, in comparison with the earlier modes, capitalism is yet young. It has been around for millennia, but it has been predominate in a world-system for less than a millennium. On the other hand, many have observed that social change in general has speeded up. The rise of tribute-taking based on institutionalized coercion took more than 100,000 years. Capitalism itself speeds up social change because it revolutionizes technology so quickly that other institutions are brought along, and people have become adjusted to more rapid reconfigurations of culture and institutions. So it is plausible that the contradictions of capitalism may lead it to reach its limits much faster than the kin-based and tributary modes did.

TRANSFORMATIONS BETWEEN MODES

For Immanuel Wallerstein (2011 [1974]), capitalism started in the long sixteenth century (1450-1640), grew larger in a series of cycles and upward trends, and is now nearing "asymptotes" (ceilings) as some of its trends create problems that it cannot solve. Thus, for Wallerstein the world-system became capitalist and then it expanded until it became completely global, and now it is coming to face a big crisis because certain long-term trends cannot be accommodated within the logic of capitalism (Wallerstein, 2003). Wallerstein's evolutionary transformations come at the beginning and at the end. In there is a focus on expansion and deepening as well as cycles and trends, but no periodization of world-system evolutionary stages of capitalism (Chase-Dunn 1998: Chapter 3). This is very different from both Arrighi's depiction of successive (and overlapping) systemic cycles of accumulation and from the older Marxist stage theories of national development. Wallerstein's emphasis is on the emergence and demise of "historical systems" with capitalism defined as "ceaseless accumulation." Some of the actors change positions but the system is basically the same as it gets larger. Its internal contradictions will eventually reach limits, and these limits are thought to be approaching within the next five decades.

According to Wallerstein (2003) the three long-term upward trends (ceiling effects) that capitalism cannot manage are:

1. The long-term rise of real wages;
2. The long-term costs of material inputs; and
3. Rising taxes.

All three upward trends cause the average rate of profit to fall. Capitalists devise strategies for combating these trends (automation, capital flight, job blackmail, attacks on the welfare state and unions), but they cannot really stop them in the long run. Deindustrialization in one place leads to industrialization and the emergence of labor movements somewhere else (Silver, 2003). The falling rate of profit means that capitalism as a logic of accumulation will face an irreconcilable structural crisis during the next 50 years, and some other system will emerge. Wallerstein calls the next five decades "The Age of Transition."

Wallerstein sees recent losses by labor unions and the poor as temporary. He assumes that workers will eventually figure out how to protect themselves against globalized market forces and the "race to the bottom". This may underestimate somewhat the difficulties of mobilizing effective labor organization in the era of globalized capitalism, but he is probably right in the long run. Global unions and political parties could give workers effective instruments for protecting their wages and working conditions from exploitation by global corporations if the North/South issues that divide workers could be overcome.

Wallerstein is intentionally vague about the organizational nature new system that will replace capitalism (as was Marx) except that he is certain that it will no longer be capitalism. He sees the declining hegemony of the United States and the crisis of neoliberal global capitalism as strong signs that capitalism can no longer adjust to its systemic contradictions. He contends that world history has now entered a period of chaotic and unpredictable historical transformation. Out of this period of chaos a new and qualitatively different non-capitalist system will emerge. It might be an authoritarian (tributary) global state that preserves the privileges of the global elite or it could be an egalitarian system in which non-profit institutions serve communities (Wallerstein, 1998).

STAGES OF WORLD CAPITALIST DEVELOPMENT: SYSTEMIC CYCLES OF ACCUMULATION

Giovanni Arrighi's (1994) evolutionary account of "systemic cycles of accumulation" has solved some of the problems of Wallerstein's notion that world capitalism started in the long sixteenth century and then went through repetitive cycles and trends. Arrighi's account is explicitly evolutionary, but rather than positing "stages of capitalism" and looking for each country to go through them (as most of the older Marxists did), he posits somewhat overlapping global cycles of accumulation in which finance capital and state power take on new forms and increasingly penetrate the whole system. This was a big improvement over both Wallerstein's world cycles and trends and the traditional Marxist national stages of capitalism approach.

Arrighi's (1994, 2006) "systemic cycles of accumulation" are more different from one another than are Wallerstein's cycles of expansion and contraction and upward secular trends. And Arrighi (2006) has made more out of the differences between the current period of U.S. hegemonic decline and the decades at the end of the nineteenth century and the early twentieth century when British hegemony was declining. The emphasis is less on the beginning and the end of the capitalist world-system and more on the evolution of new institutional forms of accumulation and the increasing incorporation of modes of control into the logic of capitalism. Arrighi (2006), taking a cue from Andre Gunder Frank (1998), saw the rise of China as portending a new systemic cycle of accumulation in which "market society" will eventually come to replace rapacious finance capital as the leading institutional form in the next phase of world history. Arrighi does not discuss the end of capitalism and the emergence of another basic logic of social reproduction and accumulation. His analysis is more in line with the "types of capitalism" and "multiple modernities" literature except that he is analyzing the whole system rather than separate national societies.

Arrighi sees the development of market society in China as a consequence of the differences between the East Asian and Europe-centered systems before their merger in the 19th century, and as an outcome of the Chinese Revolution. His discussion of Adam Smith's notions of societal control over finance capital is interesting, but he is vague as what the forces that can counter-balance the power of finance capital. In China it is obviously the Communist Party and the new class of technocratic mandarins. This is somewhat similar in form to Peter Evans's discussion of the importance of technocrats in Brazilian, Japanese and Korean national development, though Arrighi does not say so.

Arrighi also provides a more explicit analysis of how the current world situation is similar to and different from the period of declining British hegemonic power before World War I (see summary in Chase-Dunn & Lawrence, 2011: 147-151).

Wallerstein's version is more apocalyptic and more millenarian. The old world is ending. The new world is beginning. In the coming systemic bifurcation what people do may be prefigurative and causal of the world to come. Wallerstein agrees with the analysis proposed by the students of the New Left in 1968 (and large numbers of activists in the current global justice movement) that the tactic of taking state power has been shown to be futile because of the disappointing outcomes of the World Revolution of 1917 and the decolonization movements (but see below).

ECONOMIC GLOBALIZATION

Regarding the issue of whether or not the recent meltdown is itself a structural crisis or the beginning of a long process of transformation, it is relevant to examine recent trends in economic globalization. Is there yet any sign that the world economy has entered a new period of deglobalization of the kind that occurred in the first half of the twentieth century?

Immanuel Wallerstein contends that globalization has been occurring for five hundred years, and so there is little that is importantly new about the so-called stage of global capitalism that is alleged to have emerged in the last decades of the twen-

tieth century. Well before the emergence of globalization in the popular consciousness, the world-systems perspective focused on the world economy and the system of
interacting polities, rather than on single national societies. Globalization, in the sense
of the expansion and intensification of larger and larger economic, political, military
and information networks, has been increasing for millennia, albeit unevenly and in
waves. And globalization is as much a cycle as a trend (see Figure 2). The wave of
global integration that has swept the world in the decades since World War II is best
understood by studying its similarities and differences with the waves of international
trade and foreign investment expansion that have occurred in earlier centuries, especially the last half of the nineteenth century.

Wallerstein has insisted that U.S. hegemony is continuing to decline. He interpreted the U.S. unilateralism of the Bush administration as a repetition of the mistakes of
earlier declining hegemons that attempted to substitute military superiority for economic comparative advantage (Wallerstein, 2003). Most of those who denied the notion of U.S. hegemonic decline during what Giovanni Arrighi (1994) called the "belle
epoch" of financialization have now come around to Wallerstein's position in the wake
of the current global financial crisis. Wallerstein contends that once the world-system

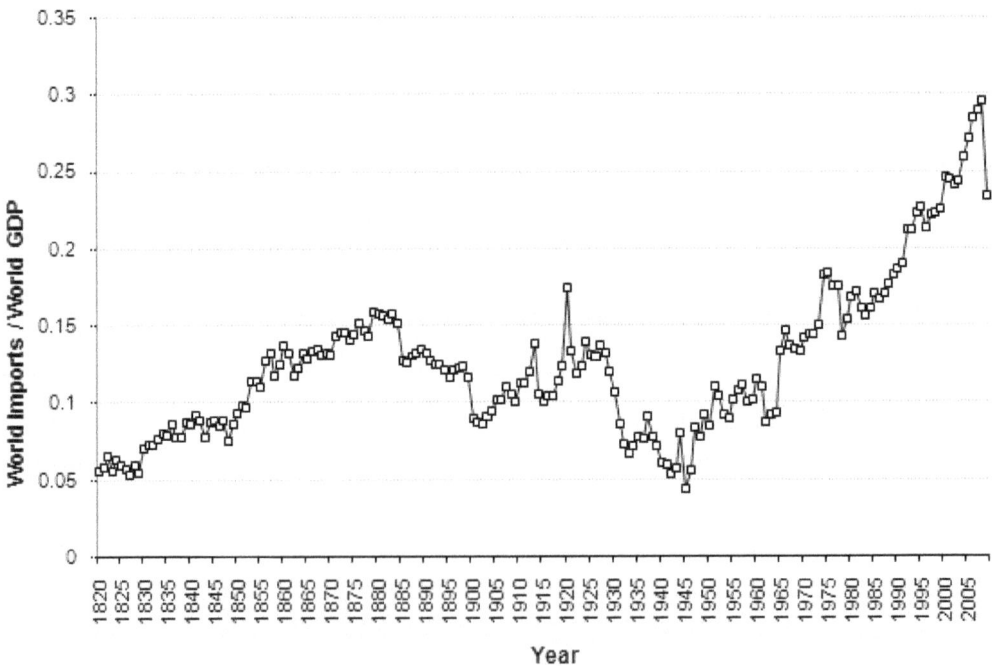

Figure 2 *Trade Globalization 1820–2009: World Imports as a Percentage of World
GDP (Sources: Chase-Dunn et al. (2000); World Bank (2011)).*

cycles and trends, and the game of musical chairs that is capitalist uneven development, are taken into account, the "new stage of global capitalism" does not seem that different from earlier periods.

Figure 2 is an updated version of the trade globalization series published in Chase-Dunn *et al.*, (2000). It shows the great nineteenth century wave of global trade integration, a short and volatile wave between 1900 and 1929, and the post-1945 upswing that is characterized as the "stage of global capitalism." The figure indicates that globalization is both a cycle and a bumpy trend. There have been significant periods of deglobalization in the late nineteenth century and in the first half of the twentieth century. Note the steep decline in the level of global trade integration in 2009.

The long-term upward trend has been bumpy, with occasional downturns such as the one shown in the 1970s. But the downturns since 1945 have all been followed by upturns that restored the overall upward trend of trade globalization. The large decrease of trade globalization in the wake of the global financial meltdown of 2008 represents a 21% decrease from the previous year, the largest reversal in trade globalization since World War II. The question is whether or not this sharp decrease represents a reversal in the long upward trend observed over the past half century. Is this the beginning of another period of deglobalization?

The Financial Meltdown of 2007-2008

The recent financial crisis has generated a huge scholarly literature and immense popular reflection about its causes and its meaning for the past and for the future of world society. This contribution is intended to place the current crisis, and the contemporary network of transnational social movements and progressive national regimes, in world historical and evolutionary perspective. The main point is to accurately determine the similarities and differences between the current crisis and responses with earlier periods of dislocation and breakdown in the modern world-system and in earlier world-systems.

This analysis is reported in Chase-Dunn and Kwon (2011). The conclusions are that financial crises are business as usual for the capitalist world-economy. The theories of a "new economy" and "network society" were mainly justifications for financialization. The big difference is the size of the bubble and the greater dependence of the rest of the world on the huge U.S. economy and the U.S. dollar sector. The somewhat successful reinflating of the global financial bubble by the government funded bail-out of Wall Street has not resolved the basic structural problems, but it has avoided (so far) a true collapse, deflation, and a wiping out of the bloated mass of paper securities that constitute the financial bubble. This is not a stable situation, but neither is it the end of capitalism.

The World Revolution of 20xx

The contemporary world revolution is similar to earlier ones, but also different. Our conceptualization of the New Global Left includes civil society entities: individuals, social movement organizations, non-governmental organizations (NGOs), but also political parties and progressive national regimes. In this chapter we will focus mainly on the relationships among the movements and the progressive populist regimes that have emerged in Latin America in the last decade and on the Arab Spring that began in Tunisia in December of 2010. We understand the Latin American "Pink Tide" regimes to be an important part of the New Global Left, though it is well-known that the relationships among the movements and the regimes are both supportive and contentious.

The boundaries of the progressive forces that have come together in the New Global Left are fuzzy and the process of inclusion and exclusion is ongoing (Santos, 2006). The rules of inclusion and exclusion that are contained in the Charter of the World Social Forum, though still debated, have not changed much since their formulation in 2001.[1]

The New Global Left has emerged as resistance to, and a critique of, global capitalism (Lindholm and Zuquete, 2010). It is a coalition of social movements that includes recent incarnations of the old social movements that emerged in the nineteenth century (labor, anarchism, socialism, communism, feminism, environmentalism, peace, human rights) and movements that emerged in the world revolutions of 1968 and 1989 (queer rights, anti-corporate, fair trade, indigenous) and even more recent movements such as the slow food/food rights, global justice/alter-globalization, anti-globalization, health-HIV and alternative media (Reese *et al.*, 2008). The explicit focus on the Global South and global justice is somewhat similar to some earlier instances of the Global Left, especially the Communist International, the Bandung Conference and the anticolonial movements. The New Global Left contains remnants and reconfigured elements of earlier Global Lefts, but it is a qualitatively different constellation of forces because:

---

1. The Transnational Social Movement Research Working Group at the University of California-Riverside has studied the movements participating in the World Social Forum since 2005. The project web page is at http://www.irows.ucr.edu/research/tsmstudy.htm.

1. There are new elements;
2. The old movements have been reshaped, and;
3. A new technology (the Internet) is being used to mobilize protests in real time and to try to resolve North/South issues within movements and contradictions among movements.

There has also been a learning process in which the earlier successes and failures of the Global Left are being taken into account in order to not repeat the mistakes of the past. Many social movements have reacted to the neoliberal globalization project by going transnational to meet the challenges that are obviously not local or national (Reitan, 2007). But some movements, especially those composing the Arab Spring, are focused mainly on regime change at home. The relations within the family of antisystemic movements and among the Latin American Pink Tide populist regimes are both cooperative and competitive. The issues that divide potential allies need to be brought out into the open and analyzed in order that cooperative efforts may be enhanced and progressive global collective action may become more effective.

The Pink Tide

The World Social Forum (WSF) is not the only political force that demonstrates the rise of the New Global Left. The WSF is embedded within a larger socio-historical context that is challenging the hegemony of global capital. It was this larger context that facilitated the founding of the WSF in 2001. The anti-IMF protests of the 1980s and the Zapatista rebellion of 1994 were early harbingers of the current world revolution that challenged the neoliberal capitalist order. And the World Social Forum was founded explicitly as a counter-hegemonic project vis-à-vis the World Economic Forum (an annual gathering of global elites founded in 1971).

World history has proceeded in a series of waves. Capitalist expansions have ebbed and flowed, and egalitarian and humanistic countermovements have emerged in a cyclical dialectical struggle. Polanyi (1944) called this the double-movement, while others have termed it a "spiral of capitalism and socialism." This spiral of capitalism and socialism describes the undulations of the global economy that have alternated between expansive commodification throughout the global economy, followed by resistance movements on behalf of workers and other oppressed groups (Boswell & Chase-Dunn, 2000). The Reagan/Thatcher neoliberal capitalist globalization project extended the power of transnational capital. This project has reached its ideological and material limits. It has increased inequality within some countries, exacerbated rapid urbanization in the Global South (so-called Planet of Slums [Davis, 2006]), at-tacked the welfare state and institutional protections for the poor, and led to global financial crisis.

A global network of countermovements has arisen to challenge neoliberalism, neoconservatism and corporate capitalism in general. This progressive network is composed of increasingly transnational social movements as well as a growing num-ber of populist government in Latin America—the so-called Pink Tide. The Pink Tide is composed of populist leftist regimes that have come to state power in Latin Amer-ica, some of which advocate dramatic structural transformation of the global political economy and world civilization.

An important difference between these and many earlier Leftist regimes in the non-core is that they have come to head up governments by means of popular elec-tions rather than by violent revolutions. This signifies an important difference from earlier world revolutions. The spread of electoral democracy to the non-core has been part of a larger political incorporation of former colonies into the European interstate system. This evolutionary development of the global political system has mainly been caused by the industrialization of the non-core and the growing size of the urban working class in non-core countries (Silver, 2003). While much of the "democratiza-tion" of the Global South has consisted mainly of the emergence of "polyarchy" in which elites manipulate elections in order to stay in control of the state (Robinson, 1996), in some countries the Pink Tide Leftist regimes have been voted into power. This is a very different form of regime formation than the road taken by earlier Left-ist regimes in the non-core. With a few exceptions earlier Left regimes came to state power by means of civil war or military coup.

The ideologies of the Latin American Pink Tide regimes have been both socialist and indigenist, with different mixes in different countries. The acknowledged leader of the Pink Tide as a distinctive brand of leftist populism is the Bolivarian Revolution led by Venezuelan President Hugo Chavez. But various other forms of progressive politi-cal ideologies are also heading up states in Latin America. Indigenist and socialist Evo Morales is president of Bolivia. The Fidelistas in Cuba remain in power. The Brazilian Workers' Party is still an important player, though its elected presidents have been pragmatic politicians rather than revolutionary leaders. In Chile social democrats are in power. Sandinistas in Nicaragua and the FMLN in El Salvador have elected national leaders. Argentina bravely and unilaterally restructured its own debt obligations in 2005. The President of Peru is a leftist. And several European-style social democrats lead some of the Caribbean islands.

Most of these regimes are supported by the mobilization of historically subordinate populations including the indigenous, poor, and women. The rise of the voiceless and the challenge to neoliberal capitalism seemed to have its epicenter in Latin America before the emergence of the Arab Spring. While there are important differences of emphasis among these Latin American regimes, they have much in common, and as a whole they constitute an important bloc of the New Global Left. We agree with William I. Robinson's (2008) assessment of the Bolivarian Revolution and its potential to lead the global working class in a renewed challenge to transnational capitalism.

The rise of the left has engulfed nearly all of South America and a considerable portion of Central America and the Caribbean. Why has Latin America been the site of both populist Leftist regimes and most of the transnational social movements that contest neoliberal capitalist globalization up until recently? We suggest that part of the explanation is that Latin America as a world region has so many semiperipheral countries. These countries have more options to pursue independent strategies than the mainly peripheral countries of Africa do. But some of the Pink Tide countries in Latin America are also peripheral. There has been a regional effect that did not seem to be operating in either Africa or Asia. The Pink Tide phenomenon and the anti-neoliberal social movements may have been concentrated in Latin America because the foremost proponent of the neoliberal policies has been the United States. Latin America has long been the neocolonial "backyard" of the United States. Most of the people of Latin America think of the United States as the "colossus of the North." The U.S. has been the titular hegemon during the period of the capitalist globalization project, and so the political challenge to neoliberalism has been strongest in that region of the world. Both Africa and Asia have a more complicated relationship with former colonial powers and with the U.S. hegemony.

President Hugo Chavez of Venezuela is perhaps the most vocal advocate of an alternative to global capitalism, and his advocacy is greatly aided by the massive Venezuelan oil reserves. The Banco del Sur (Bank of the South) that Chavez has founded, for example, has been joined by many Pink Tide nations and seeks to replace the International Monetary Fund and the World Bank in sponsoring development projects throughout the Americas. The goal is to become independent of the capitalist financial institutions headquartered in the Global North.

The early Structural Adjustment Programs imposed by the International Monetary Fund in Latin America in the 1980s were instances of "shock therapy" that emboldened domestic neoliberals to attack the "welfare state," unions and workers parties. In many countries these attacks resulted in downsizing and streamlining of urban industries, and workers in the formal sector lost their jobs and were forced into the informal economy, swelling the "planet of slums" (Davis, 2006). This is the formation of a globalized working class as described by Bill Robinson (2008). In several countries the swollen urban informal sector was mobilized by political leaders into new populist movements and parties, and in some of these, the movements were eventually successful in electing their leaders to national power, creating the Pink Tide regimes. Thus did neoliberal Structural Adjustment Programs provoke counter-movements that eventuated in the Pink Tide regimes.

The very existence of the World Social Forum owes much to the Pink Tide regime in Brazil. The Brazilian transition from authoritarian rule in the 1980s politicized and mobilized civil society, contributing to the elections of leftist presidents. One of these was Fernando Henrique Cardoso, a famous Brazilian sociologist who was one of the founders of dependency theory. The Brazilian city of Porto Alegre, where the first World Social Forum meetings were held, had been a stronghold for the Brazilian Workers' Party. The World Social Forum was born in Porto Alegre with indispensable help from the Brazilian Workers' Party and its former leader who had been elected President of Brazil, Luis Inàcio da Silva. The political trend of the Pink Tide was an important element in context and conditions that allowed for the rise of the World Social Forum.

The relations between the progressive transnational social movements and the regimes of the Pink Tide have been both collaborative and contentious. We have already noted the important role played by the Brazilian Workers' Party in the creation of the World Social Forum. But many of the activists in the movements see involvement in struggles to gain and maintain power in existing states as a trap that is likely to simply reproduce the injustices of the past. These kinds of concerns have been raised by anarchists since the nineteenth century, but autonomists from Italy, Spain, Germany and France now echo these concerns. And the Zapatista movement in Southern Mexico, one of the sparks that ignited the global justice movement against neoliberal capitalism, has steadfastly refused to participate in Mexican electoral politics. Indeed the New Left led by students in the World Revolution of 1968 championed a similar critical approach to the old parties and states of the Left as well as involvement in electoral politics. As mentioned above, Immanuel Wallerstein (1984b, 2003) agrees with this antistatist political stance. This antipolitics-as-usual has become embodied in the Charter of the World Social Forum, where representatives of parties and governments are theoretically proscribed from sending representatives to the WSF meetings.[2]

---

2. The charter of the World Social Forum does not permit participation by those who attend as representatives of organizations that are engaged in, or that advocate, armed struggle. Nor are govern-

The older Leftist organizations and movements are often depicted as hopelessly Eurocentric and undemocratic by the neo-anarchists and autonomists, who instead prefer participatory and horizontalist network forms of democracy and eschew leadership by prominent intellectuals as well as by existing heads of state. Thus when Lula, Chavez and Morales have tried to participate in the WSF, crowds have gathered to protest their presence. The organizers of the WSF have found various compromises, such as locating the speeches of Pink Tide politicians at adjacent, but separate, venues. An exception to this kind of contention is the support that European autonomists and anarchists have provided to Evo Morales's regime in Bolivia (e.g., López & Iglesias Turrión, 2006).

Latin America has been the epicenter of the contemporary world revolution. If the movements and the progressive regimes could work together this would be an energizing model for the other regions of the globe. The challenges are daunting but the majority of humankind needs organizational instruments with which to democratize global governance and the World Social Forum has been designed to be the venue from which such instruments could be organized.

The Meltdown and the Countermovements

What have been the effects of the global financial meltdown on the counter-movements and the progressive national regimes? The World Social Forum slogan that "Another World Is Possible" seems far more appealing now than when the capitalist globalization project was booming. Critical discourse has been taken more seriously by a broader audience. Marxist geographer, David Harvey, has been interviewed on the BBC. The millenarian discourses of the Pink Tide regimes and the radical social movements seem to be at least partly confirmed. The "end of history" triumphalism and theories of the "new economy" seem to have been swept into the dustbin. The world-systems perspective has found greater support, at least among earlier critics such as the more traditional Marxists. The insistence of Wallerstein, Arrighi, and others that U.S. hegemony is in long-term decline has now found wide acceptance.

On a more practical level, most of the social movement organizations and NGOs have had more difficulty raising money, but this has been counterbalanced by increased participation (Allison *et al.*, 2011). The environmental movement has received some setbacks because the issue of high unemployment has come to the fore. The Copenhagen summit was largely understood to have been a failure. The wide real-

ments, confessional institutions or political parties supposed to send representatives to the WSF. See World Social Forum Charter http://wsf2007.org/process/wsf-charter/

ization that energy costs are going to go further up has increased the numbers who support the further development of nuclear energy, despite its long-run environmental costs. But the Japanese earthquake and nuclear meltdown has led to the declaration of a non-nuclear future by the German government. And the radical alternative of indigenous environmentalism has gotten a boost (Wallerstein, 2010). The World People's Conference on Climate Change and the Rights of Mother Earth, held in Cochabamba, Bolivia in April of 2010, discussed a Universal Declaration of the Rights of Mother Earth, a World People's Referendum on Climate Change, and the establishment of a Climate Justice Tribunal. The meeting was attended by 30,000 activists from more than 100 countries, and was financially supported by the governments of Bolivia and Venezuela.

The Arab Spring

The movements that have swept the Arab world since December of 2010 are also part of the world revolution of 20xx and they may play a role in the New Global Left. As in earlier world revolutions, contagion and new technologies of communication have been important elements. And as in earlier world revolutions, rather different movements stimulated by different local conditions converge in time to challenge the powers that be. The Arab Spring movements have been rather different from the global justice movements. Their targets have mainly been authoritarian national regimes rather than global capitalism. Youthful demonstrators have used Facebook to organize mainly peaceful protests that have succeeded in causing several old entrenched regimes to step down. The countries in which these movements have succeeded are not the poorest countries in Africa and the Middle East. Rather they have been semi-peripheral countries in which a large mobilizable group of young people have access to social media. In many cases the old autocrats had been trying to implement austerity programs in order to be able to borrow more money from abroad and this set the stage for the mass movements. But the Arab Spring movements have not explicitly raised the issues of austerity and global financial dependency.[3]

The issues raised by the Arab Spring movements have mainly been about national democracy, not global justice. But the example of masses of young people rallying against unpopular regimes now seems to be spreading to the second-tier core states of Europe. Both Spain and Greece have seen large anti-austerity demonstrations that have been inspired by the successes of the Arab Spring. And in these cases the connection with the global financial crisis is even more palpable. The austerity programs

---

3. The NATO intervention in Libya mainly illustrates the illegitimacy of both Khadafy and of the nascent global state that is seeking to depose him.

are the conditions imposed by global finance capital for reinflating the accumulation structures of these countries of the European second-tier core. The popular anti-austerity rebellions might provoke an even deeper financial collapse if investors and their institutional agents lose faith in the ability of the system to reproduce the existing structures of accumulation. And anti-austerity movements have also spread to the core states, where severe fiscal crises have led to the dismantling of public services.

CONCLUSIONS

So do recent developments constitute the beginning of the terminal crisis of capitalism or another systemic cycle of accumulation. As mentioned above, predominant capitalism has not been around very long from the point of view of the succession of qualitatively different logics of social reproduction. But capitalism itself has speeded up social change and its contradictions do seem to be reaching levels that cannot be fixed. Declarations of imminent transformation may be useful for mobilizing social movements, but the real problem is the clearly specify what is really wrong with capitalism and how these deficiencies can be fixed. Whether or not we are in the midst of a qualitative transformation this task will need to be accomplished.

Regarding a new systemic cycle of accumulation, Arrighi's bet on the significance of the rise of China also needs clarification. As he has said, other countries have not experienced the trajectory that produced "market society" in China, so how can forces emerge elsewhere that can counter-balance the power of national and global finance capital. And what kind of forces could do this?

The rise of the anti-austerity movements in Spain and Greece and the Occupy Wall Street movement in the U.S. may portend the emergence of strong and effective anti-capitalist social movements in the core. The Occupy and anti-austerity movements interestingly borrowed tactics from the Arab Spring, including the use of Facebook for organizing revolt and camping in central public spaces. The Occupy movement may improve President Obama's chances of re-election and might also inspire his administration to more energetically push for re-industrialization of the U.S. This could slow or even reverse the U.S. economic decline. But the movements and the regime would have to overcome the still-strong legacy of Reaganism-Thatcherism, the political muscle of Wall Street and the Tea Party right-wing populists who call the Black President a Muslim and a socialist. Continued political stalemate in the U.S. is the most likely outcome, and this will result in the continued slow decline of U.S. hegemony. This is not surprising from the point of view of world-systemic cycles of hegemonic rise and fall.

But things seem more interesting in the semiperiphery and the Global South. So far the U.S. has not used much muscle in opposition to the rise of the Pink Tide in Latin America. Expensive U.S. military involvements in the Middle East and Central Asia have continued, and these may partly explain the relative inaction in Latin America. Can the progressive transnational social movements and the left populist regimes of the Pink Tide forge a coalition that can move toward greater global democracy? Could the emergent democratic regimes in the Arab world and protests against the austerity imposed by finance capital in the European second-tier core lead to a situation in which a strong force for global social democracy would challenge the powers that be? As in earlier world revolutions the institutions of global governance are likely to be reshaped by forces from below. Hopefully a more democratic and collectively rational global commonwealth can emerge without the violence and totalitarianism that was so prevalent in the first half of the twentieth century.

ACKNOWLEDGEMENT

Thanks to Roy Kwon and Kirk Lawrence for help with this paper.

REFERENCES

Allison, Juliann, Ian Breckenridge-Jackson, Katja M. Guenther, Ali Lairy, Elizabeth Schwarz, Hellen Reese, Miryam E. Ruvalcaba and Michael Walker 2011 "Is the economic crisis a crisis for social justice activism/" Policy Matters 5,1 (Spring) http://policymatters.ucr.eduAlexis

Alvarez, Hiroko Inoue, Kirk Lawrence, Evelyn Courtney, Edwin Elias, Tony Roberts and Christopher Chase-Dunn 2011 "Semiperipheral Development and Empire Upsweeps Since the Bronze Age" IROWS Working Paper #56 http://irows.ucr.edu/papers/irows56/irows56.htm

Amin, Samir. 1997. Capitalism in an Age of Globalization. London: Zed Books.

Arrighi, Giovanni. 1994. The Long Twentieth Century. London: Verso.

Arrighi, Giovanni. 2006. Adam Smith in Beijing. London: Verso

Arrighi, Giovanni and Terence K. Hopkins, and Immanuel Wallerstein. 1989. Antisystemic Movements. London: Verso.

Arrighi, Giovanni. and Beverly Silver. 1999. Chaos and Governance in the Modern World-System: Comparing Hegemonic Transitions. Minneapolis: University of Minnesota Press.

Barbosa, Luis C. and Thomas D. Hall 1985. "Brazilian slavery and the world economy" Western Sociological Review 15,1: 99-119.

Bergesen, Albert and Ronald Schoenberg. 1980. "Long waves of colonial expansion and contraction 1415-1969" Pp. 231-278 in Albert Bergesen (ed.) Studies of the Modern World-System. New York: Academic Press

Bornschier, Volker. 2010. "On the evolution of inequality in the world system" in Christian Suter (ed.) Inequality Beyond Globalization: Economic Changes. Social Transformations and the Dynamics of Inequality. Berlin: World Society Studies.

Bornschier, Volker and Christopher Chase-Dunn (eds.) The Future of Global Conflict. London: Sage

Boserup, Ester 1981. Population and Technological Change. Chicago: University of Chicago Press.

Boswell, Terry and Christopher Chase-Dunn 2000. The Spiral of Capitalism and Socialism: Toward Global Democracy. Boulder, CO.: Lynne Rienner.

Brenner, I Robert 2002. The Boom and the Bubble: The U.S. in the World Economy. London: Verso.

Bunker, Stephen and Paul Ciccantell 2005. Globalization and the Race for Resources. Baltimore: Johns Hopkins University Press.

Carroll, William K. 2010. The Making of a Transnational Capitalist Class. London: Zed Press.

Chase-Dunn, Christopher 1998. Global Formation: Structures of the World- Economy. Lanham, MD: Rowman and Littlefield.

Chase-Dunn, Christopher. Orfalea Lecture on the evolution of societal organization. http://www.youtube.com/watch?v=FxNgOkU6NzY

Chase-Dunn, C. and Salvatore Babones (eds.) Global Social Change: Historical and Comparative Perspectives. Baltimore, MD: Johns Hopkins University Press.

Chase-Dunn, C. and Terry Boswell 2009. "Semiperipheral development and global democracy" in Phoebe Moore and Owen Worth, Globalization and the Semiperiphery: New York: Palgrave MacMillan.

Chase-Dunn, C. Yukio Kawano and Benjamin Brewer 2000. "Trade globalization since 1795: waves of integration in the world-system" American Sociological Review 65, 1: 77-95.

Chase-Dunn, Christopher and Thomas D. Hall 1997. Rise and Demise: Comparing World-Systems. Boulder, CO: Westview.

Chase-Dunn, C. Roy Kwon, Kirk Lawrence and Hiroko Inoue 2011. "Last of the hegemons: U.S. decline and global governance" International Review of Modern Sociology 37,1:1-29

Chase-Dunn C. and Roy Kwon 2011. "Crises and Counter-Movements in World Evolutionary Perspective" in Christian Suter and Mark Herkenrath (Eds.) The Global Economic Crisis: Perceptions and Impacts (World Society Studies 2011). Wien/Berlin/Zürich: LIT Verlag

Chase-Dunn, C. and Kirk Lawrence 2011. "The Next Three Futures, Part One: Looming Crises of Global Inequality, Edcological Degradation, and a Failed System of Global Governance" Global Society 25,2:137-153 http://irows.ucr.edu/papers/irows47/irows47.htm

Chase-Dunn C. and Ellen Reese 2007. "Global political party formation in world historical perspective" in Katarina Sehm-Patomaki & Marko Ulvila eds. Global Political Parties. London: Zed Press.

Chase-Dunn, C. and R.E. Niemeyer 2009 "The world revolution of 20xx" Pp. 35-57 in Mathias, Albert, Gesa Bluhm, Han Helmig, Andreas Leutzsch, Jochen Walter (eds.) Transnational Political Spaces. Campus Verlag: Frankfurt/New York

Chase-Dunn, C. and Bruce Lerro 2008 "Democratizing Global Governance: Strategy and Tactics in Evolutionary Perspective" IROWS Working Paper #40

Chase-Dunn, Christopher and Bruce Podobnik 1995 "The Next World War: World-System Cycles and Trends" Journal of World-Systems Research,Volume 1, Number 6, http://jwsr. ucr.edu/archive/vol1/v1_n6.php

Cioffi, John W. 2010 "The Global Financial Crisis: Conflicts of Interest, Regulatory Failures, and Politics" Policy Matters 4,1 http://policymatters.ucr.edu/

Collins, Randall 2010 "Geopolitical conditions of internationalism, human rights and world law" Journal of Globalization Studies 1,1: 29-45.

Davis, Mike 2006 Planet of Slums. London: Verso

Dymski, Gary A. 2009 "Sources & Impacts of the Subprime Crisis: From Financial Exploitation to Global Economic Meltdown" UCR Program on Global Studies University of California-Riverside, 2008-2009 Colloquium Series Tuesday, January 13

Evans, Peter B. 1979 Dependent Development: the alliance of multinational, state and local capital in Brazil. Princeton Unviersity Press.

Fletcher, Jesse B; Apkarian, Jacob; Hanneman, Robert A; Inoue, Hiroko; Lawrence, Kirk; Chase-Dunn, Christopher. 2011"Demographic Regulators in Small-Scale World-Systems" Structure and Dynamics 5, 1Frank, Andre Gunder 1998 Reorient:: Global Economy in the Asian Age. Berkeley: University of California Press

Goldfrank, Walter L. 1999 "Beyond hegemony" in Volker Bornschier and Christopher Chase-Dunn (eds.) The Future of Global Conflict. London: Sage.

Goldstone, Jack A. 1991 Revolution and Rebellion in the Early Modern World. Berkeley: University of California Press.

Gramsci, Antonio 1971 Selections for the Prison Notebooks. New York: International Publishers

Grinen, Leonid and Andrey Kortayev 2010 "Will the global crisis lead to global transformations?: the global financial system—pros and cons" Journal of Globalization Studies 1,1:70-89 (May).

Hall, Thomas D. and Christopher Chase-Dunn 2006 "Global social change in the long run" Pp. 33-58 in C. Chase-Dunn and S. Babones (eds.) Global Social Change: Historical and Comparative Perspectives. Baltimore: Johns Hopkins University Press

Hamashita, Takeshi 2003 "Tribute and treaties: maritime Asia and treaty port netowrks in the era of negotiations, 1800-1900" Pp. 17-50 in Giovanni Arrighi, Takeshi Hamashita and Mark Selden (eds.) The Resurgence of East Asia. London: Routledge

Harvey, David 1982 The Limits to Capital. Cambridge, MA: Blackwell

Harvey, David 2010 The Enigma of Capital.

Harvey, David 2010 "The crisis of capitalism" http://davidharvey.org/2010/05/video-the-crises-of-capitalism-at-the-rsa/

Henige, David P. 1970 Colonial Governors from the Fifteenth Century to the Present. Madison, WI.: University of Wisconsin Press.

Henwood, Doug 1997 Wall Street : how it works and for whom London: Verso

Hilferding, Rudolf 1981 Finance Capital: A Study Of The Latest Phase Of Capitalist Development. London : Routledge & Kegan Paul.

Hobsbawm, Eric 1994 The Age of Extremes: A History of the World, 1914-1991. New York: Pantheon.

Klein, Naomi. 2007 The Shock Doctrine: The Rise of Disaster Capitalism. New York: Henry Holt and Company.

Korzeniewicz, Roberto P. and Timothy Patrick Moran. 2000. "Measuring World Income Inequalities." American Journal of Sociology 106(1): 209-221.

Korzeniewicz, Roberto P. and Timothy Patrick Moran. 2009. Unveiling Inequality: A World Historical Perspective. New York, NY: Russell Sage Foundation.

Krippner, Greta R. 2010 " The political economy of financial exuberance," Pp.141-173 in Michael Lounsbury (ed.) Markets on Trial: The Economic Sociology of the U.S. Financial Crisis: Part B (Research in the Sociology of Organizations, Volume 30) Bingley, UK: Emerald Group Publishing Limited

Kuecker, Glen 2007 "The perfect storm" International Journal of Environmental, Cultural and Social Sustainability 3

Lawrence, Kirk. 2009 "Toward a democratic and collectively rational global commonwealth: semiperipheral transformation in a post-peak world-system" in Phoebe Moore and L Owen Worth, Globalization and the Semiperiphery: New York: Palgrave MacMillan.

Lindholm, Charles and Jose Pedro Zuquete 2010 The Struggle for the World: Liberation Movements for the 21st Century. Palo Alto, CA: Stanford University Press.

López, Jesús Espasandín and Pablo Iglesias Turrión (eds.) 2006 Bolivia en movimiento. Acción colectiva y poder político http://www.nodo50.org/boliviaenmovimiento/

Mann, Michael 1986 The Sources of Social Power, Volume 1. Cambridge: Cambridge University Press.

Mann, Michael 2006 "The Recent Intensification of American Economic and Military Imperialism: Are They Connected? Presented at the annual meeting of the American Sociological Association, Montreal, August 11.

Martin, William G. et al. 2008 Making Waves: Worldwide Social Movements, 1750-2005. Boulder, CO: Paradigm

Mitchell, Brian R. 1992. International Historical Statistics: Europe 1750-1988. 3rd edition. New York: Stockton.

Mitchell, Brian R. 1993. International Historical Statistics: The Americas 1750-1988. 2nd edition. New York: Stockton.

Mitchell, Brian R. 1995. International Historical Statistics: Africa, Asia, and Oceania 1750-1988. 2nd edition. New York: Stockton.

Modelski, George and William R. Thompson. 1996. Leading Sectors and World Powers: The Coevolution of Global Politics and Economics. Columbia, SC: University of South Carolina Press.

Monbiot, George 2003 Manifesto for a New World Order. New York: New Press

Patomaki, Heikki 2008 The Political Economy of Global Security. New York: Routledge.

Pfister, Ulrich and Christian Suter 1987: "International financial relations as part of the world system." International Studies Quarterly, 31(3), 23-72.

Polanyi, Karl 1944 The great transformation. New York: Farrar & Rinehart

Podobnik, Bruce 2006 Global Energy Shifts. Philadelphia, PA: Temple University Press.

Reese, Ellen, Christopher Chase-Dunn, Kadambari Anantram, Gary Coyne, Matheu Kaneshiro, Ashley N. Koda, Roy Kwon, and Preeta Saxena. 2008. "Research Note: Surveys of World Social Forum Participants Show Influence of Place and Base in the Global Public Sphere." Mobilization: An International Journal 13(4): 431-445.

Revised version in A Handbook of the World Social Forums Editors: Jackie Smith, Scott Byrd, Ellen Reeseand Elizabeth Smythe. Paradigm Publishers

Reifer, Thomas E. 2009-10 "Lawyers, Guns and Money: Wall Street Lawyers, Investment Bankers and Global Financial Crises, Late 19th to Early 21st Century" Nexus: Chapman's Journal of Law & Policy, 15: 119-133

Reinhart, Carmen M. and Kenneth S. Rogoff 2008 "This time is different: a panoramic view of eight centuries of financial crises" NBER Working Paper 13882 http://www.nber.org/papers/w13882

Reitan, Ruth 2007 Global Activism. London: Routledge.

Robinson, William I 1996 Promoting Polyarchy: Globalization, US Intervention and Hegemony. Cambridge: Cambridge University Press.

Robinson, William I 2008. Latin America and Globalization. Baltimore: Johns Hopkins Unversity Press

Robinson, William I 2010 "The crisis of global capitalism: cyclical, structural or systemic?" Pp. 289-310 in Martijn Konings (ed.) The Great Credit Crash. London:Verso.

Santos, Boaventura de Sousa 2006 The Rise of the Global Left. London: Zed Press.

Sen, Jai, and Peter Waterman (eds.) World Social Forum: Challenging Empires. Montreal: Black Rose Books

Smith, Jackie Marina Karides, Marc Becker, Dorval Brunelle, Christopher Chase-Dunn, Donatella della Porta, Rosalba Icaza Garza, Jeffrey S. Juris, Lorenzo Mosca, Ellen Reese, Peter Jay Smith and Rolando Vazquez 2007 Global Democracy and the World Social Forums. Boulder, CO: Paradigm Publishers

Silver, Beverly J. 2003 Forces of Labor: Workers Movements and Globalization Since 1870. Cambridge: Cambridge University Press.

Suter, Christian 1987 "Long waves in core-periphery relationships within the international financial system: debt-default cycles of sovereign borrowers." Review, 10(3).

Suter, Christian 1992 Debt cycles in the world-economy : foreign loans, financial crises, and debt settlements, 1820-1990 Boulder : Westview Press

Suter, Christian 2009 "Financial crises and the institutional evolution of the global debt restructuring regime, 1820-2008" Presented at the PEWS Conference on "The Social and Natural Limits of Globalization and the Current Conjuncture" University of San Francisco, August 7.

Taylor, Peter 1996 The Way the Modern World Works: Global Hegemony to Global Impasse. New York: Wiley.

Turchin, P. 2003. Historical dynamics: why states rise and fall. Princeton University Press, Princeton, NJ.

Turchin, Peter and Sergey A. Nefedov 2009 Secular Cycles. Princeton, NJ: Princeton University Press

Turner, Jonathan H. 2010 Principles of Sociology, Volume 1 Macrodynamics. Springer Verlag

Wallerstein, Immanuel 1984a. "The three instances of hegemony in the history of the capitalist world-economy." Pp. 100-108 in Gerhard Lenski (ed.) Current Issues and Research in Macrosociology, International Studies in Sociology and Social Anthropology, Vol. 37. Leiden: E.J. Brill.

Wallerstein, Immanuel 1984b The politics of the world-economy: the states, the movements and the civilizations. Cambridge: Cambridge University Press.

Wallerstein, Immanuel 1998. Utopistics. or, historical choices of the twenty-first century. New York: The New Press.

Wallerstein, Immanuel 2003 The Decline of American Power. New York: New Press

Wallerstein, Immanuel 2010 "Contradictions in the Latin American Left" Commentary No. 287, Aug. 15, http://www.iwallerstein.com/contradictions-in-the-latin-american-left/

Wallerstein, Immanuel. 2011 [1974] The Modern World-System, Volume 1. Berkeley: University of California Press.

World Bank 2009. World Development Indicators [CD ROM]. Washington DC: World Bank

World Bank 2011. World Development Indicators [online]. Washington DC: World Bank.

Chapter 4

GEOPOLITICAL CONDITIONS OF INTERNATIONALISM, HUMAN RIGHTS, AND WORLD LAW

Randall Collins

International rule of law is not an alternative to geopolitics, but is successful only under specific geopolitical conditions. As historical sociologists in the tradition of Weber have documented, the state's existence has depended on its military power, which varies in degree of monopolization, of legitimacy, and of extent of territory controlled. Geopolitical principles (comparative resource advantage, positional or marchland advantage, logistical overextension) have determined both the Chinese dynastic cycles, and the balance of power in European history; they continue to apply to recent wars in Afghanistan, Iraq, and Pakistan. Guerrilla wars differ from conventional wars by relying especially on geopolitical principles of promoting enemy overextension. Geopolitics encompasses both war and diplomacy, the means by which coalitions among states are organized. The rule of international law depends on a dominant coalition upheld by favorable geopolitical conditions; and on the extension of bureaucracy via state penetration, but now on a world-wide scale.

Keywords: geopolitics, logistical overextension, state penetration, guerrilla war.

In the period between the fall of the USSR in 1991 and the immediate aftermath of the September 11, 2001 attacks in the United States, it was often argued that a new world order was emerging. Many politicians as well as journalists and social scientists held that we had entered in a new period where the only use of force would be international coalitions taking action against 'rogue states', 'international outlaws', and terrorist organizations. This rhetoric was shared by US President George W. Bush in setting forth the rationale for intervention both in Afghanistan in 2001, and two years later in Iraq. The first of these invasions was widely acclaimed in the West, the second—widely opposed. The ostensible terms of the debate focused on whether each of these invasions fit the rule of international law. Underneath the ideals and ideologies, however, more basic geopolitical processes have continued to be at work.

The configuration of state powers in the world has changed, of course, since 1990, and indeed again to a degree since 2001. This does not mean that the most basic principles of International Relations have changed as well. Sometimes it is true that a theory is so specific to the historical period, in which it is formulated, that when conditions change the theory no longer works. Sixty years ago much importance was given to Balance of Power theory. This theory held that when several big states struggle for power, they make alliances so that they keep up a balance of power of roughly equal strength. The theory was based on the period of European history when England, on its island off from the Continent, looked upon European struggles and always chose to fight on the side of the weaker coalition, so that no state could ever dominate the Continent. When France was strong, England allied with Germany; when Germany was strong, England allied with France. Balance of power is not a very general theory however; it does not explain why the balance of power disappeared after the end of the Second World War; and it does not explain earlier state systems such as the Roman Empire, or the dynasties of Imperial China. In fact, Balance of Power theory does not even explain England's behavior; at the same time England was maintaining balance of power politics in Europe, it was expanding an overseas empire around the rest of the world.

More recent fashions in International Relations theory include neorealism; and, on the other hand, the theory of hegemonic stability. These theories clash on the question of whether the relations among states are a realm of anarchy, where each follows its own self-interest and no laws or principles control them except their own force; or, on the contrary, that there is an international or interstate order, a framework in which the world carries on its business. In the latter theory, the strongest state or hegemon acts to enforce the rules of the international game, and thus provides stability—in this view it is functionally useful for the world to have a hegemonic power like Britain or the US to keep order. In fact, both things are possible. Under some historical conditions, the world looks like a violent confrontation of self-interested states; at other times, there is more of an appearance of international rules of the game. But this is a continuum, not an all-or-nothing choice between extremes; states exist by controlling military force, but they also tend, to lesser or greater degree, to enter into alliances and coalitions, and to make arrangements even with their enemies. We have seen these throughout history: the Roman Empire was at first a system of alliances before it became an Empire; the Holy Roman Empire (or German Empire) of the European middle ages was chiefly just a diplomatic structure, a kind of early and limited version of the United Nations. Historically, the units do not stay static; sometimes states become bigger or smaller, more independent or more amalgamated, with many possible variations in between: the history of China, Hong Kong, and other parts of East Asia provide good examples. And new state forms emerge; sometimes alliances become stronger and turn into states, as we see happening, perhaps, today in the case of the European Union, and happened 200 years ago in the federation which became the United States of America. New coalitions, including those formed under the rationale of combating terrorism, must judged as to where they will fall along the continuum from a weak decentralized alliance to a centralized structure of world government.

Another theory which is linked to a particular historical time and place is the theory of Chinese dynasties. This is the theory, held by Chinese historians for almost 2000 years, that China goes through a dynastic cycle: first, there is a strong centralized state; the emperor or state leader has high prestige and legitimacy; then the state becomes corrupt, the officials become ineffective, tax collection weakens, bandits appear inside the borders and foreign enemies outside become more troublesome. Eventually the state falls into disintegration; but then one of these small states becomes stronger; it conquers and unifies the rest, and starts a new dynasty. In some respects this is a strong theory, at least for the period from the Han dynasty up through the Qing dynasty, and some think perhaps even later. However, we may ask: does this theory apply only to China? Are there no general principles which apply equally to China and to other states? The Roman Empire, for example, expanded and then collapsed, but it never was reconstituted as a new empire; instead it broke into pieces that have never been reunited. In the Warring States period for about 500 years before the Han dynasty, there was no dynastic cycle but instead there were many states in north China which acted according to Balance of Power theory; whenever one of these states became strong enough to threaten to conquer all the others, a coalition formed against it which prevented its expansion. Why should Chinese dynastic cycles begin at a particular time in history? And does the principle of a dynastic cycle come to an end, once China becomes part of the larger global world of the 20th and 21st centuries?

Historical questions about the deep past such as cycles in China, as well as contemporary questions about the trajectory of the US military policy in its period of world hegemony, and about the prospects for international law, all hinge upon a more thorough understanding of the conditions of relations among states. Here I will summarize two theories which help explain all of these historical changes, including the situation of the world today. First is the geopolitical theory of the state; and the second is the theory of bureaucratization as basis for formal law. Both theories develop from classic analysis by Max Weber, but have been taken much further by recent historical sociologists.

THE MILITARY-CENTERED GEOPOLITICAL THEORY OF STATE POWER

A geopolitical theory of the state has developed from the implications of Weber's point that the state is fundamentally an organization of military force which claims a monopoly over the legitimate use of force upon a territory (Weber 1968 [1922]). If such a theory is to be of use, it should be treated as a set of variables, not as a constant. How much monopoly over legitimate force a state has, and how much territory it applies to, is not a constant, but changes over time as the outcome of political and military struggle. The principles which determine these changes are principles of geopolitics.

What then are the key geopolitical [GP] processes? What makes a state geopolitically stronger in its control over a geographical territory, and what makes it weaker, introducing a degree of geopolitical strain? I will summarize in a series of *ceteris paribus* principles which bring out the causes of variations in the territorial power of states; since all causes may operate simultaneously, we must combine all these principles to explain changes in the power of states.

1. *Resource advantage.* States which mobilize greater economic and population resources tend to expand at the expense of states mobilizing lesser such resources. Big states get bigger; and rich states get bigger, because they absorb smaller or poorer states on their borders—either by conquest and formal annexation, or by means of alliances, protectorates or empires absorbing their economic resources and exercising command over their military forces.

2. *Geopositional advantage.* States with potential enemies on fewer frontiers tend to expand at the expense of states with a larger number of frontiers to defend; this is sometimes referred to as the advantage of marchlands over centrally-

located states. Conversely, states in the middle of a zone of multiple states tend to be caught in a web of multiple shifting alliances and to fragment over time.

The first two principles, resource advantage and marchland advantage/interior disadvantage, cumulate over time; relatively resource-richer or geographically better positioned states grow at the expense of poorer and interior states, thereby swallowing up their resources and controlling their territory. Over long periods of time (my estimate is several centuries), a few large states consolidate. This leads to periodic showdown wars (or so-called hegemonic wars; *e.g.*, the Napoleonic wars were a hegemonic war; World War II was another). Such showdown wars are highly destructive and are fought at a high degree of ferociousness, in contrast to wars fought in balance of power situations among many small contenders, where gentlemanly rules of limited combat tend to prevail. A showdown war may end either by total victory of one side, which establishes a 'universal' empire over the accessible 'world'; or to mutual exhaustion of resources by the contenders, opening them up to disintegration and incursion from new smaller contenders on the margins.

3. *Principle of overextension or logistical overstretch.* The greater the distance from its home resource base a state extends its territorial control, the greater the logistical strain; overextension occurs at the point at which more resources are used up in transportation than can be applied in military force relative to the forces which enemies can muster at that location. Overextension not only causes military defeat and territorial loss, but is a major cause of state fiscal strain and state breakdown. The time-patterns of the growth of large states or empires, and their collapse, are quite different. Whereas the cumulative growth of resources and territorial expansion occurs gradually over a long period of time (on the order of centuries), the collapse of empires tends to occur quite rapidly (in a few crisis years).

Overextension is especially dangerous for a state because it tends to cause revolutions. Not only does the state lose territory, but also its monopoly over force, and its ruling faction or party tends to lose legitimacy; and these are crucial conditions leading towards revolution. This follows from the state breakdown theory of revolutions: the model that revolutions are never successful merely because of dissatisfaction from below, but only where popular dissatisfaction is mobilized in a situation of crisis in the state apparatus of coercion; that in turn is typically due to military strains, either directly or in their effects upon state revenues, with the situation exacerbated by conflict between propertied and state elites over who is to pay for the shortfall (Skocpol, 1979;

Goldstone, 1991; Collins, 1995). The link to external geopolitical affairs is both direct and indirect: direct because military weakness reduces the legitimacy of whoever is in control, indirect because military expenses have historically been the bulk of state expenditure and accumulated debt.

Let us see now how GP principles apply to the Chinese dynastic cycle. First: the importance of the economic resource base. Because of the geographical configuration of East Asia, any state, which unified the two great river valleys of the Yellow River and the Yangtse, would have population and economic wealth much greater than any other state in the region. Thus the central state was able to expand against enemies in almost all directions, which were certain to be smaller and poorer. Eventually the over-extension principle comes in: successful Chinese armies extend to frontiers which are very far from the home base; this produces logistical strain, and the military budget becomes increasingly expensive, at just the time that armies become less effective. This is what causes rebellions against taxation, the rise of banditry, and the corruption of officials. When the crisis occurs, China finds itself in the center defending attacks from many different directions. Thus the middle splits up, and there occurs a period of fragmentation, the recurrent warring states periods which occur in the intervals between the great dynasties. Eventually one of the smaller states located in a borderland or marchland region, begins to grow, until it attains cumulative advantage and reunifies the great population areas of the center. Now there is a strong dynasty, deriving strong legitimacy from its recent geopolitical success, and the dynastic cycle begins again.

The dynastic cycle follows from GP principles, but only as long as China was in a zone which was largely cut off from other parts of the world, especially by the inefficiency of transportation in early historical periods. Once China became part of the larger space of world power relations, the conditions for the dynastic cycle were no longer present. Thus, the cycle operated in a particular period of history, although it was the result of causal principles which are universal.

Thus although GP principles are first formulated by being abstracted from particular historical periods, it has been possible to broaden the application of such principles by reformulating them on the basis of wide historical comparisons. Classic and modern efforts to formulate GP principles, which I have drawn upon in my summary, have been based upon studies of Greco-Roman antiquity as well as early modern through contemporary Europe (Andreski, 1971; Gilpin, 1981; Kennedy, 1987; McNeill, 1963, 1982 ranged even more widely in world history). My own initial inductive formulation (Collins, 1978) was based upon analyzing historical atlases for the Middle

East and Mediterranean regions from the first ancient empires through medieval and modern times, and for China since the earliest dynasties (see also Collins 1992 for application to kinship-based societies). In other words, GP principles (resource advantage, marchlands, overextension, *etc.*) hold across the range of patrimonial and bureaucratic state forms. In addition, I was able to use GP principles successfully in 1980 to predict the strains which brought about the collapse of the Soviet empire (a continuation of the older Russian empire) (Collins, 1986, 1995). And finally, geopolitical principles fit into a coherent theory of the state, developed from scholars from Weber through Skocpol, Tilly and Mann, which as we will see, gives a well-supported picture of all major aspects of state growth, state crises, state organization, political mobilization and revolution.

THE GEOPOLITICS OF WAR IN AFGHANISTAN, IRAQ, AND PAKISTAN

The early phase of the US war in Afghanistan was not a repetition of the Vietnam war, nor a repetition of the Soviet war in Afghanistan. To apply GP principles, we must summarize the resources on each side, look at their geographical positions, and their problems of logistical extension or overextension. In the cases of the Vietnam war and the Soviet-Afghanistan war, there were two big world power blocs; hence each side in these local wars had support from much bigger chains of resources. In both cases, these were guerrilla wars. The guerrillas did not have to win the war by battlefield victories, but only to continue resistance until their opponents' supply lines became too costly—in other words, to wait until logistical overextension made their opponent withdraw. In addition, in the case of the Soviet-Afghanistan war, the Soviets had multiple military commitments on other fronts—in Eastern Europe, Northeast Asia, the long-distance nuclear weapons race, *etc.* The Soviet weakness was precisely the reverse of the marchland advantage—the USSR was in the middle extending forces in all directions. It was Gorbachev's effort to reduce these multiple military commitments that led to the Soviet policy of giving up Eastern Europe, allowing the wave of anti-Communist revolts that eventually broke up the USSR.

In contrast with this, the war in Afghanistan in October—December 2001 was an alliance of all the big powers against the supporters of the terrorist movement al-Qaeda. From the first GP principle, resource advantage, we would expect the US forces and allies to win. The second GP principle, geopositional advantage or disadvantage, posed no problem for the US forces insofar as it was not fighting multiple wars on widely separated fronts. The main GP danger was in the third principle, overextension: Afghanistan is very far from Western supply bases, and thus the war could become very costly, depending on how long it would continue. The main worry of US policy during the 1990s was to avoid logistical overextension—the so-called lesson of the Vietnam war—not to become bogged down in long and expensive wars in distant places. Thus President George W. Bush, in the early period of his administration, tried to bring the US military into a completely defensive posture, and to withdraw from international commitments. This was changed, of course, by the attack of September 11—according to the principle that external attack brings national solidarity, and widespread desire for national military action (Coser, 1956). This is what happened in the summer of 1914 in Europe, when after an assassination in Sarajevo the states of Austria, Russia, Germany, France, Italy and England began to threaten each other with war; these threats increased national solidarity in each place, and huge crowds in the streets in Vienna, Berlin, Moscow, Paris, and London demanded that their countries go to war (Scheff, 1994). After the attacks of September 11, there was a huge increase in national solidarity in the United States; the level of agreement on public policy temporarily became very high; the popularity of President Bush rose from moderate (about 50 % approval rating) to the highest ratings ever recorded (90 %) (Collins, 2004).

The question became: how long would this national solidarity last, compared to how long it would take before the problem of logistical overextension set in? According to my analysis of indicators of national solidarity—especially the display of flags on houses and cars which broke out spontaneously in the days immediately after the 9/11 attack—peaks of solidarity under attack remain high for about three months, and recede back to normal in six months (Collins, 2004). The US victory over the Taliban regime was well within this period.

One reason for the rapid victory is that the war was not a guerrilla war, but a conventional war between Taliban troops defending fixed positions, especially around the cities, and the Northern Alliance troops supported by the US. This was exactly the situation in which US superiority in air power would be most effective. A second reason was that the Taliban was not organized as a unified army but as a coalition of warlords and tribal clans, along with some ideologically-recruited troops. We must distinguish between the Taliban movement, which was concerned above all with enforcing its conservative Islamic religious policy, and the wider Taliban coalition. Hence it was very easy for the Taliban coalition of clans and warlords to unravel, once it became apparent the Taliban would lose any direct battles against superior US military resources. This is a typical case of a bandwagon effect (Marwell & Oliver, 1993).

Once the US-led coalition destroyed the Taliban regime and installed a favorable regime in office in Afghanistan, the situation did indeed shift back towards a situation more resembling the Soviet war during the 1980s. Through 2009, that war was essentially a stalemate: the Taliban guerrillas failed to overthrow the central government, the US coalition was unable to destroy the guerrillas. In addition, the guerrilla war spread across the border into Pakistan. Geopolitical principles have not as yet been applied extensively to guerrilla war; as a step in this direction, I suggest that guerrillas play for different stakes than conventional military powers. Guerrilla war is not a useful tactic for invading foreign territory, nor of holding territory; thus, the Taliban was easy to defeat when it held conventional government power. On the other hand, guerrilla war is quite efficient in denying a conventional military power full control of a territory—preventing it from establishing a Weberian state of monopoly over violence—since guerrilla war requires much fewer resources in manpower, weapons, and logistical base than conventional war. Guerrillas largely avoid problems of logistical overextension, since they operate close to their home base. In contrast, opponents of guerrillas are at a particular disadvantage the further from their home base they operate—not only are logistical lines of supply longer and more expensive, but an army from a distant land is more culturally alien and thus likely to generate more cultural resistance by its very presence.

The spread of the guerrilla war into Pakistan is explainable in part by the geopolitical pattern that power-prestige increases inflow of resources via recruitment to alliances. Power-prestige is always relative to what a military force is attempting to do; guerrillas have only to survive to build their reputation as invincible; whereas a conventional force has the goal of defeating an enemy fully and thereby establishing monopoly of force on the territory. The longer the Taliban/al-Qaeda guerrillas hold out in Afghanistan, the more it makes them appear a permanent feature of the local scene; it is this local growth in power-prestige that helps account for their successes in recruiting allies and reinforcements in nearby areas of Pakistan. And these additions to local resources need not be large—need not even be a majority of the local population—to be effective as resources for the goals of guerrilla war.

Nevertheless, although fighting against guerrillas in Afghanistan, and by proxy, in Pakistan, has increased strain on US military resources and thus the amount of overextension, the issue was not decisive for US power-prestige either at home or internationally, because the focus of attention during this period had shifted to Iraq. When the US (together with a limited number of allies) invaded Iraq in March 2003, its military success followed geopolitical principles: large resource advantages, overwhelming concentration of those resources in a single theatre of operations (the Afghanistan war having been reduced to a small scale of anti-guerrilla operations); logistical overextension would not come into play as long as the war was short. Here we must

---

elaborate on the character of geopolitical resources and how they are transformed into military resources. The US had a much bigger economic resource base and population than Iraq in 2003, while the latter was essentially isolated and without allies.

In addition gross economic and population advantages, US military organization had over the past 20 years been engaged in a so-called 'revolution in military affairs', which transformed both weaponry and organization into a high-tech, computer-centered mode. The development of aerial surveillance by global positioning satellites, of laser-guided missiles, infra-red sensors and other devices made US forces much more accurate in hitting targets with long-distance weapons. High-tech development also allowed relatively smaller armies (although still on the order of several hundred thousand troops) to carry the firepower equivalent to far larger conventional armies of the mid-twentieth century. By these means, US/coalition forces were able to destroy the command and control structure of Iraqi military organization very rapidly, and to overrun the country in a period of weeks. The relatively larger Iraqi army and its numbers of armored vehicles and artillery were immobilized by the high-tech weaponry and coordination system of the invaders. My point here is not to extol the dominance of technology; rather we must understand that technological advance is itself a means by which superior economic resources are turned into military power. Decades of US investment in military research and development, based on a substantial portion of the world-leading US economy, culminated in the high-tech military organization which won a blitzkrieg over the older and more 'low-tech' Iraqi military forces. (This is not to say Iraqi military equipment was entirely 'low-tech'—the war has been described as an army of the year 2000 fighting against an army of the 1960s.)

High tech is the mode in which geopolitical resources manifest themselves today. Indeed, this was already true at the time of the World Wars of the 20th century, but the disparity in technology between the sides was minimal in the First and Second World Wars and hence not apparent. The importance of high tech was masked during the US-Vietnam war, because the guerrillas had only to hold out, rather than actually win, until the power-prestige of the long-distance US occupation had declined and brought political pressures for settlement. Only in the brief 1991 Gulf War, and in the 2003 invasion of Iraq, was the importance of a wide disparity in military technology apparent, since these were wars which matched pairs of conventional military forces against each other.

After the successful invasion of spring 2003, the war in Iraq changed from a conventional war into a guerrilla war. Here high-tech weapons are less decisive. Guerrillas extend the trend of modern warfare in the sense that as long-distance firepower has become more lethal, military forces have dispersed instead of concentrating where they would be easy targets (Biddle, 2004). Guerrillas take this principle to the extreme, by hiding among the civilian population in very small groups, and concentrating on attacking enemy logistics lines. Guerrilla war is a war of attrition, above all aimed at increasing not just the material cost but the political cost of persisting in attempts to hold foreign territory. Nevertheless, geopolitical principles still hold in a guerrilla war. The side with large resource advantages can persist in fighting guerrillas as long as the actual rate of material attrition is not too high; and this becomes largely a matter of political will. Viewed sociologically, political will is the pattern of emotional solidarity around government leaders during a period of conflict.

As I have indicated above (in regard to the period of intense national solidarity in the months following the 9/11/01 attack), solidarity is highest at the beginning of a conflict; it also peaks at moments of victory; otherwise it gradually declines. Thus popular support for the US war in Iraq declined over the years, especially during the period of guerrilla war. These matters are always difficult to judge at the time when they are happening, especially by observers who are politically engaged and have strong feelings about the propriety of policy decisions. From an analytical viewpoint, it should be apparent that although popular enthusiasm for a war tends to decline the longer the war goes on (as was apparent in all countries during the First and Second World Wars, as well as during the long US wars in Korea, Vietnam, and Iraq), nevertheless anti-war sentiment has rarely if ever become high enough to cause a Great Power—i.e., a state with strong international power-prestige at the beginning of a conflict—to voluntarily pull out of a war when it had not yet been thoroughly defeated on the battlefield. Thus it should not be surprising that President George W. Bush was able to win re-election in 2004, despite impassioned opposition both inside the United States, and elsewhere (especially in Europe) where the rationale for attacking Iraq in 2003 was strongly criticized. By the time an ostensible anti-war candidate won the US Presidency, Barrack Obama in 2008, the war in Iraq had already been winding down, and US troops were in the process of withdrawing and turning over the task of fighting guerrillas to Iraqi government forces. In the immediate perspective of 2009, it is striking how little real difference there was between the military policy of the Bush and Obama administrations, the latter carrying over the trajectory of the former. This supports the sociological point that state leaders tend to go along with the exigencies of military power-prestige, and that the rhetoric of being a candidate in opposition is different from the actions of an elected head of state.

The political criticism of the US invasion of Iraq was largely focused on the question of international norms and rights. The Bush administration held that the US had the moral right to invade what it called a terrorist state; domestic opposition in the US (mostly from the internationalist liberal/left) and in Europe held that only an international consensus could decide such a move, and that the US unilateralism was illegitimate. The use of military power by individual states was held to be superseded now by international organization and international law. To put this in sociological perspective, we need to examine the conditions underlying diplomatic solutions to international crises and their relation to military actions.

THE GEOPOLITICS OF INTERNATIONAL COALITIONS

G P principles do not mean that states are always threatening to go to war. On the contrary, states often pursue diplomacy instead of fighting. But it is a mistake to regard GP and diplomacy as separate from each other. Diplomatic strength depends on GP strength; successful diplomacy takes account of GP principles rather than ignores them.

GP principles do not become superseded, even in a world rule of humanitarian law. It is important to emphasize that GP principles do not require the bounded independent state actor as the unit of analysis. Instead, GP analysis focuses upon the organization of force, and derives the territorial and organizational configurations into which this organization is shaped under different historical conditions. The formation of a new type of organization of force, at the level of international alliances or even world government, is compatible with these principles. GP principles were first developed by analyzing the relations of separate states, but they apply to any organization which attempts to exercise military force over a territory. It could be an international alliance, or a world government. Examples are the United Nations, which is as yet a rather weak world government, but one which nevertheless attempts to define as legitimate solely that force which it sanctions; or the European Union, which is a federation moving towards becoming a European government; it will become a such at the point at which it has an autonomous European army. There are many other kinds of international organizations and alliances, such as NATO in its recent phase of expansion, and ad hoc alliances such as the anti-terrorist alliance assembled by the US after September 11, 2001 to invade Afghanistan; and the much smaller coalition put together to support the 2003 invasion of Iraq.

To the extent that the UN, EU, NATO or any other such international alliance become effective in enforcing a new world order, it is because they have GP advantages over their potential opponents. That is to say, they must be superior in resources and in organization to mobilize those resources. They are subject to geopositional constraints, since it is easier to project force at some targets than others. It is easier to project Western forces in the Balkans than in Central Africa, which explains why there was an intervention to stop ethnic cleansing in Bosnia and Kosovo, but not in the genocide in Rwanda-Burundi. And international organizations will be in danger of logistical overextension, like all previous states. If there can be mega-states and world governments, there is also the possibility of state breakdowns in these units. There is always the possibility that international organizations may undergo revolutionary breakdowns, driven by the classic pathway of GP strain, fiscal crisis, intra-elite struggles, and coinciding popular resentments from below. Even if there is a real world government or massive world federations in the future, they will be subject to the restrictions of GP principles. The possibility that a world government might some day be established does not mean that it would necessarily be permanent; it could undergo a revolution or state breakdown, just like previous states.

Such a development remains in the hypothetical future. Thus far the transnational coalitions and their righteous crusades in favor of international law and justice look a great deal like previous alliances and federations. NATO's role in the Kosovo intervention of 1999, and the negotiations of recent years to expand NATO membership into the old Warsaw pact, can be interpreted as a project to keep the US involved in the center of European power, at a time when it has been implicitly in rivalry with the EU as alternative way of organizing military force upon the Continent. Such rival and overlapping coalitions have happened before; the geopolitics of medieval Europe was to a considerable extent a struggle between the opposing claims of Christendom unified under the papacy, as against the German (or Holy Roman) Empire; there were also some smaller confederations which battened upon the fall of the Empire to create federal states such as Switzerland and the Dutch Republic (Collins, 1999). NATO in the 1990s looks a good deal like the German Empire of the late Middle Ages, in the sense that it was mobilized for wars against external enemies (in the case of NATO this was first the Soviet bloc, then rogue states; in the case of the medieval German Empire it was mainly the Ottoman threat); this collective enterprise was always led by the strongest state (in the modern case the US, in the medieval case the Habsburg ruler) which took military command and provided the bulk of the troops.

Historically, alliances and federations have often exercised military force under strong control from its dominant member; in effect the entire alliance operates to enhance the power-prestige of its leader. In ancient Greece, the Athenian League against the Persians was also the Athenian empire coercing participation and punishing withdrawal. It is a plausible argument that whatever the surface emotions and humanitarian ideals involved, the various US-led coalitions of the post-1945 period are manifes-

---

tations of the desire of US political leaders to keep up power-prestige in the international arena. Nor is the idealism of today's transnational coalitions new; the crusades of medieval Christendom which bolstered the power of the Pope were equally idealistic, and in general every large military enterprise acts in an atmosphere of emotionally charged belief. The big test of a truly transnational political order would be if a major coalition were to go into military action against the desires of its strongest member: if the UN were to take action, for instance, against the USA.

As of today, the UN has a long way to go to become a state in the strong sense of the term. The UN assembles military forces by a feudal-like levée, in which each partner to the alliance raises and pays for its own troops and keeps them under chains of command which are largely separate, except for temporary international combinations of officers at the top. Under these conditions, the effect of warfare in galvanizing national identity is not transferred to the coalition, but reinforces the ethno-nationalism of the states identified with each body of troops. A true UN army, and thus the basis of a strongly held world-identity, would depend upon the UN being able to recruit its own soldiers from throughout its member countries, combining them into formations irrespective of origin. The state penetration of the UN (not to mention other alliances) is shallow; it does not wield coercive power to discipline its own members, but thus far has intervened only in the internal affairs of non-members. In this respect these international coalitions have operated like empires of conquest expanding their spheres of control.

STATE BUREAUCRATIZATION AS BASIS FOR RULE OF LAW

Let us return to the question: is the world of the early 21st century moving towards a new era of international rule of law to support universal human rights? Such claims have been made increasingly in recent years, and some organizational apparatus has been developed to attempt to put them into action. Nevertheless, this idealized goal in the use of force is not so new, and that it happens in accord with existing sociological principles.

Law is a set of ideals and procedures; but law always has an organizational base. Laws do not enforce themselves. Thus it was naive, on the part of some political commentators on the September 11, 2001 attacks, to say that Osama bin Laden and others responsible should be brought to trial; but at the same time to say there should be no war against the Taliban coalition. The notion that criminal justice is an alternative to war is an inaccurate extrapolation of the domestic power of the state into the realm of inter-state relations. The organizational base of law is the power of the state; and that in turn depends on geopolitical power, and on the extent and effectiveness of state organization.

In the modern ideal of the rule of law is that there should be general principles designating individual rights and responsibilities, and formal procedures for judging who has which rights, and who is responsible to be punished for violations. The organizational basis for this kind of law is the rise of the modern bureaucratic state. The rise of the modern state is a topic on which there has taken place in the last 25 years of scholarship a cumulative development of historical sociology. I will briefly summarize three points: the military revolution, state penetration into society, and the extension of bureaucracy.

The full-fledged ideal type of the force-monopolizing territorial state gradually developed since 1500 in the West, although there have been variations along this continuum elsewhere in world history. The story that we have become familiar with through the work of Mann (1986, 1993), Tilly (1990), Parker (1988) and others begins with the military revolution which drastically increased the size and expense of armed forces. State organization began to grow in order to extract resources to support current military expenses and past debts, above all by creating a revenue-extraction apparatus. This was the pathway towards bureaucratization and centralization. State penetration into society brought a series of effects in economic, political and cultural spheres. State apparatus now could increasingly regulate the economy, provide infrastructure, compel education and inscribe the population as citizens in government records. These same processes mobilized people's collective identities into social movements operating at a national level: in part because the state itself now constituted a visible target for demands from below; in part because state penetration provided the mobilizing resources of communication, transportation, and consciousness-raising. State penetration thus fostered both its own support and its domestic opposition; as Mann has demonstrated, both nationalism and class conflict were mobilized as part of the same process. The modern state became a breeding-grounds for social movements; and whenever a social movement has been successful, it has institutionalized its victories by creating new laws which are administered by the bureaucratic state.

The rise of the modern state leads directly to the theory of bureaucracy. In terms of organization, the rise of modernity is best characterized, not as a move from feudalism to capitalism, but from the *patrimonial household to bureaucratic organization*. What Weber called patrimonial organization exists where the basic unit of society is the household, and larger structures are built up as networks of links among households. It is important to note that the household mode of organization is not the same thing as the family mode of organization, although they are related. The household typically had at its core a family, the head of household with his wife (or wives) and children, perhaps with some other relatives; and thus property and authority were hereditary. But households could never be very large if the only people they included were family members. Patrimonial households were full of pseudo-familistic relationships; a household of the upper classes would include servants, retainers, guards, guests, hostages and others, all supported from the household economy, and all expected to provide some resource: work, loyalty, or military forced. An important house contained within it enough armed force to be powerful; it was a fortified household. Links to other households of lesser or greater power constituted the political structure of the society; under certain legal arrangements, these might be called properly 'feudal', but a variety of other structures were possible. The economy was also organized in patrimonial households or their linkages; the labor force consisted of servants and apprentices under familistic protection and discipline rather than independent wage relationships. To refer to a great 'house' was both literal and metaphorical; the aristocracy and the great burghers or merchants were the possessors of the largest household units with the most retainers.

The rise of bureaucracy was the dismantling of the patrimonial household. Workplace was separated from home, private force was superseded by professional military and police units belonging to the state. The physical separation among buildings where production, consumption, politics and administration took place was also the creation of the division between public and private spheres. Bureaucracy was the creation of offices separate from the persons who held them, the creation of a sphere of interaction apart from family ties and pseudo-familistic relationships of loyalty and subordination. The impersonality of bureaucratic organization depends upon paperwork, codifying activities in written rules and keeping count of performance in files and records. Bureaucracy is thus the source of modern ideologies: the rule of law, fairness, justice, impartiality; the previous practices of loyalty to the patrimonial household, and the consumption of organizational property became condemned as nepotism and corruption. Bureaucracy is the source of individualism since the unit of accounting and responsibility is the individual who can be appointed, promoted, moved from one position to another, paid, reprimanded, and dismissed, all with reference only to their personal dossier rather than their family and household connections. The shift from patrimonial households to bureaucracy promoted the ideology of individual freedom, but also the ideology of alienation from the impersonal public order; both are sides of the same coin. The shift to bureaucracy also made possible modern mass politics: ideologically, it fostered the conception of the individual's rights to democratic representation and legal status apart from the jurisdiction of the household head; structurally, it made it possible for workers, women, and youth to mobilize in their own places of assembly and their own cultural and political movements. One reason class conflict became possible in the modern era was because penetration by the revenue-extracting state created a centralized arena for political action; a complementary reason was that class and other conflicts were mobilized by being freed from the constraints of patrimonial household organization (Tilly, 1978, 1995; Mann, 1993).

The great historical transformation was the shift from patrimonialism to bureaucracy. These Weberian concepts are of course ideal types, and actual historical configurations were often mixtures. Weber used a concept of 'patrimonial bureaucracy' for intermediate forms, typically a more centralized governmental structure than feudalism or local chiefdoms ('caudillismo' in Latin America). Egypt, late Imperial Rome, many Chinese dynasties, and early modern Europe all had particular mixtures of these ideal types, which slid up and down the continuum of patrimonial and bureaucratic forms.

What caused the transition from patrimonial to bureaucratic organization? Weber's answer has usually been interpreted as a series of material preconditions (existence of writing, long-distance transportation, a monetary system, *etc.*) or as a functionalist argument that bureaucracy arises because it is the most efficient way to coordinate large-scale and complex activities. For the grand historical transition we are concerned with, there is a more directly political answer. Recall that we are considering the state processually, as a struggle to monopolize legitimate force upon a territory. The state is a project, an attempt to control and coordinate force in as definite a manner as possible; under particular historical conditions, what is possible along that line may be quite limited. How then do organizations move along that continuum towards increasing monopolistic control? Weber sees the shift from kinship alliance politics towards patrimonial household domination as one move towards centralization and monopolization; the shift to the bureaucratic state is a much stronger move higher up the continuum. What enabled some states to make that move earlier or to a greater degree than others?

Bureaucratization was a move in the struggle between whoever was the paramount lord at any particular moment and his allies and rivals among the other great patrimonial households. A crucial condition was the geopolitical configuration. Decentralized chiefdoms and hereditary feudal lineages raised less military resources for their paramount lords and thus tended to be conquered, or were forced to imitate the bureaucratizing manners of the more successful states. Dynastic states proved geopolitically weak because far-flung marriage ties produced scattered states, in effect subject to the effects of logistical overextension. History of course is more complicated than a simple winnowing out of non-bureaucratic states by bureaucratic ones; resource advantage is not the only GP principle, and some states favored by marchland positions might survive with more quasi-patrimonial structures (as Britain did down through the 19th century); and bureaucratizing states might nevertheless fail to expand their territorial power because of logistical overextension. Nevertheless, the long-run trend is towards the victory of the bureaucratizers. The successive waves of the military revolution were steps in the development of bureaucracy, first within the military itself (especially logistically-intensive branches such as artillery), then in the revenue-extraction service. State penetration was largely bureaucratization at the expense of the patrimonial household. Extensive market capitalism and especially its industrial form prospered under particular versions of state penetration and military mobilization; in this way bureaucracy spread from government into the economic sector; and this in turn fed back into still further government bureaucracy.

I have sketched a theoretical perspective of causality from the outside in: the various ramifications of the military revolution and the revenue-extracting state. In important ways, geopolitical processes are prime movers, even as they play into a multi-causal situation. Not to say that states cannot take alternative pathways, but they do so at a risk: if they are too weak geopolitically *vis-à-vis* their neighbors, they become swallowed up into an expanding state which has successfully negotiated the military revolution and thereby have state-penetrating structures imposed upon them.

Bureaucratization underlies both the positive and negative features of modern societies. In contemporary discourse, the term bureaucracy is a negative one: it implies inefficiency, paperwork, impersonality, and endless complexity. In some parts of the world, the term bureaucracy also has the connotation of corruption, a regime of bribery; but this is not a sociological use of the term; it would be more accurate to describe corruption as a form of patrimonial organization—the rule of personal connections—which reemerges inside the framework of bureaucracy. The cure for bureaucratic corruption is more rule of law, which is to say bureaucratic administration in the strict sense of the term. Structurally, bureaucracy is the basis of the rule of law; and hence the question of a new world order is a question of the future of bureaucracy.

SOCIAL CONDITIONS FOR EXPANSION OF WORLD LAW

The transition now being proposed at the beginning of the 21st century, to a world rule of law and universal human rights, is an extension of bureaucratic organization and its ideological ethos. The rule of law and the focus upon individual rights are central to the way bureaucratic organization functions. What may be afoot now is not a transition beyond bureaucracy but an expansion of legalistic bureaucratic organization from the national to a global scale. To put this more precisely, there have long been in existence networks organized on bureaucratic principles which have overlapped the boundaries of national states; what is happening today is that the sheer quantity of such transnational organizations has increased, and they have moved more intensively into attempting to regulate human behavior everywhere in the world according to an explicit formal code. We are seeing efforts which are analogous to the state penetration which took place earlier at the national level, both in conjunction with fledgling international government, and in international business, charitable, and social movement organizations whose networks overlap even wider than today's international alliances. What determines whether this movement to spread universal law will succeed?

The rule of law developed first inside those states which became bureaucratic and penetrated deeply into their own societies, so that every individual became subject to the law. For there to be a world law of human rights, there must be an organization which carries out an analogous penetration into every society around the world. This could be some kind of international organization or coalition. But—and this is my main point—its degree of success depends on its geopolitical strength. And that is to say that the expansion of universal rights and protection of those rights must go through a phase where the organizations upholding world law are geopolitically stronger than those who oppose it. This extension could be diplomatic, but it is bound to be at least partly military. International organizations will sometimes have to fight and win to establish world law. This may be accompanied by some peaceful extension, if the power-prestige of the international coalition grows stronger, attracting other societies who want to join, in another bandwagon effect.

The ideal of world law is where individuals are held responsible for crimes against human rights. But in order to get to that point, world bureaucratic organization has to penetrate all societies; and the struggle against this penetration is carried on by groups, not individuals. Struggles are bound to produce group animosities—following the principle that external attack increases group solidarity—so there are always processes like Islamic groups supporting al-Qaeda because it is perceived as a form of

---

loyalty to embattled Islam. And when conflicts are violent, there are always individual members of groups who are caught in conflicts for which they as individuals are not responsible. This is particularly true in war, where some civilians and noncombatants always get killed—since warfare is a very crude and dangerous instrument. But there seems to be no escape from this on the pathway to world law. On the opposing side, the crimes against humanity which some people are attempting to control—genocide, murderous ethnic cleansing, terrorist attacks—are by their very nature attacks on groups, not on individuals, and largely on civilian populations. It is only at the end of this process—in a territory where the rule of law prevails, and there is an organization to enforce it, which people consider legitimate—that law can successfully treat conflicts as crimes for which individuals are to be held responsible.

Finally, let us ask: where does the opposition to universal human rights come from? Much international ideological conflict of the last 20 years has pitted those regions with bureaucratic ideals against parts of the world which are still relatively more patrimonial. Interventions against ethnic cleansing and genocide are attempts to impose the universalism of bureaucratic regions upon the patrimonial ethics of noncosmopolitan, relatively closed communities whose structure fosters ethnic particularism and reinforces the bloody ritualism of group vendetta. Conflicts over the rights of women in the Islamic world also have this character: the bureaucratic part of the world pitted against patrimonial households that Islamic conservatives struggle to preserve. The conflict over international terrorism is a struggle between these two organizational forms. We see this organizational conflict in Afghanistan. 'Taliban' means students of a madrasa teacher, which is to say a traditional Islamic school in which the teacher acts like head of household for his students; and they are bound to him throughout their lives by ties of patrimonial and religious obligation. The Taliban was thus based on explicitly patrimonial organization, although it has to take on some bureaucratic elements as it attempts to administer the state. Fundamentalist or conservative Islam is a form of religious organization which is both patrimonial in its own church structure, and which sees itself in a violent struggle to maintain itself against the threat of the outside world based on bureaucratic organizational principles.

Over the long run of history, modern bureaucratic organization has everywhere prevailed over the patrimonial household. Much of international terrorism today is an attempt to defend the patrimonial structures remaining in parts of the world, against the structures and ideologies of bureaucratic organization. If world law and rights for individuals are based on bureaucratic organization, it is realistic to expect that the organizational procedures claiming to protect human rights will increase during future history. This will not be a smooth and continuous trend, since the international organizations for administering and enforcing rights are part of the struggle for geopolitical power, and are subject to geopolitical tensions and possibilities for breakdown. Human rights will become an increasingly widespread ideological theme, but their realization will depend on the contingencies of organized state power. And that has always been a process of ongoing tensions and conflict.

REFERENCES

Andreski, S. 1971. Military Organization and Society. Berkeley, CA: University of California Press.

Biddle, S. 2004. Military Power. Explaining Victory and Defeat in Modern Battle. Princeton, NJ: Princeton University Press.

Collins, R. 1978. Long-term Social Change and the Territorial Power of States. In Kriesberg, L. (ed.), Research in Social Movements, Conflicts, and Change. Vol. 1 (pp. 1–34). Greenwich, CT: JAI Press.

Collins, R. 1986. Weberian Sociological Theory. New York: Cambridge University Press.

Collins, R. 1992. The Geopolitical and Economic World-systems of Kinship-based and Agrarian-coercive Societies. Review 15 (summer): 373–388.

Collins, R. 1995. Prediction in Macro-sociology: The Case of the Soviet Collapse. American Journal of Sociology 100: 1552–1593.

Collins, R. 1999. Macro-History. Essays in Sociology of the Long Run. Stanford University Press.

Collins, R. 2004. Rituals of Solidarity and Security in the Wake of Terrorist Attack. Sociological Theory 22: 53–87.

Coser, L. 1956. The Functions of Social Conflict. New York: Free Press.

Gilpin, R. 1981. War and Change in World Politics. New York: Cambridge University Press.

Goldstone, J. A. 1991. Revolution and Rebellion in the Modern World. Berkeley, CA: University of California Press.

Kennedy, P. 1987. The Rise and Fall of the Great Powers: Economic Change and Military Conflict from 1500 to 2000. New York: Random House.

Mann, M. 1986, 1993. The Sources of Social Power. Vols 1, 2. New York: Cambridge University Press.

McNeill, W. H. 1963. The Rise of the West. A History of the Human Community. Chicago, IL: University of Chicago Press.

McNeill, W. H. 1982. The Pursuit of Power. Chicago, IL: University of Chicago Press.

Marwell, G., and Oliver, P. 1993. The Critical Mass in Collective Action. A Micro-Social Theory. New York: Cambridge University Press.

Parker, G. 1988. The Military Revolution: Military Innovation and the Rise of the West. New York: Cambridge University Press.

Scheff, Th. J. 1994. Bloody Revenge. Emotions, Nationalism and War. Boulder, CO: Westview Press.

Skocpol, Th. 1979. States and Social Revolutions. New York: Cambridge University Press.

Tilly, Ch. 1978. From Mobilization to Revolution. Reading, MA: Addison-Wesley.

Tilly, Ch. 1990. Coercion, Capital, and European States. A.D. 990–1990. Oxford: Blackwell.

Tilly, Ch. 1995. Popular Contention in Great Britain, 1758–1834. Cambridge: Harvard University Press.

Weber, M. 1968 [1922]. Economy and Society. New York: Bedminster Press.

Chapter 5

CONTEMPORARY GLOBALIZATION AND NEW CIVILIZATIONAL FORMATIONS

Shmuel N. Eisenstadt

In this article I would like to examine some specific aspects of contemporary globalization as they bear on the crystallization of new distinct civilizational formations. The new very intensive processes of contemporary globalization are characterized by growing interconnectedness between economic, cultural and political processes of globalization. The full impact of the processes can be understood only in the new historical context, especially against the background of changes in the international arenas which have been closely connected with processes of globalization during this period. Among different contemporary cultural and civilization forms we note a very important component of contemporary civilization attesting to the fact that different religions are now acting in a common civilizational setting. In this context competition and struggles between religions often became vicious—yet at the same time there developed strong tendencies toward the development of common encouraging interfaith meetings and encounters which focused on their relations in terms of some of the premises of the new civilizational framework rooted in the original program of modernity. These premises implied the possibility of cooperation between them—indeed, even going beyond that. Such attempts at the reformulation of civilizational premises have been taking place in some movements and in new institutional formations such as the European Union, in different local and regional frameworks, as well as in the various attempts by the different 'peripheries'.

Keywords: globalization, hegemonic center, contemporary civilization, civilizational formations, global confrontations, non-Western societies.

INTRODUCTION

The new very intensive processes of contemporary globalization are characterized by growing interconnectedness between economic, cultural and political processes of globalization. Each of these processes entails continuous encounters between different societies and their respective sectors. In the cultural arena the processes of globalization were closely connected with the expansion especially through the major media that were often conceived in many parts of the world as uniform, hegemonic and Western, above all American, cultural programs or visions, giving rise to strong tendencies for global cultural homogenization and, what has been referred to as 'de-traditionalization'.

These processes of globalization have been characterized by continual growing mutual impingement of different societies and social sectors throughout the world. This process gives rise to the possibility of more intensive confrontations between them. These processes entail the continual movements of hitherto peripheral, 'local' non-hegemonic groups and sectors to the centers of their respective national and internal systems. The movement from periphery into existing centers and also into emerging hegemonic centers often bypasses the trans-local institutions and public arenas; concomitantly there is a closely related movement of non-Western societies or sectors thereof into the hitherto mostly Western centers of modernity.

The movements of many 'peripheral', be they national or international, sectors into the very hegemonic centres of globalization, were connected first with the continual development of new modes of resistance to globalization, of various 'counter'-globalization tendencies and movements; these forms of resistance include the intensification of terrorist activities and associated tendencies to appropriate conventions of modernity thus leading to the development of new visions of civilization.

Second, such incorporation entailed continual intensive encounters and confrontations between different civilizational traditions and the respective hegemonic centres—encounters and confrontations which were intensified by the multiple movements of migration and by the impact of the media.

Third, the incorporation of multiple social sectors, indeed of entire societies into the global framework was closely interwoven with far-reaching processes of dislocation of large sectors of population of many societies and their push, as it were, into states of insecurity and anomie.

Fourth, there emerge growing discrepancies in economic, political and social processes between the hegemonic centres and the more peripheral sectors. Such discrepancies were of course characteristic both of 'traditional' pre-modern globalization, as well as of the processes of globalization of early modern period and in the era of the hegemonies of the nation and revolutionary states and of capitalist market economies. In contrast to such discrepancies in the earlier periods, contemporary discrepancies develop against the background of the homogenizing and centralizing tendencies and ideologies of the nation, revolutionary states, and more contemporary forces. These discrepancies entail the possibility for the continual mutual impingement of these different societies and social sectors.

Of special importance in this context is the combination of discrepancies between those social sectors which were incorporated into the hegemonic financial and 'high-tech' frameworks and those which were left out. The closely connected far-reaching dislocation of many of the people who comprise the latter sectors, suffered a decline in their standard of living and, as a result, gave rise to acute feelings of dislocation and dispossession. Most visible among such dislocated or dispossessed groups were not necessarily—and certainly not only—those from the lowest economic echelons—poor peasants, or urban lumpen-proletariat, important as they were in those situations. Rather, most prominent among such dislocated sectors were, first, groups from the middle or lower echelons of the more traditional sectors. Those sectors comprise people who were hitherto embedded in relatively stable, even if not very affluent, social, cultural and economic frameworks or niches. These sectors (and the people they comprise) were transferred into the mostly lower echelons of new urban centers. Secondly, large social sectors which were put out from the work force; and third, various highly mobile, 'modern' educated groups—professionals, graduates of modern universities and the like who were denied autonomous access to the new political centers or participation in them—find themselves dispossessed from access to the centres of their respective societies or from their cultural programs. Thus, for instance, it was not only the dislocation of the Shia clergy from strong positions in the cultural centre or close to it that was important in the success of the Khomeini revolution. Of no less importance was the fact that highly mobile modernized occupational and professional groups, which developed, to no small extent, as a result of the processes of modernization, and which were controlled by the Shah, were barred from any autonomous access to the new political center or participation in it—very much against the premises inherent in these processes. Such groups were especially visible in Turkey, India and Pakistan, and in many of the Muslim Diasporas in Europe—but they were also important in other Muslim or South Asian societies.

These groups often find themselves in a situation of social anomie in which old ways of life have lost their traditional standing. They are caught in the pressure of globalization and of international markets for greater efficiency and are losing their security nets and for whom the programs promulgated by the existing modernizing regimes, are not able to provide meaningful interpretations of the new reality. A very important group which may be highly susceptible to communal-religious or fundamentalist messages are younger generation of seemingly hitherto well-established urban classes who distance themselves from the more secular style of life of their relatively successful parents. But even more important are the relatively recent members of second-generation immigrants to the larger cities from provincial urban and even some rural centres (Eisenstadt, 1999).

CHANGES IN THE INTERNATIONAL ARENAS AND IN THE CONSTITUTION OF HEGEMONIES

The full impact of the processes analyzed above can be understood only in the new historical context, especially in the changes in the international arenas which have been closely connected with processes of globalization that have been taking place in this period.

The most important aspects of the new international scene were: first, shifts in hegemonies in the international order; second, the development of new power relations between different states; third, the emergence of new actors, institutions and new regulatory arenas and rules in the international arena. All of these changes attest to the continual disintegration of the 'Westphalian' international order with far-reaching implications for the transformation of political arenas, especially those of the national and revolutionary states.

In the continuous shifts in the relative hegemonic standing of different centres there developed the concomitant growing competitions or contestations between such centres about their presumed hegemonic standing. Second, there developed continual contestations between different societies and sectors about their place in the international order—and the concomitant increasing destabilization of many state structures—above all but not only in the different peripheries—all of them contributing greatly to the development of the 'New World Disorder' (Jowitt, 1993). The development of such a disorder was intensified with the demise of the Soviet Union, the disappearance of the bipolar order of the 'Cold War' and the relative stability it entailed, and of the disappearance of the ideological confrontation between Communism and the West. These developments—with only one Superpower, the US, remaining—gave rise to greater autonomy of many regional and trans-state frameworks and within these frameworks to new combinations of geopolitical, cultural and ideological conflicts and struggles over their relations standing and hegemony, including indeed those between major global powers—the US, the European Union, post-Soviet Russia and China.

Further, far-reaching transformations in the power relations in the international order took place around the last decade of the twentieth century. During the first two decades after the fall of the Soviet Union, the United States was not only the single superpower but also the almost non-contested hegemon, in both military and economic terms, of the neo-liberal economic order. This status was epitomized by the Washington Consensus being aggressively pursued by the major international agencies such as the International Monetary Fund and the World Bank. But the situation has greatly changed with the onset of the second post-Soviet decade. In this decade, post-Soviet Russia, China and to a smaller extent India and Brazil, became much more independent players in the international economic order, pursuing more independent policies, pursuing their own geopolitical as well as economic interests, generating changes in the balance of regional geopolitical and economic formations challenging the American hegemony, as well as the premises of the Washington Consensus. All these tendencies were intensified attendant on the international financial crises which developed from 2008 on—which shattered and transformed most of the hitherto predominant arrangements for regulating power relations in the international economic and political arenas.

INTERCIVILIZATIONAL SETTINGS— ANTI-GLOBALIZATION MOVEMENTS AND TRANSFORMATION OF MOVEMENTS AND IDEOLOGIES OF PROTEST

All these processes provided the background for the crystallization of new civilizational frameworks. One of the most important manifestations of the new civilizational framework that developed attendant on all the processes analyzed above has indeed been the close interweaving between the numerous anti-globalization movements and the new types of orientations and movements of protest that developed from the late sixties of the twentieth century. While intercivilizational 'anti-globalization' or anti-hegemonic tendencies combined with an ambivalent attitude to the cosmopolitan centres of globalization developed in most historical cases of globalization—be it in the Hellenistic, Roman, the Chinese Confucian or Hinduistic, in 'classical' Islamic, as well as early modern ones—yet on the contemporary scene they become intensified and transformed. First, they became widespread especially via the media throughout the world. Second, they became highly politicized, interwoven with fierce contestations formulated in highly political ideological terms. Third, they entailed a continual reconstitution in a new global context, of collective identities and contestations between them. Fourth, the reinterpretations and appropriations of modernity (giving rise to new inter-civilizational orientations and relations) were attempts by these actors to decouple radical modernity from Westernization, and to take away from the 'West', from the original Western 'Enlightenment'—and even Romantic programs—the monopoly of modernity; to appropriate modernity and to define it in their own terms, often above all in highly transformed civilizational terms. A central component of this discourse was a highly ambivalent attitude to the West, above all to the US, its predominance and hegemony most fully manifested in the worldwide expansion (including many European countries) of strong anti-American movements.

All these developments were perhaps most clearly visible in the various new Diasporas and virtual communities and networks. It was indeed within these virtual communities and networks that there developed extensive and highly transformed intensified 'reactions' to the processes of globalization, especially to the hegemonic claims of the different, often competing centers of globalization, attesting, to follow Arjun Appadurai's felicitous expression, 'the power of small numbers' (Appadurai, 2006) and constituting one of the most volatile and highly inflammatory components on the global scene; as well as an important factor in the transformation of inter-civilizational relations in the contemporary scene, often promulgating visions of clashes of civilizations.

One of the most important manifestations of the new civilizational framework that developed attendant on all the processes analyzed above has indeed been the close interweaving between these processes and the new types of orientations and movements of protest that have developed since the late sixties of the twentieth century (Eisenstadt 2006).

Movements and symbols of protest continued indeed to play a very important central role in the political and cultural arenas—as they did in the constitution and development of modern states—but their structure, as well as their goals of visions have been continually reinforced by the processes of globalization. The most important among these movements were the new student and anti-(Vietnam) movements of the late 1960s—the famous 'movements of 1968', which continued in highly transformed way in the great variety of movements that have developed since then. These movements and orientations went beyond the 'classical' model of the nation state and of the 'classical' or liberal, national and socialist movements, and they developed in two seemingly opposite but in fact often overlapping or cross-cutting directions. On the one hand, there developed various 'post-modern', 'post-materialist' movements such as the women's, ecological and anti-globalization movements; on the other hand, many movements promoted very particularistic local, regional, ethnic cultural autonomous movements that were very aggressive and ideological in spirit. Among different sectors of the dispossessed there also blossomed various religious-fundamentalist and religious-communal movements that promulgated conceptions of which identity was supreme above all others.

The themes promulgated by these movements were often presented or perceived as the harbingers of far-reaching changes being spawned by the contemporary cultural and institutional scene, indeed possibly also of the exhaustion of the entire classical program of modernity entailed far-reaching transformations, both in internal state and international arenas. In turn these themes of protest spawned the revolutionary imagination and thus were constitutive of the development of the modern social order and above all indeed of the modern and revolutionary states.

The common core of the distinctive characteristics of these new movements, attesting to their difference from the 'classical' ones, has been first the transfer of the central focus of protest orientations from the centers of the nation and revolutionary states and from the constitution of 'national' and revolutionary collectivities as the charismatic bearers of the vision of modernity into various diversified arenas of which the by now transformed nation states was only one; second, the concomitant weakening of the 'classical' revolutionary imaginaire as a major component of protest; third, the development of new institutional frameworks in which these options were exercised; and fourth, the development of new visions of inter-civilizational relations.

Contrary to the basic orientations of the earlier, 'classical' movements, the new movements of protest, were oriented to what one scholar has defined as the extension of the systemic range of social life and participation, manifest in demands for growing participation in work, in different communal frameworks, citizen movements, and the like. Perhaps the initial simplest manifestation of change in these orientations was the shift from the emphasis on the increase in the standard of life which was so characteristic of the 1950s as the epitome of continuous technological-economic progress to that of 'quality of life'—a transformation, which has been designated in the 1970s as one from materialist to post-materialist values. In Habermas' (1989) words these movements moved from focusing on problems of distributions to an emphasis on the 'grammar of life' (Taylor, 2007: 299–505). One central aspect of these movements was the growing emphasis, especially within those which developed among sectors dispossessed by processes of globalization, on the politics of identity; on the constitution of new religious, ethnic and local collectivities promulgating in narrow, particularist themes often in terms of exclusivist cultural identity—often formulated in highly aggressive terms.

Closely related to these processes was the transformation of the utopian, especially transcendental, orientations whether of the totalistic 'Jacobin' utopian ones that were characteristic of many of the revolutionary movements, or the more static utopian visions which promulgated a flight from various constraints and tensions of modern society. The focus of the transcendental utopian orientations shifted from the centers of the nation state and overall political-national collectivities to more heterogeneous or dispersed arenas, to different 'authentic' forms of life-worlds, often in various 'multicultural' and 'post-modern' directions.

In the discourse attendant these developments, above all in the West, but spreading very quickly beyond it, there developed a strong emphasis on multiculturalism as a possible supplement or substitute to that of the hegemony of the homogeneous modern nation-state model and as possibly displacing it.

New Intercivilizational Relations, Anti-Globalization Tendencies and Movements, Global Confrontations, Attempts at Appropriation of Modernity

The crucial differences from the point of view of civilizational orientations between, the major 'classical' national and religious, especially reformist, movements, and the new contemporary communal, religious and above all fundamentalist movements,—all of which were closely connected with the constitution of the new virtual communities—stand out above all with respect to their attitude to the premises of the cultural and political program of modernity and to the West. They constitute part of a set of much wider developments which have been taking place throughout the world, in Muslim, Indian and Buddhist societies, seemingly continuing, yet indeed in a markedly transformed way, the contestations between different earlier reformist and traditional religious movements that developed throughout non-Western societies.

These developments signaled far-reaching changes from the earlier reformist and religious movements that developed throughout non-Western societies from the nineteenth century to the present. Within these contemporary anti-global movements confrontation with the West does not take the form of searching to become incorporated into the modern hegemonic civilization on its terms, but rather to appropriate the new international global scene and modernity for themselves, in their own terms, in terms of their traditions.

These movements do indeed promulgate a markedly confrontational attitude to the West, to what is conceived as Western, and attempts to appropriate modernity and the global system on their own non-Western, often anti-Western, terms. This highly confrontational attitude to the West, to what is conceived as Western, is in these movements closely related either to the attempts to decouple radically modernity from Westernization or to take away from the West the monopoly of modernity, and to appropriate the contemporary scene, contemporary modernity in terms of visions grounded in their own traditions.

They aim to take over as it were the modern program in terms of their own civilizational premises, which are rooted, according to them, in the basic, indeed highly reformulated images and symbols of civilizational and religious identity—very often formulated by them as the universalistic premises of their respective religions or civilizations, and aiming to transform the global scene along such terms.

At the same time, however, the vistas grounded in these traditions have been continually reconstituted under the impact of 'modern' programs and couched paradoxically enough in terms of the discourse of modernity in the contemporary scene. Indeed these discourses and the discussions around them resemble in many ways the discourse of modernity as it developed from its very beginning in the very centres of the modernities in Europe, including far-reaching criticisms of the predominant Enlightenment program of modernity and its tensions and antinomies. Thus, for instance, many of the criticisms of the Enlightenment project as made by Sayyid Qutb, possibly the most eminent fundamentalist Islamic theologian, are in many ways very similar to the major religious and 'secular' critics of Enlightenment from de Maistre, the romantics, the many populist Slavophiles in Central and Eastern Europe, and in general those who, in Charles Taylor's words emphasized the 'expressivist dimension of human experience', then moving, of course, through Nietzsche up to Heidegger. Or, in other words, these different antiglobal and anti-Western movements and ideologies reinforce in their own terms the basic tensions and antinomies of modernity, attesting—perhaps in a paradoxical way—that they constitute components of a new common global civilizational framework rooted in the program of modernity, but also going beyond it.

Another very important component of the contemporary civilizational scene attesting to the fact that different religions are now acting in a common civilizational setting is the changes in the relations between the different—especially the 'major'—religions. Competition and struggles between religions became very often vicious—yet at the same time there developed strong tendencies to the development of common encouraging interfaith meetings and encounters focused on their relations to some of the premises of the new civilizational framework rooted in the original program of modernity and on the possibility of cooperation between them – but indeed going beyond it.

Such attempts at the reformulation of civilizational premises have been taking place not only in these movements, but also—even if perhaps in less dramatic forms—in new institutional formations such as the European Union, in different local and regional frameworks, as well as in the various attempts by the different 'peripheries'—as for instance in the discourse on Asian values, to contest the Western, especially American, hegemony, as well as to forge their own constitutive modernities. These reformulations of rules and premises have also been taken up by many developments in the 'popular' cultural arenas challenging the seeming predominance of the American vision. Thus giving rise to distinct new trans-state Indian and East Asian media productions and regional, diasporic and even global spheres of influence.

The debates and confrontations in which these movements or actors engage and confront each other may often be formulated in 'civilizational' terms, but these very terms—indeed the very term 'civilization' as constructed in such a discourse—are already couched in the language of modernity, in totalistic, very often essentialistic, and absolutizing terms derived from the basic premises of the discourse of modernity, its tensions and antinomies, even if it can often draw on older religious traditions. When such clashes or contestations are combined with political, military or economic struggles and conflicts they can indeed become very violent.

Indeed, at the same time, the combination of the far-reaching changes in the international arena and the distinct characteristics of the contemporary processes of globalization with the changes in the structure of the international arena has given rise to the multiplication and intensification of aggressive movements and inter-civilizational contestations and encounters.

Indeed among various anti-global movements, of special importance was the multiplication, extension and intensification of highly aggressive terrorist movements, which became closely interwoven with international and intercivilizational contestations and encounters. Already in the first period of the post (Second) World War era, a central component of the international scene was the growth of revolutionary and terrorist groups and this component became even more central being interwoven with the crystallization of new international and intercivilizational orientations, new patterns of intercivilizational relations. When these transformations became connected with increasing confrontations in many societies, both in local, as well as in global scenes and arenas, and with political, military or economic struggles and conflicts they can indeed become very violent; they may become a central player in connection with movements of independence of different regional contestations, what G. Münkler (2003) has defined as non-symmetric wars, in contrast with the symmetric wars between nation-states in the framework of the Westphalian order, which became a continual component of the international order and in which such movements played a central role.

REFERENCES

Appadurai, A. 2006. Fear of Small Numbers: An Essay on the Geography of Anger. Durham: Duke University Press.

Eisenstadt, S. N. 1999. Fundamentalism, Sectarianism and Revolution: The Jacobin Dimension of Modernity. Cambridge: Cambridge University Press.

Eisenstadt, S. 2006. The Great Revolutions and the Civilizations of Modernity. Leiden: Brill.

Habermas, J. 1989. The Structural Transformation of the Public Sphere. Oxford: Polity Press.

Jowitt, K. 1993. New World Disorder: The Leninist Extinction. Berkeley, CA: University of California Press.

Münkler, H. 2003. Über den Krieg: Stationen der Kriegstgeschichte in Spiegel ihrer theoretischen Reflexion. Weilerwist: Velbrück.

Taylor, Ch. 2007. A Secular Age. Cambridge, MA: Harvard University Press.

Chapter 6

THE 'RETURN' OF RELIGION AND THE CONFLICTED CONDITION OF WORLD ORDER

Roland Robertson

The question of the return of religion to the study of world politics and international relations is considered in terms of the neglect of religion since the Peace of Westphalia. This neglect has largely occurred because of the primacy given to changes and events in the West, particularly since the formal separation of church and state and its imposition on or emulation by Eastern societies. The recent concern with globalization has provided the opportunity to undertake historical discussion in new perspectives which overcome the Western 'normality' of the absence of religion from Realpolitik. Moreover, it is argued that much of the neglect of religion in work on world affairs has largely been the product of the inaccurate perception of ongoing secularization. The overall discussion is framed by some objections to the limiting consequences of disciplinarity.

Keywords: religion, globalization, disciplinarity, Realpolitik, international relations.

INTRODUCTION

While this paper is primarily concerned with the conditions that are giving rise to the conspicuousness of religion in contemporary international politics, it should be said at the outset that the recent controversy surrounding the alleged evils of religion—notably in the UK and the USA—is less than marginal to this focus. This is because much of the polemical 'shouting' that has issued from the anti-religious, or anti-God, camps has undoubtedly been much influenced by the overlapping presence of religion in intranational, transnational, and international politics. At the same time the militancy of, for example, Richard Dawkins and Christopher Hitchens has certainly contributed significantly to the presence of religion in the minds of contemporary politicians, journalists and academics (Dawkins, 2006; Hitchens, 2007). Another big controversy has also played a part in subduing the significance of religion in international affairs—namely, the significance of religion in the American policy toward Israel. It has become very clear in recent years that this is a subject which many avoid, for fear of arousing ethnic passion that can effectively damage academic careers, at least in the UK and the USA.

However, there is currently a strong move within sociology, philosophy and related disciplines away from atheistic secularism. This shift against the latter, as well as proliferating critiques of the idea of rampant secularization, is of great consequence for the general comprehension of global trends and circumstances (Robertson 2007; McLennan, 2006; Robertson & Chirico, 1985).

What follows is divided into two sections. The first deals with what can be called the 'disciplinary' world, while the second may be called the 'real' world. There are most certainly oversimplifications involved in this delineation, not least because what was once a matter of disciplinarity then becomes central to reality. Put another way, while disciplinarity is a constructed, 'artificial' way of comprehending reality, at the same time reality is partly constituted by disciplinarity. The complexity of this problem cannot be pursued here—not merely for reasons of space, but also because it has been, and will continue to be, an intractable one in all of the sciences, both natural and human. Many philosophers have sought over the centuries, in different civilizational contexts, to solve this epistemological and/or ontological problem and many have claimed to have resolved it. In full recognition of various contributions of the latter kind, in this paper the author will simply take the problem for granted and deal with it in a very simple way.

THE PROBLEM OF DISCIPLINARITY

At least since the late 18th or early 19th centuries interpretation and/or analysis of the world have, for the greater part, been undertaken from increasingly specialized and compartmentalized perspective. A vast amount has been written about the origins, the histories and the genealogies of various disciplines, as well as variations in such from society to society, region to region, and civilization to civilization. Nonetheless, it should be stipulated here that the present focus is primarily a Western one and that it involves no systematic attempt to be specific about the civilizational structuring of particular academic disciplines; nor of their trajectories or configurations within different societies. What has to be firmly stated is that each discipline in the western academy, as well as in the primary and secondary sectors of school systems, has rested upon rhetorical constructions and academic contingencies. Thus the idea that disciplines reflect the natural condition of life is without any foundation. One has to make this point strongly, precisely because it seems that many academics and intellectuals—and not least their bureaucratic administrators—do believe that disciplines reflect or grasp reality, although some of these may also grant that so-called reality is partly constituted by disciplinarity.

In spite of these considerations it should be said that throughout the last century and a half or so various individuals and schools of thought have attempted to overcome or lay out the preconditions and sustaining infrastructures of the disciplines on a universalistic basis. For example, Comte made an extended attempt to connect systematically all disciplines, Marx also approached the same issue (but, of course, from a very different perspective), as did John Stuart Mill. The same might be said of Freud and certainly this is true of the rise of General Systems Theory in the 1930s and also of the work of Talcott Parsons during the mid-twentieth century. Foucault explored rather thoroughly the basis and forms of disciplinarity in the broadest possible sense—which led in his work to the casting of academic discipline as similar to discipline in the penal sense.

Increasingly, during the past twenty years or so, there has been much disciplinary mutation, particularly around the theme of globalization. Much of the study of the latter, in spite of its enormous fashionability, has unfortunately been centred upon the idea of interdisciplinarity. This has been very counterproductive and has served more the bureaucratic interests of academic administrators and power-seekers within academic professions than it has the enhancement of substantive intellectual progress. Specifically, interdisciplinarity has consolidated, rather than overcome, disciplinary and professional distinctiveness. For example, interdisciplinary collaboration often involves the practitioners of two or more disciplines getting together and seeing what each can contribute to a particular topic from their own disciplinary standpoint. What, on the other hand, ideally ought to occur is a direct concern with the substantive issue as opposed to a rehearsal of the identity of particular disciplines. Many enterprises of a so-called interdisciplinary nature have entailed little more than each disciplinary representative pronouncing what her or his discipline could/should contribute to the topic in question. Thus, we should turn in the direction of what preferably should be called either cross-disciplinarity or trans-disciplinarity (although cogent claims could and have been made on behalf of 'counter-disciplinarity' and 'post-disciplinarity').

In the case in hand—namely the study of international relations, or world politics, in connection with the study of religion—such reflections on the limitations of disciplinary approaches have contributed mightily to the relative absence of attention to the interpenetration of religion and IR in historical terms. This means that rather than trying to account for the great significance of religion in world politics at the present time—as if religion had suddenly erupted onto the world-political scene—we would be much better advised to try and account for why the relationship between religion and IR has been grossly neglected for many years. Indeed, International Relations as a discipline or sub-discipline was professionally established without any reference to the significance of religion. While economics has often been called the dull science, that label might well be equally applied to IR, at least until fairly recently.

From the standpoint of those who have been mainly concerned with religion, the obsession with the secularization thesis has served both to insulate the sub-discipline of the sociology of religion from other disciplinary perspectives and subdue its influence in the wider society. Indeed, for many decades, sociologists of religion have, not unironically, expressed much regret that their sub-discipline is marginal to the wider discipline of sociology and have complained in so doing that the findings of the sociology of religion are not taken seriously by political elites and the more intellectual elements in the mass media. Thus, since the 1960s individual practitioners of the sociology of religion have only recently been invited to contribute to discussions of political events, circumstances and trends. For much of this period students of religion have been mainly involved in public discussion in relation to controversies surrounding religious sects (sometimes called 'cults'). However, with the advent of religions of violence sociologists have been called upon increasingly to participate in public debate and give advice to governments, particularly since 9/11. Similar, but greater, neglect of religion can be said of the study of world politics, although there has been an increasing concern with the relation between religion and societal politics during the same period. Thus the neglect of religion has been most evident in the study of international relations.

Another significant factor in the neglect of the involvement of religion in international relations is the way in which assumptions have been formed concerning distances between particular disciplines or subdisciplines. For example, at the beginning of the twentieth century it was possible for historians and sociologists to say that it was strange to connect the study of religion to the study of economics. Within a few years, however, the relationship between religion and the development of capitalism had assumed the status of the obvious. Much of this was a consequence of Max Weber's The Protestant Ethic and the Spirit of Capitalism which was first published in 1904/1905 (Weber, 1930). In the particular case at hand, many practitioners on the IR side would have asked, only a few years ago, what religion could have to do with their own domain of analysis? Now, in the early years of the 21st century very few would be so daring—perhaps, one might say foolish—as to ask this same question. Undoubtedly this has a great deal to do with the present so-called war on terror (a term which is, in fact, fast retreating) and, more specifically, with the problematic thesis as to the clash of civilizations (which is also in retreat). To be more precise, it is the centrality of jihadist, or caliphate, Islam and its opponents—not to speak of its targets—that has been so crucial in the attention to the subject of the present paper. The surprise among the relevant disciplines as to the apparent eruption of Islam onto the world scene as symbolised and expressed by the events of 9/11 now seems rather difficult to comprehend (Lincoln, 2006). Even most of those who have been studying religion and regretting its marginality within and without the academy seem to have been amazed by 9/11. This can, in significant part, be attributed to the insulation of IR from the study of religion and vice versa. On the other hand, it should be said that the study of the politics/religion connection had been expanding in the last quarter of the 20th century.

This expanding interest almost certainly had much to do with the increasing conspicuousness of religion within and without nation-states since the late 1970s. At that time such events as the coming to power in Nicuragua of the Sandinistas, the complex connection between those opposed to the latter, Iran and the US Republican government (the so-called Iran–Contra affair); the injection of theocratic ideas into the global arena in the aftermath of the Iranian revolution of 1979; and the rise of the Solidarity movement—heavily backed by the Catholic Church—in Poland raised, so to speak religion, above the parapet for systematic attention. The spread and intensification of tensions between 'church' and state constituted the end of a long era that had begun following the Peace of Westphalia in 1648 which had marked the termination of religious wars within the West. Prior to the Westphalian settlement, the sacred and the profane were seen to have coexisted—although often problematically. Westphalia marked the end of such coexistence, in such a way as to largely separate religion from politics.

The consummation of that trend was the Declaration of Independence in the nascent American Republic in 1776, with its commitment to the constitutional separation of church and state. This rapidly produced globe-wide implications, even more important than in the USA itself (Armitage, 2007). Moreover, it was not a coincidence that it was in this same period that Jeremy Bentham pronounced, in 1789, the need for a specialised focus on international relations. Few scholars have recognized the significance of this conjunction. However, this was a Western phenomenon which was, nonetheless, imposed upon, or emulated by, a number of Asian societies during the late 19th and the first half of the 20th centuries. The variety of political orientations to attempts to disentangle religion and politics—or church and state—cannot be explored here. Suffice it to say that in East Asia one finds that whereas in China the demise of religion was taken to be a prerequisite of a modern society, in Japan there was a serious attempt to emulate the American separation of 'church' from state. In Japan State Shinto was established in the Meiji period by denying that it constituted a religion in the Western sense of the word. In contrast, the Chinese political elite and leading intellectuals took the lead from such Western philosophers as Bertrand Russell and insisted that there was no significant place for religion in a modern society (Robertson, 1992: esp. 115–128, 146–163; also Gong, 1984).

The areas of the world which most strongly resisted both of these trends were, overwhelmingly, Islamic. In view of this it is not surprising that the 'return' of religion to the international arena should have come in the form of a conflict between Islam and much of the rest of the world, particularly those parts of the latter that were seen to be particularly responsible either for the separation of religion and politics and church and state, or the imposition of state organized atheism, as in Communist regimes.

The considerable interest in the theme of globalization has undoubtedly drawn attention to the significance of religion in world politics and international relations. In arguing this I am emphasizing strongly the multidimensionality of globalization. Rather than conceiving of the latter in the form of neoliberalism, thus giving it a distinctively economic gloss, I regard it as having political, social, and cultural dimensions. This type of broad conception of globalization has constituted the basis of an ever-expanding interest in global, or world, history. This revival in the study of world history is significantly different from the kind of West-skewed interest in the latter that thrived at the end of the 19th and the beginning of the 20th centuries. The new global history—at least as it is practised in the West—is not anywhere near so Eurocentric. In fact has not infrequently been anti-Eurocentric. This means that in many societies and world regions different, often competing, paradigms and images of global history are being presented and promoted. Many, if not most, of these involve situating a particular society or region at the centre of world history. Clearly, this has a great deal to do with the present globe-wide concern with national identities.

In the frame of globalization this has come about for two main reasons. On the one hand, globalization involves the increasing connectivity of the global whole—sometimes expressed as a compression of the entire world, producing a circumstance in which each society, region or civilization is under constraint to identify and proclaim its own uniqueness. On the other hand, globalization also involves increasing global consciousness—better, self-consciousness, in the sense that, with periodic interruptions, the world as it increasingly has become 'one place'. This frequently neglected feature of globalization enhances, problematically, the sense of humanity being one. Needless to say, in recent times, the actuality of pandemics, epidemics, climate change—as well as the rise of religions concerned with 'the end-time'—has greatly consolidated this heavily contested oneness. In fact, much of the contemporary globe-wide concern with religious and civilizational conflict is centred upon the issue of religio-cultural hegemony.

In the wake of the rise of a new form of global history there has also arisen a fast-growing interest in the subject of imperialism and its great relevance to the theme of globalization. A good example of this is John Darwin's book, After Tamerlane: The Global History of Empire. Darwin argues that 'Tamerlane was the last of the series of "world-conquerors" in the tradition of Attila and Genghis Khan, who strove to bring the whole of Eurasia—the "world island"—under the rule of a single vast empire' (Darwin, 2007; cf. Bayly, 2002). After 1405 there soon began the exploration of the sea routes that became what Darwin calls 'the nerves and arteries of great maritime empires' (Darwin, 2006: x; Fernandez-Armesto, 2006). The European expansion after Tamerlane led to 'the rise of the West', but when the European empire dissolved—in the period lasting from the beginning of World War Two until the mid-twentieth century—the story of world history began to be retold, particularly with the rise of the so-called Third World. As Darwin says, this retelling cannot be written without a fully global view of the past. He cogently quotes Teggart, who in his Rome and China argued that 'the study of the past can become effective only when it is fully realized that all peoples have histories, that these histories run concurrently and in the same

---

world and that the act of comparing them is the beginning of knowledge' (Darwin, 2007: xi; Teggart, 1939; Robertson, 1998). This suggestion of the need to coordinate inter-unit relations with comparative analysis is, perhaps, the most important step forward that we must make in the study of international relations. This has been the main methodological consequence of the widespread concern with globalization. Much has recently been written about the need for new approaches to the latter, but virtually none of this has dealt with this analytic desideratum, or with the substantive relevance of religion and culture. Undoubtedly the 'terror wars' that were, in a sense, 'scripted' by Huntington's Clash of Civilizations, have brought religion—via radically politicized Islam—into a central, but highly problematic position, in world affairs. But little has been seen of the necessary analytic readjustments (Huntington, 1996). On the IR side, this has much to do with the so-called positivism of the discipline, which has largely eschewed any concern with such matters, particularly in the USA. Nye's concept of soft power is a rather meagre acknowledgement of these kinds of consideration (Nye, 2004).

Even though IR has continued since its inception in the early 1920s to display continuing controversies about Realpolitik it has nonetheless been overwhelmingly centred upon 'realistic' motifs. Many would, perhaps, contest this strong argument, but it is here claimed that—at least until recently—that has been the case. The rising attention to international society, global civil society, and global society by what Buzan calls 'the English School' contrasts with the emphasis on Realpolitik. Buzan seeks to establish a view of world society as 'a concept to capture the non-state side of the international system' or, to put it more elaborately, to 'create a synthesis between the structural elements of the Bull/Vincent side of English school theory about international and world society, and Wendt's... social theory of international politics'. In so doing Buzan speaks disparagingly about 'the analytical vacuousness of "the 'G' word"' (Buzan, 2004: 3; see also Wendt, 1999). (Wendt, of course, refers to the concept of globalization.) However, despite some praiseworthy attempts to bring back the social into IR, Buzan dangerously simplifies the concept of globalization. This is so, largely because he treats the latter concept in primarily political terms.

The unidimensional tendencies of many contributions to globalization theory have severely limited its analytical and empirical purchase, even though Buzan himself displays considerable interest in some sociological conceptions of world society outside conventional IR. For example, he attends, appropriately, to the work of the so-called Stanford school (led by John Meyer) which has promoted an important extra-IR view of the world as a whole. In the process, on the other hand, he has entirely neglected the major contributions of the Stanford school to the study of religion. Undoubtedly, there is a failure in the meeting of minds in so far as he rightly accuses members of the Stanford school of being either unaware or unwilling to consider the work of such people as Martin Wight and Hedley Bull in the English school of IR. Buzan rightly emphasizes that one—if not the—central concerns of the work of the Stanford sociological school is that of global culture. However, he overlooks the fact that a close relation of the Stanford school has been what used to be called the Pittsburgh school, whose major figures included Roland Robertson, Frank Lechner, Peter Beyer and Victor Roudometof. In the works of such sociologists religion has been absolutely central. Another lacuna in Buzan's approach is the neglect of the fact that some representatives of the sociological approach to globalization deny that, that process is greatly concerned with what has conventionally been called micro-sociological aspects of what Robertson has conceptualized as the global field (Robertson, 1992: 25–31). In articulating his ideas about the latter, Robertson has typologically divided the world into four major elements: individual states (national or otherwise), the system of states (or nation-states), humanity; and, not least, individual selves. The principal reason for including the latter within the frame of globalization is that it is completely impossible—when one seriously thinks about it—to exclude individuals from the world! Nonetheless, the idea that globalization is primarily a macro topic continues, in spite of anthropologists and sociologists insisting that globalization occurs interpersonally, that personal interaction can have very large consequences and that globalization occurs on the street, in the supermarket, in marital and other relationships, among but a few examples.

A great deal has recently been written in the millennial genre. This can be seen in both utopian and dystopian forms. For the most part, the present global millennial concern is more of the latter than the former kind, certainly in the Western portion of the world. It is in terms of this standpoint that it is particularly necessary to consider the relationship between religion and IR (Robertson, 2007).

The millennial and apocalyptic view of the 'terror wars', is at the centre of what may be called the religiocultural turn in world politics, specifically the relationship between radically politicized Islam and the 'modern West'. Indeed, the degree to which this global conflict between the two major actors on either side—namely al-Qaeda and the Bush regime in the USA—has assumed heavily religious terms cannot responsibly be questioned. However, there are those who still cling implausibly to the contention that this conflict is 'really' about oil, water and other material factors. The insistence on reducing all phenomena to a single factor is, however—it should be noted—a distinctively Western disposition. The failure to recognize that all human phenomena and interaction are—to put it in 'Western' terms—multidimensional, or

---

multifactorial has been, throughout the course of Occidental history, an egregious limitation. Looked at from another angle, we should not now be misled by the fact that communist regimes, for example, have claimed that they have considered international relations and world politics in 'atheistic' terms. Needless to say, virtually all communist or neo-communist regimes have claimed to be committed to either the complete elimination of religion or its totalitarian control. But, from a sufficiently sophisticated perspective, one can surely see that such ideological commitments have been framed historically by ancient religious traditions. In the most obvious case of Marxist Communism the religious or theological context of such is well documented. In any case, within forms of orthodox Marxism this embeddedness has been clearly acknowledged—for example, by Engels in his writing about European peasant utopianism as a forerunner of working class militancy and in Marx's contention that theology provides the basic categories for theoretical struggles (Burleigh, 2006a, 2006b).

At the same time, it has not been sufficiently recognized that the major opponent of Communism also has had a very strong millennial culture (Gray, 1998: 157; 2007; Harrington, 1986; Reynolds, 2002: 243–260). Or, at least, the millennial thrust of American culture—at least since the late 18th century—has rarely been analysed and represented from an international affairs standpoint. In this specific sense, IR, as well as the sociology of religion, have both 'developed' as forms of false consciousness. It should be reiterated that the obsession with the secularization thesis among a (declining) majority of sociologists of religion has been as responsible for the neglect of politics / IR as has the neglect of religion from the latter side.

CONCLUSION

The main concern in this article has been with the way in which religion has appeared in recent years to be a crucial theme in world politics and international relations, not least because religion appears to be at the centre of some of the world's most formidable global conflicts. It has been argued that the new global history that has developed in the context of the disputed concern with globalization provides us with an opportunity to comprehend how, on the one hand, religion has been greatly overlooked in the interrogation of world politics and how, on the other hand, the study of religion—particularly in its sociological form—has similarly neglected international relations because of its continuous and misplaced concern with secularization. In sum, on both sides of the equation there has been much mutual neglect. This has been largely attributed to the structure of academic disciplines, particularly in the Western world.

REFERENCES

Armitage, D. 2007. The Declaration of Independence: A Global History. London: Harvard University Press.

Bayly, C. A. 2002. 'Archaic' and 'Modern' Globalization in the Eurasian and African Arena, c. 1750–1850. In Hopkins, A. G. (ed.), Globalization in World History (pp. 47–73). London: Pimlico.

Burleigh, M. 2006a. Earthly Powers: Religion and Politics in Europe from the French Revolution to the Great War. London: HarperCollins.

Burleigh, M. 2006b. Sacred Causes: Religion and Politics from the European Dictators to Al Qaeda. London: HarperCollins.

Buzan, B. 2004. From International to World Society? English School Theory and the Social Structure of Globalization. Cambridge: Cambridge University Press.

Darwin, J. 2007. After Tamerlane: The Global History of Empire. London: Allen Lane.

Dawkins, R. 2006. The God Delusion. London: Bantam.

Fernandez-Armesto, F. 2006. Pathfinders: A Global History of Exploration. Oxford: Oxford University Press.

Gong, G. 1984. The Standard of 'Civilization' in International Society. Oxford: Clarendon Press.

Gray, J. 1998. Global Utopias and Clashing Civilizations: Misunderstanding the Present. International Affairs 74: 149–164.

Gray, J. 2007. Black Mass: Apocalyptic Religion and the Death of Utopia. London: Allen Lane.

Harrington, M. 1986. The Dream of Deliverance in American Politics. New York: Knopf.

Hitchens, Ch. 2007. God is Not Great. London: Allen and Unwin.

Huntington, S. P. 1996. The Clash of Civilizations and the Remaking of World Order. New York: Simon Schuster.

McLennan, G. 2006. Towards Post-secular Sociology? Sociology 41: 857–870.

Lincoln, B. 2006. Holy Terrors: Thinking about Religion after September 11. Chicago: University of Chicago Press.

Nye, J. S., Jr. 2004. Soft Power: The Means to Success in World Politics. New York: Public Affairs.

Reynolds, D. 2002. American Globalism: Mass, Motion and the Multiplier Effect. In Hopkins, A. G. (ed.), Globalization in World History (pp. 243–260). London: Pimlico.

Robertson, R. 1992. Globalization: Social Theory and Global Culture. London: Sage.

Robertson, R. 1998. The New Global History: History in a Global Age. Cultural Values 2: 368–384.

Robertson, R. 2007. Global Millennialism: A Post-mortem on Secularization. In Beyer, P., and Beaman, L. (eds.), Globalization, Culture and Religion (pp. 9–34). Leiden: Brill.

Robertson, R., and Chirico, J. 1985. Humanity, Globalization and Worldwide Religious Resurgence: A Theoretical Exploration. Sociological Analysis 46: 219–242.

Teggart, F. 1939. Rome and China. Berkeley, CA: University of California Press.

Weber, M. 1930. The Protestant Ethic and the Spirit of Capitalism. London: Allen and Unwin.

Wendt, A. 1999. Social Theory of International Politics. Cambridge: Cambridge University Press.

Chapter 7

CULTURE AND THE SUSTAINABILITY OF THE GLOBAL SYSTEM

Ervin Laszlo

The values and associated behaviors of the dominant culture of the contemporary world gave rise to a globally extended system that is not sustainable in its present form. If a cataclysmic breakdown is to be averted, the influential culture that shapes today's world must change. Humanity can no longer afford to be dominated by a narrowly materialist and manipulative culture focused on ego-centered, company-centered, or nation-centered short-term benefit, with no regard to the wider system that frames existence on this planet. Consciously moving toward a harmonious system of cooperative societies focused on the shared objective of sustaining the systems of life on the planet is an urgent necessity. To this end a mutation is needed in the cultures of the contemporary world, so as to create the values and aspirations that would bring together today's individually diverse and largely self-centered societies in the shared mission of ensuring the sustainability of the global system of humanity in the framework of the biosphere.

The global system is highly diverse today, but it is insufficiently coordinated. Creating a higher level of unity within its diversity is intrinsically feasible: it calls for system-maintaining cooperation among the diverse societies that make up the system.

Keywords: sustainability, cultural mutation, global warming, diversity, cooperation.

THE CULTURAL ROOTS OF THE UNSUSTAINABILITY OF THE CONTEMPORARY WORLD

Today's socioeconomic and ecological world system is structurally unstable and dynamically unsustainable. This condition has been created by practices oriented by the values and perceptions of a dominant layer of society. These values and perceptions have now become largely obsolete. For example:

Nature is inexhaustible. The long-standing belief that the Earth is an inexhaustible source of resources and an inexhaustible sink of wastes leads to the over-mining of natural resources and overloading of the biosphere's regenerative cycles.

The biosphere is a mechanism. The belief that we can engineer the biosphere like a building or a bridge is producing a plethora of unforeseen and vexing side-effects, such as the destruction of natural balances and the disappearance of myriad living species.

Life is a struggle where the fittest survives. This application of Darwin's theory of natural selection to society is mistaken in principle (Darwin did not mean by the 'fittest' the strongest and most aggressive, but the most adaptive and cooperative), and it is dangerous: it produces a growing gap between rich and poor, and legitimates the use of force on the premise that the possession of power is the natural attribute of a species that is fit to survive.

The market distributes benefits. The free market, governed by Adam Smith's principle of the 'invisible hand', is believed to distribute the benefits of economic activity in society. However, the poverty and marginalization of nearly half of the world population indicates that under current conditions trust in this belief is unfounded. The invisible hand does not operate: the holders of wealth and power garner for themselves a disproportionate share of the material benefits resulting from economic activity.

Some of the current beliefs produce paradoxical conditions.

- Millions are suffering from overeating and obesity, while a thousand million go hungry;
- Six million children die annually of starvation, and 155 million are overweight;
- There are millions of intelligent women ready to play a responsible role in society, but they do not get a fair chance in education, business, politics, and civic life;

- In order to save on the cost of labor, millions are put out of work, wasting human capital that would be essential to tackle the social, economic, and environmental problems now faced by humanity;

- Vast herds of animals are brought into the world for the sole purpose of being slaughtered for meat, something that, apart from its questionable ethical and health implications, is wasting an enormous amount of water and grain, resources urgently needed to ensure nutrition for human populations;

- The problems of the human community call for long-term solutions, but the criterion of success in the business world is the bottom line in annual or semi-annual corporate profit-and-loss statements;

- The planet is bathed in solar energy, and technologies are on-line to tap the energy of wind, tides, hot subsurface rocks, biomass, and animal waste and side-products, yet the world continues to run on polluting and finite fossil fuels and inherently dangerous nuclear power;

- Hi-tech weapons that are more dangerous than the conflicts they are intended to cope with are developed and stockpiled at vast investment of money and human and natural resources, and;

- The ineffectiveness of military force to achieve economic and political objectives has been demonstrated over and over again, yet the world's governments spend over $1.2 trillion dollars a year on arms, wars and military establishments, and similar amounts on empire-building objectives often disguised as projects of national defense and homeland security.

Such values and beliefs, and the conditions, to which they give rise, produce multiple strands and forms of unsustainability. They are manifest in the contemporary world in the sphere of society, in that of the economy, as well as in the domain of the ecology.

The Strands of Unsustainability

1. Unsustainable conditions in society

In the rich countries job security is disappearing, competition is intensifying, and family life is suffering. More and more men and women find satisfaction and companionship outside rather than within the home. And in the home, many of the functions of family life are atrophying, taken over by outside interest groups. Child rearing is increasingly entrusted to kindergartens and company or community day-care centers. The provision of daily nourishment is shifting from the family kitchen to supermarkets, prepared food industries and fast food chains. Leisure-time activities are colored by the marketing and public relations campaigns of commercial enterprises. Children's media exposure to TV, video games, and 'adult' themes is increasing, and it motivates violent and sexually exploitative behavior. In the United States the rate for first marriages ending in divorce is fifty percent, and about forty percent of children grow up in single-parent families for at least part of their childhood.

Social structures are breaking down in both the rich and the poor countries. In poor countries the struggle for economic survival destroys the traditional extended family. Women are extensively exploited, given menial jobs for low pay; often they are obliged to leave the home in search of work. Fewer and fewer women have remunerated jobs and more and more are forced to make ends meet in the socially and economically marginal informal sector. According to the International Labour Organization, fifty million children, for the most part in Africa, Asia, and Latin America, are employed for a pittance in factories, mines, and on the land. In some countries destitute children are recruited as soldiers and forced into prostitution, or are forced to venture into the streets as beggars.

2. Unsustainability in the economy

The human community is economically polarized: there is a large and in some regions still growing gap between diverse layers of population. The gap depresses the quality of life of hundreds of millions, and reduces the chances of survival of the poorest and most severely marginalized populations.

a) Wealth distribution. Wealth and income differences have reached staggering proportions. The combined wealth of the world's billionaires equals the income of three billion people, nearly half of the world's population. Eighty percent of the global domestic product belongs to one billion, and the remaining twenty percent is shared by six billion.

Poverty has not diminished in absolute numbers. In the poorest countries seventy-eight percent of the urban population subsists under life-threatening circumstances—one in three urban dwellers lives in slums, shanty towns, and urban ghettoes, and nearly one billion are classified as slum-dwellers. Of the seven billion people who now share the planet, 1.4 billion subsist on the equivalent of less than 1.25 dollars a day and an additional 1.6 billion live on less than 2.50 dollars.

b) Resource use. The rich-poor gap shows up in food and energy consumption as well as in the load placed on natural resources. People in North America, Western Europe, and Japan consume 140 per cent of their daily caloric requirement, while populations in countries such as Madagascar, Guyana, and Laos live on 70 per cent. The average amount of commercial electrical energy consumed by the Africans is half a kilowatt-hour (kWh) per person; the corresponding average for the Asians and Latin Americans is 2 to 3 kWh, and for the Americans, Europeans, Australians, and Japanese it is 8 kWh. The average American burns five tons of fossil fuel per year, in contrast with the 2.9 tons of the average German and places twice the environmental load of the average Swede on the planet, three times that of the Italian, thirteen times the Brazilian, thirty-five times the Indian, and two hundred and eighty times the Haitian.

Reducing excessive resource use is made urgent by the rapid growth of the population. World population has increased from about five billion twenty-two years ago to about seven billion today. Today, for the first time in history, in regard to a number of natural resources the rising curve of human demand exceeds the descending curve of natural supply. Since the end of World War II, more of the planet's resources have been consumed than in all of history until then. Global consumption is nearing, and in some cases has already surpassed, planetary maxima. The production of oil, fish, lumber, and other major resources has already peaked; forty percent of the world's coral reefs are gone, and annually about 23 million acres of forest are lost. The per capita availability of land for meeting human requirements has shrunk from 19.5 acres per person in 1900 to less than 5 acres today. Ecologists also speak of 'peak water', since the quantity of water suited for human use in the biosphere is rapidly diminishing.

The Fourth *Global Environment Outlook* of the UN Environment Programme estimated that satisfying the average resource demand in the world calls for the use of around 8.9 acres of land per person. (This figure masks great disparities between rich and poor economies: resource availability drops to 1.23 acres in the poorest countries such as Bangladesh, and mounts to 25.5 acres in the United States and the oil-rich Arab states.) However, 8.9 acres is more than twice the amount of land that could respond to human use on a sustainable basis: the sustainable 'Earth-share' of every man, woman and child on the planet is 4.2 acres (UNEP n.d.).

c) The financial system. The precarious structure of the world's financial system is a major factor in the unsustainability of the world's economy. Instability in the financial sector is not a new phenomenon, but it was not widely recognized prior to the credit crunch of 2008. The bubble that burst at that time has led to the loss of over two million jobs in the United States alone, and resulted in a global reduction of wealth estimated at 2.8 trillion dollars.

The structural unsustainability of the world's financial system is not uniquely due to the creation and burst of speculative bubbles: it is rooted in the imbalance of international trade. Already in 2005, the IMF's *Economic Outlook* (IMF, 2005) noted that it is no longer a question of *whether* the world's economies will adjust, only *how* they will adjust. If measures are further delayed, the adjustment could be 'abrupt', with hazardous consequences for global trade, economic development, and international security.

3.   Unsustainability in the ecology

Social, economic, and financial unsustainability is exacerbated by damages produced by human activity in the environment, resulting in a diminution of the resources effectively available for social and industrial use.

a) Water. The amount of water available for per capita consumption is diminishing. In 1950 there was a potential reserve of nearly 17,000 m³ of freshwater for every person then living. Since then the rate of water withdrawal has been more than double the rate of population growth, and in consequence in 1999 the per capita world water reserves decreased to 7,300 m³. Today about one-third of the world's population does not have access to adequate supplies of clean water, and by 2025 two-thirds of the population will live under conditions of critical water scarcity. By then there may be only 4,800 m³ of water reserves per person.

b) Productive land. There is a progressive loss of productive land. The Food and Agriculture Organization estimates that there are 7,490 million acres of high quality cropland available globally, seventy-one percent of it in the developing world. This quantity is decreasing due to soil erosion, destructuring, compaction, impoverishment, excessive desiccation, accumulation of toxic salts, leaching of nutritious elements, and inorganic and organic pollution owing to urban and industrial wastes.

Worldwide, 12 to 17 million acres of cropland are lost per year. At this rate 741 million acres will be lost by mid-century, leaving 6.67 billion acres to support 8 to 9 billion people. (This figure may still be overly optimistic, since the amount of available land will be further reduced by flooding due to a progressive rise in sea levels.) The remaining 0.74 acres of productive land could only produce food at the bare subsistence level.

c) Air. Changes in the chemical composition of the atmosphere reduce the availability of air capable of supporting adequate health levels. Since the middle of the nineteenth century oxygen has decreased mainly due to the burning of coal, and it now dips to nineteen percent of total volume over impacted areas and twelve to seventeen percent over major cities. At six or seven percent of total volume, life can no longer be sustained. At the same time, the share of greenhouse gases is growing. Two hundred years of burning fossil fuels and cutting down large tracts of forest has increased the atmosphere's carbon dioxide content from about 280 parts per million to over 350 parts per million.

At the same time, carbon dioxide is accumulating in the atmosphere. During the 20th century human activity has injected one terraton of CO_2 into the biosphere, and is currently injecting another terraton in less than two decades. The speed with which carbon dioxide is introduced makes it impossible for natural ecosystems to adjust. In the oceans, the explosive growth of CO_2 at the surface makes the water too acid for the survival of shell-forming organisms, the basis of the marine chain of life. On land, carbon dioxide absorption is reduced by the destruction of the ecosystems that had previously absorbed this gas. As much as 40 per cent of the world's forest cover has disappeared, due to acid rain, urban sprawl, and the injection of a variety of toxins into the soil.

The influx of greenhouse gases generated by human activity is matched by an influx from nature that is also largely catalyzed by human activity: the warming of the atmosphere. In Siberia a million square kilometer area of permafrost formed 11,000 years ago at the end of the last ice age is now melting. The area, the world's largest peat bog, is releasing as much methane into the atmosphere as all of human activity put together.

d) Global warming. The cumulative effect of the changes induced by human activity produces a greenhouse effect.

In recent years average temperatures have risen significantly, and the warming trend is accelerating. Conservative elements claim that global warming is due primarily to natural causes, at the most exacerbated by human activity: a new cycle in the fusion-processes that generate heat in the Sun sends an increasing amount of solar radiation to Earth, and this heats up the atmosphere. However, the injection of carbon dioxide, together with methane and other greenhouse gases into the atmosphere is likely to be a significant factor in creating and accelerating the global warming trend. The historical record of the past million years shows that the amount of CO_2 in the air correlates with variations in temperature: even if with some time delay, more carbon dioxide correlates with higher temperatures. A humanly generated shield in the upper atmosphere is now preventing heat generated at the surface from escaping into surrounding space.

Climate models show that even relatively minor changes in the composition of the atmosphere can produce major effects, including widespread harvest failures, water shortages, increased spread of diseases, the rise of the sea level, and the die-out of large tracts of forest. Global warming is already producing persistent drought in various parts of the world. In Northern China, for example, prolonged aridity has prompted the government to generate rainfall through artificial cloud-seeding.

By reducing the yield of productive lands, drought is creating a global food shortage. It is exacerbated by falling world food reserves: the current stocks are not sufficient to cover the needs of the newly food-deficit countries.

THE NEED FOR CULTURAL MUTATION

The practices that characterize human activity have their roots in the dominant values and perceptions of people. These values and perceptions are now obsolete. Allowing them to inspire action is strongly counterproductive; it produces growing crises and could issue in a world-scale breakdown.

The values and practices that inspire the dominant practices of the contemporary world need to change. We need a conscious and well focused cultural mutation.

The needed cultural mutation does not require people and societies to reject and discard their cultural heritage or disown their cultural preferences. It only requires a positive change in regard to those values and beliefs that reduce the sustainability of the system that frames human life on the planet.

Diversity is a positive attribute of the world system; a significant reduction would impair its resilience. Monocultures are inherently unstable, in society the same as in nature. Diversity, however, needs to be balanced by unity. Viable systems manifest unity within diversity: their diverse parts or elements are cooperatively focused on the attainment of shared goals, above all, that ensuring the continued persistence of the whole system.

GROUND RULES FOR HARMONIZING THE DIVERSITY OF THE CONTEMPORARY WORLD

The ground rule for achieving a higher level of unity in the contemporary world is simple and basic: maintain the diversity of the cultures and societies that compose the system, but join it with a higher level of harmony among them. A global-level harmonization of the system's diverse elements would allow the pursuit of a variety of goals and objectives as long as they do not damage that vital balances and processes that maintain the whole system. Achieving a higher level of dynamic stability in the world system is in the best interest of all people and societies, since without an adequate level of viability in the whole system, the viability of its parts is compromised.

The basic ground rule is both simple and evident:

Allow diversity to flourish among the cultures and societies that make up the contemporary socioeconomic and ecological world system, but do not allow this diversity to damage or destroy the harmony required to ensure the overall system's viability.

Additional precepts are required to ensure the effective application of the basic rule:

- Every society has an equal right to access and use the resources of the planet, but it also has equal responsibility to sustain the world system on the planet.
- Every society is free to live in accordance with the values and beliefs that accord with its historical heritage and its current wisdom, as long as these values and beliefs do not result in action that constrains the freedom of other societies to live in accordance with their own values and beliefs.
- All societies have a legitimate obligation to safeguard the freedom, physical security, and territorial integrity of their population, and to this end maintain an armed force, but no society has the right to produce and stockpile weapons that threaten the freedom, physical security, and territorial integrity of any other society.
- All societies forego technologies that waste essential resources, produce dangerous levels of pollution, or pose a threat to the health and wellbeing of their own people and the people of other societies.

Embracing these and related ground-rules would allow the world system to achieve the unity required to balance its diversity and thereby create and sustain conditions necessary to ensure the flowering of human life and wellbeing. Motivating and promoting the cultural mutation that would inspire and motivate this vital development is the moral obligation of all conscious and rational members of the human family.

REFERENCES

IMF—International Monetary Fund

2005. World Economic Outlook: Globalization and External Imbalances. URL: http:// www.imf. org/external/pubs/ft/weo/2005/01/index.htm.

UNEP—United Nations Environment Programme

n.d. GEO 4. Global Environment Outlook: Environment for Development. Nairobi: United Nations Environment Programme. URL: http://www.unep.org/geo/geo4.asp.

Chapter 8

MEASURING GLOBALIZATION— OPENING THE BLACK BOX: A CRITICAL ANALYSIS OF GLOBALIZATION INDICES

Axel Dreher, Noel Gaston, Pim Martens, and Lotte Van Boxem[1]

Indices of globalization are employed in various ways. This paper discusses the measurement of globalization with a view to advancing the understanding of globalization indices. Our assessment is that a true understanding of globalization must be an interdisciplinary enterprise. Moreover, it would be fruitful if academics, both quantitative experts and theoreticians, can work together on this challenge. Despite the different methodologies, choice of variables and weights, in order to study and measure globalization meaningfully, new cooperative frameworks are needed.

Keywords: globalization, measurement, globalization indices.

---

1. The authors contributed equal shares to this article; the order of names is chosen alphabetically.

INTRODUCTION

The objective assessment of both the causes and consequences of globalization is an essential agenda for contemporary societies. Positive economic, social and political analyses require data and globalization indices are a most promising means for providing objective data. Existing indices of globalization are employed in various ways. Apart from academic analysis, globalization indices are used in business analysis, mass and specialized media, as well as policy circles.

In business analysis, indices can be employed for gaining insight into the investment climate, the current developments of growth, and for helping business understand the global environment in which it now operates. In the mass media, the latest release of a globalization index can be the subject of a short news item or a feature article. It can also serve as an illustration for news coverage on related topics, such as technological developments. In policy circles, globalization indices provide a world view which reinforce the global context that policy makers work within.

This paper discusses the measurement of globalization with a view to advancing the understanding of globalization indices. Can globalization be better understood by measuring it? What are the intellectual and political implications of the existing globalization indices? We will discuss the attributes and limitations of globalization indices. A central theme of our argument is what we perceive to be the considerable gap between the quantitative and the qualitative analysis of globalization.

We critically analyze the types of index that can contribute to the debate on globalization. By the 'globalization debate' we mean the different viewpoints and facts about globalization that circulate between citizens, academics, scientists, politicians, media and business institutions. We argue that if globalization indices are to make a substantive contribution, they ought to bridge some existing gaps in our understanding of globalization. For example, if cultural transformation is important to globalization, can we include indicators of this transformation in the measurement of globalization? Obviously, the indices need to make a transparent and significant contribution to the debate. Finally, we look at the fields in which indices of globalization can be used. Stepping outside the realm of the indices, and considering the contribution to the wider debate, is a useful step to better understanding of the (im-)possibility of measuring globalization. Next, we discuss the most prominent indices of globalization.

GLOBALIZATION INDICES

In what follows, we discuss two indices of globalization developed by two of the authors.[2] The Maastricht Globalization Index, or MGI, developed by Martens and Zywietz (2006), and Martens and Raza (2009) refers to a cross-section of 117 countries, while the 2002 KOF Index of Globalization constructed in Dreher (2006) covers 122 countries for the period from 1970 to 2002. We also present the most recent KOF index that is based on the 2002 KOF Index of Globalization, covering 158 countries. Decisions are made concerning which variables should focus on the extensity, intensity, velocity or impact of the measured aspect as well as whether to adjust the variables for the geographical characteristics of a country, among others (Held *et al.*, 1999). While the MGI and KOF indices are very similar in many respects, there are notable methodological differences. For example, the MGI explicitly includes an environmental dimension. The latter is outcome-based and therefore excluded from the KOF Index. These differences partly reflect disagreements about the relative merit of various methodological options. Differences have also arisen due to the simultaneous and independent development of the indices. However, the resulting rankings do not crucially depend on the specific methodological choices made.

Another major difference is the adjustment of variables included in the indices for the geographical characteristics of countries. Controlling for these factors might improve the understanding of the other, more subtle determinants of globalization (e.g., past and present policy choices) that might ultimately be more interesting. Given the geographical characteristics of a country, these policy choices also affect economic development (e.g., GDP per capita). 'Stripping out the effects of economic development from the various measures of globalization would in fact be removing valuable information from these measures' (Lockwood, 2004), which is why they should be included. Pritchett (1996) argues that, when comparing countries' trade intensity, account needs to be taken of obvious structural features of the economy, such as the size and differences in transportation costs. Intuitively, these factors will also affect the other measures of globalization. For example, the trade intensity of Panama of 201.6 % in 1998 was more than eight times higher than the 24.4 % of the United

---

2. Arguably, the best-known indices of globalization are the ATKearney/Foreign Policy globalization index, which we abbreviate as 'ATK/FP'; the Maastricht Globalization Index, the 'MGI'; the World Market Research Centre G-index; and the KOF index of globalization produced by the KOF Swiss Economic Institute. The latter index is extensively used in academic analysis. Dreher *et al.* (2008: 75–78) list 36 journal articles published between 2003 and 2008 that employ the KOF index in statistical analyses. Some of the material in this section is drawn from Dreher *et al.* (2008); readers requiring greater detail are referred there. More information on the MGI, including its related publications, can be found on www.globalizationindex.info; more details on the KOF Index of Globalization are provided at http:// globalization.kof.ethz.ch/

States according to ATK/FP (2002). Whether Panama is eight times more economically globalized than the United States is debatable. The geographical location of Panama at one of the major crossroads of international trade, its size and its history are likely to be primary factors in its openness. However, one could equally well argue that the reasons for a country's openness should not matter for its globalization score. Put differently, the fact that Panama is more open than the United States because it is at one of the major crossroads of international trade does not change the fact that it is indeed more open and—by definition—more globalized. Whether correcting for such exogenous factors is a priori desirable is an open question. Correcting some variables included in globalization indices while not correcting others makes the results hard to interpret. The preferable option might be to control for these factors statistically when analyzing the causes and consequences of globalization rather than correcting the index a priori. While the MGI opts to correct for such exogenous factors, the KOF Index does not.

The construction of an index requires that the measures be normalized. If this were not done, then relatively small variations in one component or its distribution might completely swamp relatively larger variations in others. However, different methods for normalizing the data have significantly different impacts on the outcome, that is why the choice is important. On the one hand, when normalizing data from several years at the same time, termed panel normalization, the results are well-behaved in terms of sensitivity to extreme values. On the other hand, changes in one year could affect the ranking of countries in another year—a decidedly undesirable property. For this reason Lockwood (2004) proposes annual normalization, that is, the data are normalized for each year. Normalization with different parameters (mean, variance, extreme values) for each year can have the effect of 'moving the goal posts'; in effect letting a country slip in the rankings despite absolute gains in integration. However, Noorbakhsh (1998a: 522) argues that 'in an international context the goal posts are in fact moving'. If the extant rest of the world is becoming more globalized, a country whose integration is less than the rest of the world is being left behind. Different scales, means and distributions will alter any weights that are assigned to the different index components and therefore change the relative composition of the index. As described in more detail below, the KOF Index uses panel normalization. The MGI uses a cross-section of data, so panel normalization is not an issue. Both indices normalize the original variables before including them in the respective indices.

Another issue refers to how the variables included in the index should be weighted. There are several options for assigning these weights, all with their advantages in certain situations. For human development, for example, there might be subjective reasons for assigning a priori weights (e.g., the belief that education is equally important as life expectancy). For globalization, however, the case is less clear-cut. Since there is no universal agreement on what globalization is, and even less agreement on the relative importance of its components, some authors have advocated the use of statistical methods to derive weights for the index components (e.g., Noorbakhsh, 1998b; Lockwood, 2004; Dreher, 2006). They evaluate the impact of using statistically optimal weights instead of a priori weights as significant but small in absolute terms. The modification adds considerable complexity to the index. It is possible that the cost in terms of complexity may fall short of the benefit. While the MGI simply adds the individual dimensions, the KOF Index uses statistical analysis to derive the weights.

The MGI: Many previous indices have a decidedly neo-liberal focus on the economic dimensions of globalization. This may stem from the definition of globalization used. As argued earlier, the definition of globalization should refer to the process in its current state, including social, cultural and environmental factors. Hence, contemporary globalization is defined as the intensification of cross-national interactions that promote the establishment of trans-national structures and the global integration of cultural, economic, environmental, political, technological and social processes on global, supra-national, national, regional and local levels (Rennen & Martens, 2003). Another objective of the MGI is to broaden existing analyses of globalization by including coverage of sustainable development.

Components of the MGI: Reflecting the need for a balance between broad coverage, data availability and quality motivated the following choice of indicators, with data for 117 countries.

Global Politics: First among the indicators of political integration are the diplomatic relations that constitute a historical basis for communication between countries. Logically, the more important are the links to the outside world, the more diplomatic links will be established by countries to stay informed, protect their interests and facilitate communication. Since no aggregated statistics on diplomatic relations are available at the global level, the number of in-country embassies and high commissions listed in the Europe World Yearbook are used. The data are available for nearly all countries world-wide, but are corrected for country size, since very small countries can rarely afford the expense of maintaining multiple embassies and often accredit one representative for several countries. Membership in international organizations is a similar measure of the extensity of the international relations and involvement of a country. Moreover, since such memberships do not necessarily entail the need to maintain expensive representations abroad, this measure is less dependent on country size.

Organized Violence: This indicator measures the involvement of a country's military-industrial complex with the rest of the world. While the quality of the data is low, they nevertheless offer an insight into weapons proliferation, international military aid and the reasons and results of international peace-keeping operations. As this dimension has not previously appeared in other globalization indices, no comparison is possible with those indices. Of the quantitative military indicators proposed by Held *et al.* (1999), trade in conventional arms, compiled by the Stockholm International Peace Research Institute (SIPRI), is the only variable available for a reasonable number of countries. To make the data internationally comparable, a country's trade in conventional arms is related to its military expenditure. Since a large share of the trade is in 'big-ticket' items and programmes that are approved and recorded in one year may actually take several years to deliver and service, a moving three-year average is used. The period is arbitrary but offers a reasonable compromise between data availability and the need to smooth the data for infrequent, large purchases.

Global Trade: Like other globalization indices, trade intensity is included as a measure of the intensity of economic globalization. Trade intensity is the sum of a country's exports and imports of goods and services as a share of GDP. The data in this domain have been documented thoroughly over an extended period, in many cases extending back to the nineteenth century. Trade in services has brought new challenges to the statistical process, as it is far easier to value goods physically crossing border checkpoints than, for example, data processing or telecommunications, or even outsourced management consultancy services. Nevertheless, the data are widely available and generally reliable.

Global Finance: Foreign direct investment (FDI), representing financial enmeshment, is the primary indicator. Gross FDI, used here, is the sum of the absolute values of inflows and outflows of FDI recorded in the balance of payments financial accounts. It includes equity capital, reinvestment of earnings, as well as other long-term and short-term capital. This indicator differs from the standard measure of FDI, which captures only inward investment. For the measurement of globalization, however, the direction of the flow is less important than the volume. FDI is the long-term involvement of a foreign firm in a country and has cascading effects throughout an entire economy. It exposes local companies to foreign technical innovations, management styles, techniques as well as increased competition. Because of these long-term effects and the high volatility of the flows in the face of changing economic conditions, a trailing three-year average instead of single-year figures is used.

The second measure of financial interdependence used is gross private capital flows (as a percentage of GDP). This is the sum of the absolute values of direct, portfolio and other investment inflows and outflows recorded in the balance of payments financial accounts, excluding changes in the assets and liabilities of monetary authorities and the government. It measures the wider involvement of international capital in an economy and complements the FDI data. Once again, trailing three-year averages are employed.

People on the Move: This measure encapsulates migration and the international linkages that come with the movement of populations between different countries. Newly-arrived immigrants often maintain close connections to their home countries based on family ties and cultural similarities, often sending money home to their relatives and economic dependents. While a detailed analysis of migrant stocks and flows, specified by type and reason of migration would certainly be instructive, again only limited data are available on a global scale. As immigration and naturalization policies vary widely internationally and illegal immigration is widespread, the share of foreign-born residents of a given country has to suffice as a measure of the intensity of this increasingly controversial dimension of globalization.

Tourism brings people in contact with each other. It changes attitudes and promotes understanding between cultures that would otherwise have little contact. As a major economic activity, it can bring prosperity to regions with no resources other than the natural beauty of the surroundings or the cultural value of historic sites. Tourism has grown steadily in the last century, the major impetus being cheaper air travel. It represents an important part of globalization and is therefore included in the index. The World Tourism Organization, the source of the data, provides the sum of international inbound and outbound tourists, that is, the number of visitors who travel to a country other than their usual residence for a period not exceeding twelve months and whose main purpose in visiting is not employment related.

Technology: Although strongly related to GDP (with a Pearson correlation coefficient of 0.88), the share of a country's population that uses the internet still adds detail to the picture of the intensity of the technological aspect of globalization. Whether informing the international community about human rights abuses in reclusive countries or giving farmers access to commodity prices on the world's exchanges, as a global medium that transmits information cheaply over large distances it is an important factor.

The second component, international telephone traffic (again measuring intensity), can be used with fewer reservations, as the technology is older and therefore more widespread and less dependent on a country's income. International telephone traffic is defined as the sum of incoming and outgoing phone calls for a country, measured in minutes per capita (the original data are from the International Telecommunication Union).

The Environment: Overlooked by existing indices are environmental indicators, that is, measures of the intensity of globalization in the ecological domain. Held *et al.* (1999: 376–378) investigate global environmental degradation and the corresponding political and societal responses. These responses, however, are very difficult to track on a country-by-country basis. A more promising approach is to measure international linkages in terms of trade of goods that have a strong environmental impact, if not a high monetary one. Trade in software, for example, will generally have a far smaller impact on the environment than trade in tropical hardwoods, hazardous waste or water-intensive agricultural products.

Ecological footprint data offer a summary for many of these components since production and trade of these kinds of goods are included in a single measure. An ecological deficit (a footprint greater than the bio-capacity) indicates that a country must either 'import space' from somewhere (or stop 'exporting' it) or face rapid ecological degradation. Similarly, an ecological surplus offers opportunities to 'export space' by trade in space-intensive goods and services. The World Wide Fund for Nature's (WWF) Living Planet Reports provide ecological footprint and bio-capacity data in several categories (cropland, grazing land, forest, fishing grounds, energy lands and built-up land) and aggregate them into a single index, the ecological deficit. While a country with neither an ecological deficit nor surplus could be either completely autarchic or a major trader, by definition there is less dependence on outside linkages. A higher ranking according to this indicator therefore denotes more involvement with the outside world and, accordingly, a more globalized country along this dimension.

Method of Calculation: The MGI is constructed in a four-stage process (see UNDP, 2002; Martens & Zywietz, 2006). The first stage is conceptual and choices are made about which variables are most relevant and should be included in the index. In the second stage, suitable quantitative measures are identified for these variables. In the third stage, following Dreher (2006), each variable is transformed to an index with a 0 to 100 scale (this differs from earlier calculations constructing the MGI, see Martens & Zywietz, 2006). Higher values denote greater globalization. The data are transformed—on the domain level—according to the percentiles of the base year (2000) distribution (using the formula $((V_i-V_{min})/(V_{max}-V_{min})*100)$. In the last and final stage, a weighted sum of the measures is calculated to produce the final score, which is then used to rank and compare countries. The 'most globalized' country has the highest score. Within each domain, every variable is equally weighted. The MGI scores are simply added, that is all domains receive the same weight. The MGI is calculated for 2000 and 2008.

Underlying assumptions: Since there are missing data on the share of international linkages that are regional rather than global, it is impossible to distinguish globalization from internationalization and regionalization with complete certainty. Therefore, there is an assumption that countries with many international links have a correspondingly greater number of global linkages.

As expected, international statistics on 11 different indicators ranging from politics and military to the environment have widely varying degrees of data quality, reflecting the different capabilities and priorities of the organizations collecting the data. Of particular concern are the domains in which the underlying data have not been collected by official international bodies like the World Bank, IMF or UN, but by private or semi-public organizations. In addition, many countries are reluctant to share information about activities related to their national security, which creates data gaps that are not easily filled.

The fact that countries with fewer international linkages tend to publish less data and are less likely to be included in international statistics biases against states that are less globalized (see Rosendorff & Vreeland, 2006). Additionally, despite being members of the UN and most other international bodies, countries with totalitarian or communist economic systems (e.g., North Korea, Cuba) are often excluded in international financial statistics. Therefore, this also leads to their exclusion due to lack of data. Finally, yet importantly, countries that are too small to collect internationally coherent statistics and/or are strongly integrated into the economies of large neighbors (e.g., Luxembourg, Monaco and Swaziland) are also missing from the statistics and therefore excluded from the MGI.

The results: The world's most globalized country is Ireland with a score of over 70. This result is driven by a top 5 score on most of the indicators. On the other hand, Ireland ranks only 67th when it comes to political integration (and also has a relatively low ranking when it comes to the ecological integration). France has the highest political integration with the rest of the world, followed by the United Kingdom, Russia and Germany. According to the political integration index, Turkmenistan is the country with the lowest score. The sociocultural globalization ranking is headed by Kuwait, Austria, and Ireland, while Mali, Madagascar and India place at the bottom of the ranking. From a technological perspective, next to Ireland, Switzerland, New Zealand, the Netherlands, and Sweden complete the top 5 (with Bangladesh, Cambodia and Madagascar being the bottom 3). Kuwait ranks 1 on the (non-normalized) ecological index, followed by Belgium and Israel. Least ecologically integrated are Gabon and Bolivia. While Panama scores in the top 5 in terms of economic globalization, overall, they are ranked much lower. This is mainly due to their lower integration within the other domains with the rest of the world. Ireland, Belgium and the Netherlands compose the top-3 in this domain. Haiti is the country least integrated in economic terms. The world's least globalized country in 2008 is Madagascar, with an index of less than 15.

Figure 1 shows a globalization world map, where the more globalized countries are in darker colors. Western European and North American Countries are usually the most globalized, while countries in Sub-Saharan Africa are the least globalized.

As for the evolution of globalization, the overall MGI rose continuously, starting from a value of about 25 in 2000 to almost 32 in 2008. The increase is largely driven by technological and political integration. Economic and social-cultural globalization evolved similarly over time, while ecological globalization changed less (or decreased

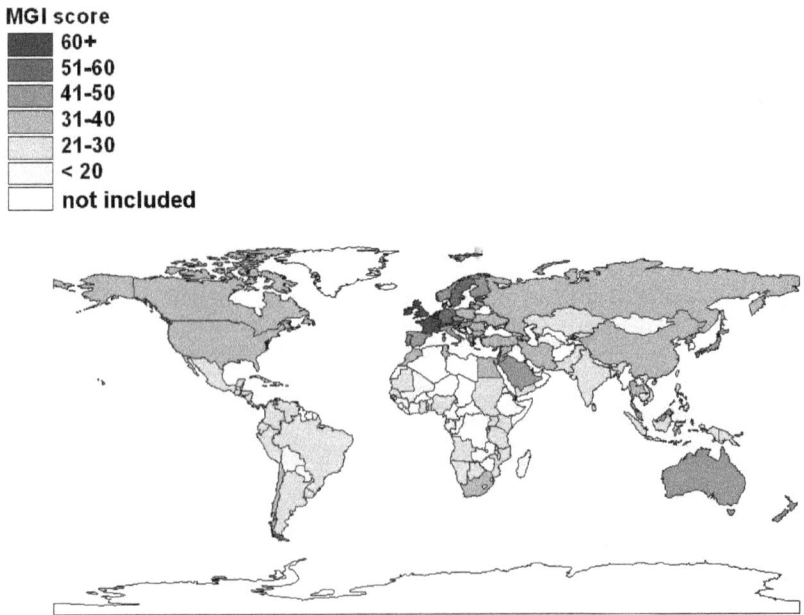

MGI score
- 60+
- 51-60
- 41-50
- 31-40
- 21-30
- < 20
- not included

Figure 1 *Map of the MGI, 2008*

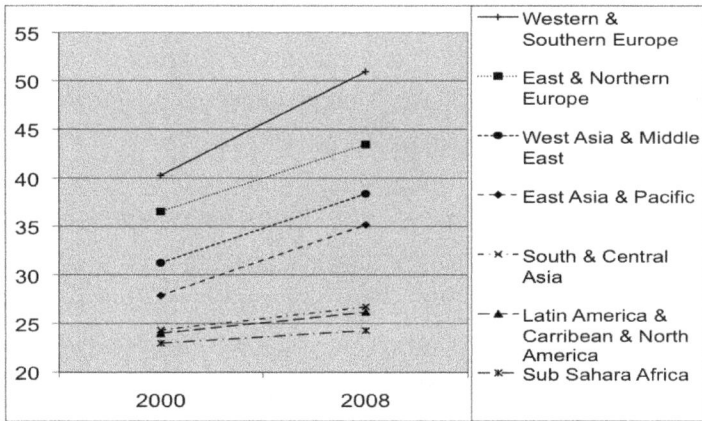

Figure 2 *Development of globalization across regions*

in the case of East and Northern Europe). For most countries, globalization increased. In some cases, the increases were substantial. The biggest increase was experienced by Ireland (+20.2), followed by the Netherlands (+19.7) and Belgium (+18.5), while globalization decreased most in Turkmenistan (–3.6) and Uruguay (–5.6).

Figure 2 displays the pattern of the overall globalization index by region.[3] Globalization has been relatively independent of region, even though the degree of globalization varies considerably. Overall, the index suggests that some countries are systematically more globalized than others. While in the last eight years globalization has been pronounced in all regions, some regions are more globalized than others. In particular, Western European and other industrialized countries display the greatest integration, South Asia and Sub-Saharan Africa are the regions least globalized.

The MGI has been linked with sustainability indices to analyze if more globalized countries are doing better in terms of sustainable development and its dimensions. The results suggest that the process of globalization may render world development more sustainable (Martens & Raza, 2010).

The KOF Index: The KOF globalization index was first published in 2002 (Dreher, 2006). It covers a large number of countries and has a long time span. The KOF Index also adds neglected dimensions of globalization.

The 2002 KOF Index covers 123 countries and includes 23 variables. The overall index covers the economic, social and political dimensions of globalization. Globalization is conceptualized as the process of creating networks among actors at multi-

---

3. The regions are based on http://www.un.org/depts/dhl/maplib/worldregions.htm

continental distances, mediated through a variety of flows including people, information and ideas, capital and goods. It is a process that erodes national boundaries, integrates national economies, cultures, technologies and governance, and produces complex relations of mutual interdependence.

More specifically, the three dimensions of globalization are defined as: economic globalization, characterized by the long distance flows of goods, capital and services as well as information and perceptions that accompany market exchanges; political globalization, characterized by the diffusion of government policies; and social globalization, expressed as the spread of ideas, information, images and people.

Economic Globalization: Economic globalization has two dimensions. First, actual economic flows are usually taken to be measures of globalization. Second, the previous literature employs proxies for restrictions on trade and capital. Consequently, two indices are constructed which include individual components suggested as proxies for globalization.

Actual flows: The sub-index on actual economic flows includes data on trade, FDI and portfolio investment. Trade is defined as the sum of a country's exports and imports and portfolio investment is the sum of a country's assets and liabilities; each measure is normalized by GDP. Included are the sum of gross inflows and outflows of FDI (again, normalized by GDP). While these variables are straightforward, income payments to foreign nationals and capital are also included to proxy for the extent to which a country employs foreign people and capital in its production processes.

International trade and investment restrictions: The second sub-index refers to restrictions on trade and capital flows using hidden import barriers, mean tariff rates, taxes on international trade (as a share of current revenue) and an index of capital controls. Given a certain level of trade, a country with higher revenues from tariffs is less globalized. To proxy restrictions on the capital account, an index constructed by Gwartney and Lawson (2002) is employed. Mean tariff rates are obtained from various sources. Gwartney and Lawson allocate a rating of 10 to countries that do not impose any tariffs. As the mean tariff rate increases, countries are assigned lower ratings. The rating declines toward zero as the mean tariff rate approaches 50 % (a threshold not generally exceeded by most countries in their sample). The original source for hidden import barriers is various issues of the World Economic Forum's Global Competitiveness Report, based on the survey question 'Hidden import barriers—no barriers other than published tariffs and quotas [are used]'.

Social Globalization: The KOF Index classifies social globalization in three categories. The first covers personal contacts, the second includes data on information flows and the third measures cultural proximity.

Personal Contacts: This index is intended to capture the direct interaction among people living in different countries. It includes international telecom traffic (outgoing traffic in minutes per subscriber), the average cost of a call to the United States and the degree of tourism (incoming and outgoing) a country's population is exposed to. Government and workers' transfers received and paid (as a percentage of GDP) measure whether and to what extent countries interact, while the stock of foreign population is included to capture existing interactions with people from other countries.

Information flows: While personal contact data are meant to capture measurable interactions among people from different countries, the sub-index on information flows is meant to measure the potential flow of ideas and images. It includes the number of internet hosts and users, cable television subscribers, number of telephone mainlines, number of radios (all per 1,000 people) and daily newspapers (per 1,000 people). To some extent, all these variables proxy the potential for receiving news from other countries and thus contribute to the global spread of ideas.

Cultural Proximity: Cultural proximity is arguably the dimension of globalization most difficult to grasp. According to Saich (2000: 209), cultural globalization to a large degree refers to the domination of U.S. cultural products. Arguably, the United States is the trend-setter in much of the global sociocultural realm (Rosendorff, 2000: 111). As proxy for cultural proximity, the number of McDonald's restaurants located in a country is included. For many people, the global spread of McDonald's is synonymous with globalization itself.

Political Globalization: To proxy the degree of political globalization, the number of embassies and high commissions in a country, the number of international organizations in which the country is a member and the number of UN peace missions a country participated in are used.

Method of calculation: In constructing the indices of globalization, each variable is transformed to an index with a 0 to 10 scale. Higher values denote more globalization. When higher values of the original variable indicate higher globalization, the formula $((V_i-V_{min})/(V_{max}-V_{min})*100)$ is used for transformation. Conversely, when higher values indicate less globalization, the formula is $((V_{max}-V_i)/(V_{max}-V_{min})*10)$. The weights for the sub-indices are calculated using principal components analysis. The year 2000 is used as the base year. For this year, the analysis partitions the variance of the variables used. The weights are then determined in a way that maximizes the variation of the resulting principal component. Therefore, the index captures the variation as fully as possible. As Gwartney and Lawson (2001: 7) emphasize, this procedure is particularly appropriate when several sub-components measure different aspects of a principal component. The same procedure is applied to the overall index. If possible, the weights determined for the base year are then used to calculate the indices for each single year back to 1970. Where no data are available, the weights are readjusted to correct for this. All yearly indices are averaged over five years to avoid huge fluctuations due to changes in yearly data.

2009 KOF Index of Globalization: An updated version of the original index is presented below. In most cases, the updating simply involves using more recent data. The costs of a telephone call to the United States are no longer included in the index, however. This was done to avoid the criticism of this variable being overly-centred on the United States. The update also excludes the number of telephone mainlines, as nowadays these are not the best measure of international flows of information. Similarly, to enhance the international focus of the index, the number of newspapers sold is replaced by the number of newspapers imported and exported. In addition, a number of proxies for globalization that are not included in the original 2002 index are included: FDI stocks, international letters sent and received, the number of Ikea outlets located in a country and trade in books and pamphlets. The number of international letters sent and received measure direct interaction among people living in different countries. Imported and exported books (relative to GDP) are used as a measure, as suggested by Kluver and Fu (2004). Traded books are intended to proxy the extent to which beliefs and values move across national borders. The number of Ikea outlets per country is motivated in a similar fashion to the number of McDonald's restaurants. The political dimension now also includes the number of treaties signed between two or more states since 1945 (as provided in the United Nations Treaties Collection).

The 2009 index introduces a number of methodological improvements over earlier versions. Each of the variables introduced above is transformed to an index on a scale of 1 to 100, where 100 is the maximum value for a specific variable over the period 1970 to 2006 and 1 is the minimum value. Once again, higher values denote greater globalization. The data are transformed according to the percentiles of the original distribution. Compared to the previous method, this has the advantage that a variable's actual weight in the index is not overly affected by its distribution. Consequently, the results are no longer driven by extreme outlying observations and missing values. The weights for calculating the sub-indices are determined using principal components analysis for the entire sample of countries and years. This is a methodological change compared with the construction of the 2002 Index, where the weights were determined using data for the most recent period. Employing data for the whole period yields better comparability over time. As discussed, one drawback is that the resulting globalization index is affected by the inclusion of additional countries. The analysis again partitions the variance of the variables used in each sub-group and determines the weights in a way that maximizes the variation of the resulting principal component. However, compared to the 2002 index, the weights are calculated using all data currently available instead of calculating them for the base year 2000. The same procedure is applied to the sub-indices in order to derive the overall index of globalization.

Data for the 2009 index are calculated on a yearly basis. However, not all data are available for all countries and for all years. In calculating the indices, all variables are linearly interpolated before applying the weighting procedure. Instead of linear extrapolation, missing values at the border of the sample are substituted by the latest data available. When data are missing over the entire sample period, the weights are readjusted to correct for this. As observations with value 0 do not represent missing data, they enter the index with weight 0. Data for sub-indices and the overall index of globalization are not calculated if they rely on a small range of variables in a specific year and country. Observations for the index are reported as missing if more than 40 % of the underlying data are missing or at least two out of the three sub-indices cannot be calculated. The indices on economic, social and political globalization as well as the overall index are calculated employing the weighted individual data series instead of using the aggregated lower-level globalization indices. This has the advantage that the data enter the higher levels of the index even if the value of a sub-index is not reported due to missing data.

The results: The methodological changes, new variables and data update do not substantially affect the weights of the individual dimensions of globalization. This is an indication of the robustness of the KOF index vis-à-vis the choice of method and data. Economic and social integration obtain approximately equal weights (38 % and, respectively, 39 % in the 2009 index), while political globalization has a substantially smaller weight in the overall index (23 % in the 2009 index).

According to the 2009 KOF Indices (which refer to data for the year 2006), the world's most globalized country is Belgium with a score of almost 92. This result is driven by high economic and political integration with the rest of the world. On the other hand, Belgium ranks only tenth when it comes to social integration. France has the highest political integration with the rest of the world, followed by Italy, Belgium and Austria. Other countries ranking high on the overall index include Ireland and the Netherlands. While Singapore and Luxembourg are ranked first and second, respectively, in terms of economic globalization, they are ranked considerably lower overall. This is mainly due to their low political integration with the rest of the world. According to the political integration index, the Channel Islands, the Isle of Man and Mayotte are the countries with the lowest score. Overall, the world's least globalized country is Myanmar with an index of less than 24. The country least integrated in economic terms is Rwanda, while Myanmar has the lowest social globalization score. Figure 3 shows the more globalized countries in a darker color. Once again, Western European and North American countries have usually been the most globalized, while countries in Sub-Saharan Africa are the least globalized.

The evolution of globalization as measured by the KOF index has been more pronounced in the later decades. The overall index rose continuously, starting from a value of about 37 to more than 60 in 2006. Economic globalization evolved similarly over time, while social and political globalization rose less steadily.

Figure 4 displays the pattern of the overall globalization index by income. In the last 30 years globalization has been pronounced in all income groups, however, some groups are clearly more globalized than others. As can be seen, high income OECD countries are, on average, the most globalized, while low income countries are the least globalized.

Overall, the index suggests that some countries are systematically more globalized than others. In particular, richer countries seem to be, on average, more global-

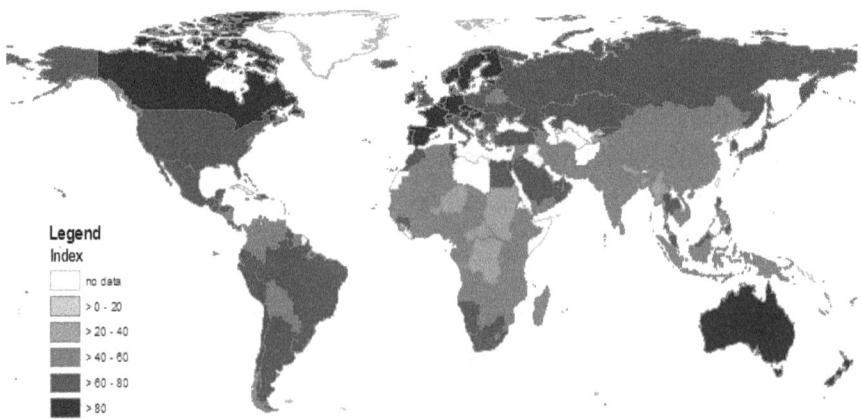

Figure 3 *Map of the KOF Index of Globalization, 2009*

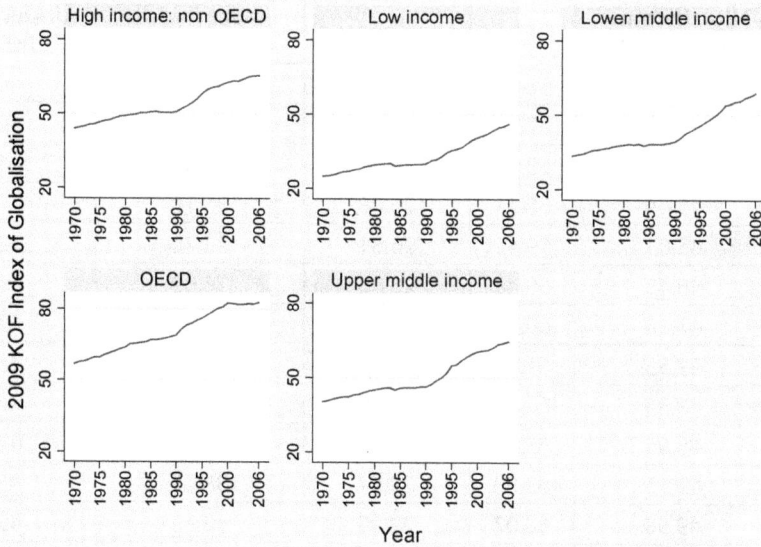

Figure 4 *Development of globalization, by income group*

ized than poorer ones. Western industrialized countries are also more globalized than the average country. The average OECD country is far more globalized than the average non-OECD country. Table 1 displays the corresponding data on a yearly basis.

THE RELEVANCE OF GLOBALIZATION INDICES

Any assessment of the relevance of the existing indices must consider the different definitions of globalization used. To facilitate comparison, the key globalization indices appear side-by-side in Table 2 from Dreher *et al.* (2008). As the Table indicates, the WMRC's G-index includes primarily economic factors; the ATK/FP index does so as well by an a priori weighting scheme that heavily favours economic factors. Unfortunately, with these indices, globalization is indistinguishable from internationalization and liberalization. This is not to say that data collected with the country as the relevant unit of analysis have no value. However, the assumptions made and the limitations of using these data for the measurement of globalization should be clearly stated—something which both indices fail to do.

Many authors examining the measurement of globalization concur with the view that 'culture is the most visible manifestation of globalization' (Kluver & Fu, 2004). However, despite culture's importance to globalization, no index provides an adequate solution to its measurement. Martens and Zywietz (2006) side-step the issue

year	High income: non OECD	OECD	Low income	Lower middle income	Upper middle income
1970	44.02	56.91	25.02	33.63	40.38
1971	44.63	57.74	25.23	34.08	40.68
1972	44.85	58.03	25.50	34.43	41.36
1973	45.49	58.71	26.08	35.16	41.97
1974	45.81	59.49	26.81	35.76	42.31
1975	46.45	59.49	26.86	36.09	42.30
1976	47.16	60.51	27.60	36.45	43.02
1977	47.53	61.47	27.95	36.77	43.32
1978	48.35	62.26	28.95	37.52	44.13
1979	48.95	63.13	29.26	37.88	44.66
1980	49.15	63.80	29.77	38.02	45.02
1981	49.58	65.07	29.99	38.33	45.49
1982	49.59	65.26	30.19	38.24	45.83
1983	50.36	65.82	30.28	38.62	46.09
1984	50.25	65.98	29.27	37.88	45.08
1985	50.60	66.95	29.52	38.33	45.78
1986	50.90	67.10	29.70	38.37	46.15
1987	50.51	67.20	29.78	38.32	45.98
1988	50.42	67.59	29.86	38.54	45.99
1989	50.50	68.21	30.00	39.06	46.37
1990	50.70	68.95	30.15	39.48	46.41
1991	52.15	71.56	31.40	40.82	47.34
1992	52.99	72.97	31.93	42.61	48.92
1993	54.51	73.92	32.99	43.62	50.10
1994	55.75	75.02	34.64	44.99	51.78
1995	58.23	76.08	35.58	46.00	54.89
1996	59.53	77.26	35.98	47.25	55.36
1997	60.60	78.76	36.87	48.47	57.10
1998	61.02	80.11	38.33	50.08	58.56
1999	61.82	80.92	39.88	51.57	59.61
2000	62.45	82.48	40.85	53.97	60.51
2001	63.19	82.26	41.39	54.70	60.92
2002	63.11	81.80	42.46	55.36	61.10
2003	63.88	81.90	43.61	55.77	62.12
2004	64.78	82.21	44.71	57.21	63.53
2005	65.13	82.13	45.21	57.88	63.83
2006	65.29	82.61	46.07	58.99	64.61

Table 1 *Development of globalization, by income group*

by stating that the concepts of culture and communication are inherently intractable and difficult to quantify. Kluver and Fu (2004) construct a Cultural Globalization Index. They argue that it is impossible to directly measure the diffusion of cultural values and ideas across national borders. So they use cultural proxies: 'the conduits by which ideas, beliefs and values are transmitted'. Although cultural globalization is adequately conceptualized, the available empirical measures once again fall short. The authors use the imports and exports of books and brochures, newspapers and periodicals because all other possible indicators lack systematic data sources. Countries at the top of the cultural rankings are generally affluent and English-speaking. One danger of the failure to measure cultural factors is the risk of dismissing the importance of culture. In our opinion, we should be asking why it is that we know so little about what should be discussed. Clearly, it would be useful if the publication of the indices include some discussion of cultural globalization.

The KOF Index includes some cultural indicators in the 'social globalization' sub-index. The indicators that have been included are the number of McDonald's restaurants per capita, the number of Ikea outlets per capita and the number of books traded (as a percentage of GDP). This sub-index can indicate the extent to which cultural globalization matters for economic and social phenomena.

Rather inevitably, the 'top 10' countries in the leading indices are usually lauded. An exception to this is the MGI because it has integrated two variables—the environment and organized violence—that change the meaning of the overall outcome. Notwithstanding, it is useful to consider what it means to be at the top, middle or bottom of a globalization ranking.

The inclusion of new indicators, that cannot be considered 'positive', changes the discussion about a country's ranking according to an index. For example, if the Netherlands ranks highly in every index of globalization is that something to be applauded? It does imply, of course, that this country has many linkages with the world outside its national borders. According to the MGI, the Netherlands, for example, ranks fourth in both the overall rank and in the environmental rank. It is placed fortieth in the 'organized violence' rank. This implies that the Netherlands has a large ecological footprint and relatively intense trade in conventional arms. It also scores well in other areas such as capital flows, trade, and telephone traffic.

A large ecological footprint implies a large ecological deficit, which needs to be compensated for by 'space' outside the country's territory. In this way, the growth in transport is connected to the exploitation of natural resources (Martens & Rotmans,

Category	Sub-category	WMRC (Randolph 2001)	ATK (A.T. Kearney / Foreign Policy 2007)	MGI (Martens and Raza 2009)	KOF (Dreher 2006)
Relevance	Definition of globalization used	Very narrow, only economic	Medium	Very broad	Very broad
	Differentiation of globalization from internationalization	No differentiation	No differentiation	No differentiation	No differentiation
	Type of change measured	Extensity, intensity	Extensity, intensity	Extensity, intensity	Extensity, intensity
	Geographical adjustment	No	No	Yes	No
	Coverage	185 countries	72 countries	117 countries	122 countries
	Correlation with economic development	Low	High	High	High
Robustness	Sensitivity to extreme values	Method not published	High (cross-panel normalization)	Low	Low
	Sensitivity to year-to-year data variations	Very high (exclusive use of strongly fluctuating indicators)	High (some indicators with lower fluctuation)	Low (indicators with high fluctuations are averaged)	High (some indicators with lower fluctuation)
	Method for determining weights	A priori, with normative discussion	A priori, with normative discussion	Equal weights	Principal components analysis
	Weight distortion	Method not published	Some distortion	No distortion	Some distortion
Added value	Correlation with own components	High	Low	Some	Some
	Correlation among components	Not published	Not published	Moderate	Moderate
Transparency	Transparency of methodology	Moderate	High	High	High
	Data published	Partially	Yes	Yes	Yes

Note: Relevance is concerned with whether the index is really measuring globalization (instead of, for example, internationalization).

Robustness is concerned with the reliability of the measurement under adverse circumstances; how sensitive to extreme values and year-to-year variations is the index.

To add value, the index should help us understand globalization better than we could by just looking at its components.

Transparency helps others to judge how valuable the index is for their purposes; whether the index, based on readily available data and literature, is reproducible; and whether the underlying assumptions are made explicit.

Table 2 *Existing globalization indices and criteria for good composite indices*

2005), for instance. So while this helps to elevate the Netherlands to the top ranking of this index, it also raises questions about the relationship between globalization, economic growth and the environment. Unlike the other variables in the index, this environmental factor appears to be a consequence of globalization rather than a driving force. However, as the globalizing processes intensify over time, the 'indirect impacts of human-induced disruption of global biogeochemical cycles and global climate change start to become apparent' (Martens & Rotmans, 2005).

If consumerism and global economic processes have polluting side-effects, it needs to be asked which direction these dynamics need to take for a sustainable future. With the environment integrated into the index, the long-existing 'environment versus growth' tension can be exposed, for which the term 'sustainable development' is often used (Ibid.). The demands for environmental protection and economic development are said to be competing. Some claim an eternal competition, while others emphasise a possible win-win situation (Van Kasteren, 2002).

Since globalization implies inter-connectedness and complexity, its various aspects need to be considered. The environment cannot be treated separately from everything else that is global. Moreover, an integrated index of globalization can stimulate a new framework of analysis for the market system, recognizing the need to integrate ecological costs in trade and consumption (Ibid.).

The inclusion of trade in conventional arms in the MGI also serves to highlight such trade. Do global mechanisms promote production and open gateways to trade in arms? Clearly the issue is complicated as it involves economic costs and benefits, political risks, social tensions and ethical values. While such issues are far from being resolved, the way the addition of such indicators influence the relevance of a measurement of globalization needs to be emphasized.

An important criticism of many indices, such as the MGI and the ATK/FP, is that, strictly speaking, they measure internationalization and regionalization rather than globalization. For example, the MGI's 'top 10' is composed of European nations which reinforces an impression of increased regionalization.

All indices have component indicators and data that fail to distinguish between globalization and internationalization (or liberalization) to some degree. They also fail to include supra-territorial indicators. For example, while the number of embassies a country has abroad may mirror increasing cooperation and even integration, these data have a territorial base.

Even leaving the problem of 'methodological territorialism' to one side, the epistemology of globalization makes one doubt the possibility of measuring it. Globalization occurs at levels that make measurement difficult, for example, trans-border environmental issues, cultural transformations and a so-called 'global consciousness'. Those features of globalization are obviously interesting and new to us which, in turn, is one reason why they are so difficult to capture.

The 'qualitative' side of research generally focuses on multi-dimensional analyses of globalization by constructing frameworks and concepts. This is useful, but does not provide a solid scientific footing with which to evaluate the over-arching phenomenon of globalization. On the other hand, the 'quantitative' side of research, with its focus on data, statistics and indices, runs the risk of over-simplification.

As we have argued, to confront new questions on the essential nature of globalization requires an interdisciplinary approach. Sociologists, critics of science and technology, and economists and others need to work on dimensions of the same questions. A composite index of globalization can reconcile multi-facetted approaches. An index needs to be conceptually analyzed and formulated and this leads to the issue of measurement. Instead of questioning the adequacy of measuring globalization, a certain degree of optimism is vital for making the improvements in measurement, which are necessary to advance an understanding of the globalization phenomenon.

CAN WE REALLY MEASURE GLOBALIZATION?

As we have discussed, the measurement of globalization should try to include the essential features of contemporary globalization. However, when we think about a possible methodology, we face a greater problem which applies to existing indices of globalization—classic or modified. Even if we could manage to find suitable supra-territorial indicators and indicators that portray cultural and other complex global features, how could such measures fit in with the rest of the existing measures, since the end result is still country-based? This dead end in the measurement of globalization is well described by Caselli (2006).

Given this situation, it is paradoxical and misconceived to insist on studying reality in general, and globalization all the more so, with instruments that take the nation-state as their unit of analysis. It is at most possible to study internationalization in this way, but not globalization. In other words, the globalization measures currently available are vitiated by what has been variously called methodological nationalism (Beck, 2004), embedded statism (Sassen, 2000), or methodological territorialism (Scholte,

2000)—a perspective which distorts the essence of globalization precisely when its study begins, and which yields data that 'in the best of cases are irrelevant and in the worse misleading, or even false' (Beck-Gernsheim, 2004, as cited by Caselli, 2006: 20).

Those features of globalization that are essentially new to us are those which are most difficult to measure by means of data collection and index construction. If the current epistemological basis of measuring globalization is so theoretically unsatisfactory and empirically problematic, we need to question why we should pursue the construction and maintenance of globalization indices which may be too narrow to understand globalization.

A possible solution to these issues is to assess globalization by thematic order. For example, we can measure how globalized our worldwide politics are. Bauman's (1998) idea of a new class division between the globalized upper classes and the localized lower classes may also be promising. This leads to the proposal to measure globalization along individual lines, or along the lines of demographic groups. We could also measure the amount of supra-territorial institutions, both formal and informal. However, once again the problem rises of fitting in these trans-border results with a country-based index.

IS THE MEASUREMENT OF GLOBALIZATION A DEAD END?

The measurement of globalization contains so many pitfalls that it is tempting to retreat to purely qualitative analyses. However, this would burn the fragile bridge between the qualitative and quantitative analysis of globalization. The qualitative side of research generally focuses on a multi-dimensional analysis of globalization, by constructing frameworks and concepts through which to understand it. This provides some tools, but not a solid scientific footing which can fully comprehend the entire phenomenon of globalization. It is simply theory without measurement; running the risk of unsubstantiated and unscientific speculation. The quantitative side of research assesses the state of play about globalization using data, statistics and indices. While this approach runs the risk of oversimplification and may take on an overly enthusiastic air of truth, its transparent use of the available data is its ultimate salvation.

There is a possibility to bridge the gap between theory and measurement. Composite indices of globalization can provide the meeting place or forum for both approaches. Composite indices need matters to be conceptually analyzed and continu-

---

ally reformulated. Instead of rejecting the possibility of measuring globalization adequately, the measurement of globalization needs to be, and can be, improved upon. A new mode of thinking, such as supra-territoriality, can trigger new ideas on both the analysis and quantification of globalization.

The confrontation with new questions on the essential nature of globalization needs to be an interdisciplinary cooperation. It would be fruitful for academics from the quantitative side (modeling, conclusive statements, certainty and proofs) and qualitative side (analysis, discussion, conceptual revision, background and textual form) to sit together and work on the challenges. Despite the different methodologies, choice of variables and weights, and so on, they need to recognize that in order to study globalization concisely, new cooperative frameworks are needed.

Sociologists, critics of science and technology and economists need to work on dimensions of the same questions. For instance, an interdisciplinary review of science and technology analyses different lines of approach and formulates conceptual criticism to technical problems. It provides an overview of possible solutions and elaborates upon quantitative issues. Rather than handing over responsibility from discipline to discipline, what is required is tackling collectively the measurement of globalization. In this case, the whole is greater than the sum of the individual parts. The study and ultimate understanding of globalization requires academics and professionals alike to step outside their own narrow disciplinary boundaries.

ACKNOWLEDGEMENT

We thank Mohsin Raza for his help in analyzing the MGI.

REFERENCES

Kearney, A. T. / Foreign Policy. 2002. Globalization Index. Washington, D.C.: Carnegie Endowment for International Peace.

Kearney, A. T. 2007. Globalization Index. Washington, D.C.: Carnegie Endowment for International Peace.

Bauman, Z. 1998. Globalization: The Human Consequences. New York: Columbia University Press.

Beck, U. 2004. Der kosmopolitische Blick oder: Krieg ist Frieden. Frankfurt am Main: Suhrkamp Verlag.

Beck-Gernsheim, E. 2004. Wir und die Anderen. Frankfurt am Main: Suhrkamp Verlag.

Caselli, M. 2006. On the Nature of Globalization and its Measurement. UNU-CRIS Occasional Papers. Milano: Università Cattolica del Sacro Cuore. URL: http://www.cris. unu.edu/ admin/documents/20060220113557.O-2006-3.pdf.

Dreher, A. 2006. Does Globalization Affect Growth? Evidence from a New Index of Globalization. Applied Economics 38(10): 1091–1110.

Dreher, A., Gaston, N. et al. 2008. Measuring Globalization: Gauging its Consequences. New York: Springer.

Gwartney, J., and Lawson, R. 2001, 2002. Economic Freedom of the World: Annual Report. URL: http://www.freetheworld.org.

Held, D., McGrew, A. G. et al. 1999. Global Transformations: Politics, Economics and Culture. Stanford, CA: Stanford University Press.

Kluver, R., and Fu, W. 2004. The Cultural Globalization Index. Foreign Policy online Web Exclusive. URL: http://www.foreignpolicy.com/story/cms.php?story_ id=2494.

Lockwood, B. 2004. How Robust is the Kearney / Foreign Policy Globalization Index? The World Economy 27: 507–524.

Martens, P., and Raza, M. 2009. Globalization in the 21st Century: Measuring Regional Changes in Multiple Domains. The Integrated Assessment Journal 9(1): 1–18.

Martens, P., and Raza, M. 2010. Is Globalization sustainable? Sustainability 2: 280–293.

Martens, P., and Rotmans, J. 2005. Transitions in a Globalizing World. Futures 37: 1133–1144.

Martens, P., and Zywietz, D. 2006. Rethinking Globalization: A Modified Globalization Index. Journal of International Development 18: 331–350.

Noorbakhsh, F. 1998a. A Modified Human Development Index. World Development 26(3): 517–528.

Noorbakhsh, F. 1998b. The Human Development Index: Some Technological Issues and Alternative Indices. Journal of International Development 10(5): 589–605.

Pritchett, L. 1996. Measuring Outward Orientation in LDCs: Can it be Done? Journal of Development Economics 49(2): 307–335.

Randolph, J. 2001. G-Index: Globalization Measured. URL: http://www.wmrc.com/.

Rennen, W., and Martens, P. 2003. The Globalization Timeline. Integrated Assessment 4(3): 137–144.

Rosendorf, N. M. 2000. Social and Cultural Globalization: Concepts, History and America's Role. In Nye, J. S., and Donahue, J. D. (eds.), Governance in a Globalizing World (pp. 109–134). Washington, D.C.: Brookings Institution Press.

Rosendorff, B. P., and Vreeland, J. R. 2006. Democracy and Data Dissemination: The Effect of Political Regime on Transparency. Memphis, TN: Mimeo.

Saich, T. 2000. Globalization, Governance, and the Authoritarian State: China. In Nye, J. S., and Donahue, J. D. (eds.), Governance in a Globalizing World (pp. 208–228). Washington, D.C.: Brookings Institution Press.

Sassen, S. 2000. New Frontiers Facing Urban Sociology at the Millennium. British Journal of Sociology 51(1): 143–159.

Scholte, J. A. 2000. Globalization—a Critical Introduction. Hampshire; New York: Palgrave.

United Nations Development Programme (UNDP). 2002. Human Development Report 2002. New York: Oxford University Press.

Van Kasteren, J. 2002. Duurzame Technologie. Amsterdam: Uitgeverij Natuur & Techniek.

Chapter 9

ON FREE TRADE, CLIMATE CHANGE, AND THE W.T.O.

Rafael Reuveny

This article focuses on the argument that a free global market benefits the environment. I explore the link between climate change, which has recently emerged as the greatest environmental threat, and world trade, which has grown continuously since WWII. The growth of world trade, facilitated by the GATT-WTO regime, evokes an important question. Is this regime good for the environment, or has it contributed to the increase of greenhouse gases, the primary driver of climate change? While this question cannot be fully ans-wered in this paper alone, it is important to consider it now because many of the expected damages caused by climate change may be considerable and nonreversible. After discussing the state of knowledge on the effects of trade on the environment, we evaluate whether the biosphere can accommodate perpetual economic growth. The purpose of this paper is to integrate the insights gained by outlining a proposed research program focusing on the WTO and the environment in the context of climate change.

Keywords: environment, economic growth, World Environmental Protection Agency.

INTRODUCTION

The ideology of liberalism can be generally categorized into two interrelated categories, republican and commercial liberalism. Republican liberalism focuses on the causes and consequences of democracy, as opposed to autocracy. Commercial liberalism focuses on the causes and particularly the consequences of free domestic and international markets, as opposed to central governmental control of economic activities. Both types of liberalism link political and economic freedoms to many socio-political-economic forces, including international relations, war propensity, income distribution, standard of living, economic growth, quality and performance of institutions, and the state of the environment. A common thread shared by both classes of liberalism is the argument that political and economic freedom, or democracy and the free market, are superior across the board, promoting peace, prosperity, and political stability.

According to a derivative of this argument, free domestic and global economic markets also promote environmental quality and reduce environmental degradation within national and domestic systems. The argument that free global markets promote global environmental quality stands at the center of this paper.

In recent decades, climate change has emerged as the largest threat to the global environment. During the 1980s and early 1990s there was still some uncertainty as to whether climate change was occurring, particularly whether it was human-induced or natural. Today there is a general scientific consensus that climate change is occurring and human activity, particularly the burning of fossil fuels, is the cause (IPCC, 2007, 2001a).

The global market involves a number of international economic interactions, including trade flows, foreign direct investments, financial capital movements, currency exchanges, labor flows or migration, technological transfers, and movements of physical capital. Of these interactions, this article focuses on international trade flows for two reasons. First, most people identify international trade as the impetus behind a free global market. Second, and perhaps more importantly, the global policymaking community has focused more on international trade than any other subject since World War II.

A number of authors have reviewed the evolution of the international trade regime after World War II, including GATT (1991), Cole (2000), and Salvatore (2006). In 1948 several countries led by the U.S. created the General Agreement on Tariffs and Trade (GATT). The evolution of the GATT reflected the liberal view that free trade benefits everyone. In the following decades several multilateral trade negotiations took place; the Kennedy Round (1963–1967), Tokyo Round (1973–1979), and the Uruguay Round (1986–1994) removed many trade barriers. In 1994 the GATT was replaced by a newly created global institution, the World Trade Organization (WTO), which was given more powers in promoting free trade. Today, almost every country in the world has joined the WTO.

Under the GATT-WTO regime, world trade has continuously expanded. Before the 1960s it was concentrated among industrialized countries. Today it involves all the countries in the world to a greater degree, and developing countries such as China and India have become major traders. Naturally, this trade growth would not be possible without the liberalization of trade barriers. This move was spawned, nurtured, supervised, and enforced first by the GATT and then by the WTO. Today, the WTO is one of the strongest international organizations. It has jurisdiction to decide on international trade disputes, rendered by the member countries, and can also impose penalties on members that break its laws. WTO members, in turn, agree to follow the decisions of the WTO court system, as well as implement all of their contractual responsibilities according to the WTO body of law.

This paper addresses the relationship between trade liberalization and activities under the GATT-WTO regime and the global environment; particularly the risk of climate change. I specifically address the following research question: Is this regime good for the environment, or has trade liberalization under this regime contributed to the increase of greenhouse gases, the primary driver of climate change? The results obtained by answering these questions can serve as a basis for evaluating the need and possibility to include climate change concerns in future WTO policies and laws.

My question is not easy to answer since climate change is an evolving and complex phenomenon whose primary effects are still not fully manifested, nor fully understood. An investigation of this research question is complex and can yield several outcomes. We may find that free trade has nothing to do with environmental degradation, or even promotes environmental quality, thus there is no need to bring climate change concerns into the WTO. We may also conclude that even though trade has promoted environmental degradation, the WTO has defended the environment, thus we should enlarge its responsibilities and powers in this regard. Alternatively, we may find free trade causes environmental degradation, including climate change, and the WTO has not addressed environmental concerns. We may even find that the WTO has made things worse, promoting environmental degradation in its pursuit of free trade.

Even though the research question is complex and cannot be fully answered within the scope of one article, it is important to start discussions now. Time is critical because many of the expected adverse damages caused by climate change, including rising sea levels, inundation of low-lying areas, seasonal changes such as lengthening of heat waves, land degradation, intensification of storms and other weather events, drying of fresh water sources, and melting glaciers, tundra, and ice-poles may be considerable and irreversible. We must therefore attempt to gain as many insights as possible on the research question today and not postpone the discussion until the time when these damages are fully manifested.

I will approach the question in three stages. First, I will discuss the state of theoretical and empirical knowledge on the effects of trade on the environment. As we shall see, trade sometimes affects the environment through the channel of economic growth. Second, this observation suggests that we could gain insights by discussing whether the global biosphere can accommodate a situation of perpetual global economic growth. Third, I will integrate the insights gained into the last section by outlining a proposed research agenda focusing on two interrelated topics: the connection between the WTO trade regime and the environment, and the public policy implications for the current design of the WTO and, more generally, trade liberalization with the goal of slowing the rate of global climate change. My research findings may perhaps suggest that attempts to bring environmental considerations into the WTO would require the design of a new international trading system.

THE EFFECTS OF TRADE ON THE ENVIRONMENT

International trade can affect the environment through two mechanisms. One mechanism directly influences human economic activities that affect the environment and works regardless of whether the economy grows. The second mechanism affects the environment indirectly because it affects the rate of economic growth which, in turn, affects the environment.

Mechanism One: Direct Effects

As detailed in Pugel (2007), Harris (2006), OECD (1994) and others, the total direct effects of international trade on the environment are the result of several competing channels. Each of these channels may promote or reduce environmental degradation, depending on the strength of the competing effects they represent. We can classify these effects by their types: compositional, structural, regulatory, and technological.

The compositional effect of trade can promote or reduce environmental degradation by changing the composition of traded goods. Consider, for example, a nation that produces a labor-intensive good whose production does not affect the environment and a capital-intensive good whose production damages the environment. Assume the country is capital-abundant, or has more capital relative to labor compared with other countries. This country, then, has comparative advantage in capital-intensive goods, or can produce them cheaper than other countries. Market logic implies that this country would specialize in producing capital-intensive goods, or produce more of them relative to no trade, exporting them to others. Consequently, it will also produce less of the labor-intensive goods, relative to no trade, importing them from others. Heavier production of the environmentally damaging capital-intensive good will obviously increase damage to the environment. If, in contrast, the country is labor-abundant, trade will increase production and export of the labor-intensive good and reduce production of the capital-intensive good, thereby reducing relative damage to the environment.

The structural effect of trade involves changes in the structure of the local economy due to changes in the location of consumption, investment, and production. For example, consider a country that grows chemical-intensive crops, and the chemicals employed (e.g., pesticides, fertilizers) damage the environment. As the country opens for trade, it may decrease production of chemical-intensive crops, importing them from countries producing them at lower costs. This country will see a change in the structure of its economy since it will employ fewer chemicals, all other things being equal. As a result, environmental quality will rise. If, however, another country increased production of these chemical-intensive crops to satisfy greater global demand, it could face greater environmental degradation due to chemical application.

The regulatory effect of trade works by promoting certain policies. Some trade agreements, for example, require countries to keep environmental damage in check, calling for environmentally-friendly regulations. Another example involves a large and influential country pushing others to take a pro-environment approach in order to be able to sell in its markets. This effect, however, may also work in the opposite direction. If the influential country is not environmentally conscious, others may follow its lead, ignoring the degradation. In a third example, consider countries with parochial trade interests pushing to relax environmental regulations in order to employ cheaper production methods that are also less environmentally-friendly. If other countries adopt this course of action, environmental degradation may rise globally, as the relaxing of environmental regulations becomes a 'race to the bottom'.

Finally, the technological effect of international trade can raise or reduce environmental degradation by promoting changes in production methods. For example, countries may be required to reduce the quantity of fertilizers or pesticides they use in agriculture since foreign consumers may seek to consume organically grown edible plants and crops. By opening domestic societies to new ideas and innovations, international trade may also promote a move toward environmentally cleaner technologies and production methods. However, the technological effect of international trade could also globally propagate the use of environmentally damaging methods and technologies (e.g., fossil fuel-based methods). Countries may use these technologies and production methods because they are cheaper to employ and legal according to extant environmental laws. This outcome may also lead to a 'race to the bottom', as countries seek to reduce their production costs by relaxing pro-environment laws and existing regulations.

Mechanism Two: Indirect Effect

Since the indirect effect of international trade on the environment works through the channel of economic growth, we need to first discuss the effect of trade on the economy. Commercial liberalism assumes that people want to maximize consumption. Economic growth, it is argued, ensures continuously rising consumption. Free markets are argued to be the best social mechanism to promote economic growth because they allocate inputs of production to their most efficient uses, and they provide incentives for innovation by granting large profits to the innovators until others learn to imitate the innovation.

The liberal argument for free international trade is an application of the general argument for free markets. Expanding trade enables national specialization in producing goods according to the principle of comparative advantage, increasing production and promoting economic growth. Nationality is not a variable in the assumptions describing the behavior of people in commercial liberalism. To put it differently, classical and neoclassical economics do not distinguish between the intrastate interactions of American producers from Philadelphia and consumers from Baltimore, for example, or producers from India and consumers from Italy. Neoclassical economists, then, implicitly make the connection that since free markets make sense domestically, they also make sense internationally.

In principle, we could end the discussion here, yet commercial liberals elaborate further. Export, they argue, promotes fuller utilization of underemployed domestic inputs since it provides new outlets for domestic production. Imports can stimulate domestic demand, ultimately enabling larger domestic production. By expanding overall production, free trade promotes more efficient division of labor between production activities and enables economies of scale, which reduces average costs and increases profits, thus providing incentives for growth. Trade also transmits new ideas and technologies across national boundaries. When countries restrict trade, they also curtail flows of technologies and improved products, which harms growth. Finally, by increasing the number of producers in the market place, trade pushes domestic producers to become more efficient, which accelerates economic growth.

The indirect effect of trade on the environment works through the 'environmental Kuznets curve' (EKC). The theory behind the EKC is discussed in a number of sources, including Thompson and Strohm (1997), Perman *et al.* (2003), Dinda (2004), and Li and Reuveny (2007). As argued in the preceding paragraphs, international trade promotes economic growth. This, however, is said to affect the environment. Up to some threshold, damage to the environment is said to rise as income per capita rises. Above this threshold of income per capita, environmental damage is said to decline as income per capita rises. The plot of environmental degradation as a function of income per capita thus takes the shape of an inverted U. The name EKC is given by analogy to the original Kuznets curve proposed by Nobel Prize-winning economist, Simon Kuznets (Kuznets 1955). The original curve plots income inequality in a country as a function of income per capita and also takes the shape of an inverted U (see Figure 1 for an illustration).

The shape of the EKC is driven by two competing forces, the scale and the income effects. With current technology, larger production and consumption generates more environmental degradation (e.g., pollution, waste), denoted as the scale effect of economic growth. However, as income per capita rises, human preferences arguably

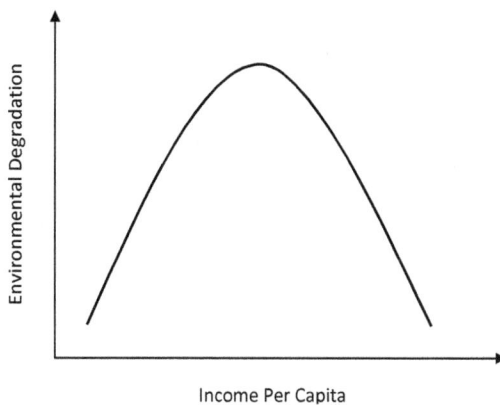

Figure 1 *A Generic Enviromental Kuznets Curve*

shift toward consuming and producing goods that generate less environmental damage. Essentially, richer people are not only more willing to pay more for environmental-friendly goods and environmental protection, but are also able to pay for these goods. This is known as the income effect of economic growth. The scale effect, then, is positive: environmental degradation rises with income per capita. The income effect is negative: environmental degradation falls with income per capita.

According to the EKC theory, as income per capita rises, the income effect will dominate the scale effect, generating the inverted U shape that indicates a decline in environmental degradation with income. Whether the U shape exists empirically is a question of interest for us. I also seek to discover whether the EKC holds true for environmental degradation, since the EKC is primary rationale supporting the position that free trade raises environmental quality. This view sees no need for policy intervention; the system can fix itself, provided that markets are set free. However, the EKC is not a hypothesis to be tested here. Rather it is an issue into which we can gain insight by discussing existing results.

The empirical literature on the EKC effect is substantial and cannot be fully discussed here. Extensive reviews are available, for example, in Panayotou (2000, 2003), Dinda (2004), and Stern (2004). In general, the obtained empirical results are inconclusive. Some studies find that EKCs exist for some air pollutants, but not for others. Other studies dispute the results. EKC results for carbon dioxide emissions and deforestation, the primary drivers of climate change (emissions on the source side and deforestation on the sink side, as forests absorb carbon dioxide), are also inconclusive. Even if the EKC effect exists, the estimated turning points of the inverted U curve, beyond which the damage arguably declines, range from about $5000–$30,000 in real terms, depending on the particular environmental indicator, statistical model specification, estimator, and sample. Given that real income per capita of most developing countries is much smaller than $5000, even if the EKC effect exists, we would have to wait many years before it materializes.

The number of empirical studies on the effect of trade on environmental degradation is comparatively small. Lucas et al. (1992) conclude that the growth rate of toxic intensity declines with openness to trade. Grossman and Krueger (1993) find that trade openness (ratio of export plus import to gross domestic product) reduces sulfur dioxide emissions but has no effect on smoke and suspended particulate matter. Suri and Chapman (1998) report a negative effect of the ratio of import to GDP on energy consumption per capita, interpreted to indicate that air pollution falls with trade. Antweiler et al. (2001) find that trade liberalization reduces sulfur dioxide emissions, but the effect is very small. Barbier (2001) finds that agricultural export promotes agricultural land expansion, concluding that trade intensifies environmental pressure. Dean (2002) finds that trade liberalization promotes water pollution in China. Li and Reuveny (2007) find that trade openness promotes deforestation and does not affect land degradation.

Taken together, the results presented in this and the previous subsections are inconclusive. However, the problem of trade and the environment is in fact even more complex than has been suggested by these results. At stake is yet a bigger question: can the biosphere accommodate a constantly growing global economic system?

PERPETUAL ECONOMIC GROWTH AND THE ENVIRONMENT

For environmental damages that arguably exhibit the EKC effect, the income per capita turning points found in empirical analyses are almost always much higher than current per capita incomes of developing countries. Since the large majority of global population lives in developing countries, even if the EKC effect exists for some damages, global environmental degradation may not decline autonomously with free trade and economic growth in the foreseeable future. In no area is this issue more important than in the area of climate change.

According to the Intergovernmental Panel on Climate Change (IPCC), there were systematic patterns of climate change consistent with a tendency of global warming in the 20th century, including an increase in the frequency and duration of warm periods, glacial retreat, an approximately 20 centimeter rise in sea level, an approximately 0.10 C per decade rise in average global temperature, a 10 % decline in winter snow covers, a 40 % decline in northern sea ice thickness, a 15 % decline in summer northern sea ice coverage, and a considerable rise in the frequency and intensity of extreme weather events. These variations are attributed primarily to greenhouse gas emissions generated by man-made fossil fuel burning (IPCC, 2007).

The predicted effects of climate change in this century vary, depending on assumptions about energy use, population growth, technological progress, and economic growth. However, all forecasts predict that the sea level and intensity and frequency of extreme weather events will rise. Existing predictions on the effects of a one meter sea-level rise on land and population, assuming no protective measures are taken, suggest that hundreds of millions of people will be displaced. Several small island-states in the Pacific may be completely submerged and other countries may suffer significant land loss, including Gambia, Bangladesh, Vietnam, and the Netherlands. Agriculture, forestry, fresh water, and coastal infrastructures are expected to be particularly sensitive to climate change. Forecasts suggest that lesser developed countries (LDCs) are the most vulnerable to climate change due to their limited adaptive capacity and large dependence on the environment for generating livelihoods (IPCC, 2001b).

Recalling that trade promotes economic growth, with the current state of technology and energy generation, it is apparent that as free trade expands under the auspices of the WTO, at least the scale effect of trade will intensify climate change in the coming decade. What about the income effect and the EKC effect as a whole? And what about the possibly positive direct effects of trade on the environment? Let us assume that these effects together will benefit the environment in general and mitigate climate change in particular. Does this mean that free international trade is ultimately the best policy to combat environmental degradation?

These are complex questions. To gain insight, let us assume that the EKC effect and free trade are the answers to environmental degradation. Hence, we should focus on promoting economic growth and free trade. For example, we should aid LDCs in attaining the standard of living in developed countries (DCs), and strengthen the WTO to better monitor, report, litigate and punish countries that deviate from free trade. Before we jump to this conclusion, we must ask yet another question: can the biosphere accommodate the standard of living in DCs for all people in the world? If the answer is no, even if trade and growth promote environmental quality, policies promoting these forces may prove to be counterproductive.

The English economist Thomas Malthus (1798) believed there were limits to economic growth. In the long run, he argued, the growth of food would fall below population growth and society would converge in a state of poverty and conflict. Neoclassical economists have criticized Malthus for ignoring the role of technological progress in alleviating environmental pressures, and his ideas subsequently lost favor. If Malthus was wrong, then either there are no limits to growth, or technological progress can expand them forever. One way to approach these issues is to first evaluate whether it is possible for all nations in the world to attain the current United States standard of living with current technology, then consider the possible effects of technological progress.

Existing results suggest that the current per capita ecological footprint of the United States (land and water areas required to sustain its actual production, waste, and pollution) is about five times larger than the world's per capita bio-capacity (available biologically productive land and water area). By mid century, the world's per capita bio-capacity is expected to fall by about fifty percent due to population growth (Wackernagel *et al.*, 1999; Reuveny, 2002, 2005; Harris, 2006). Reviewing studies on the number of people the Earth can carry, Cohen (1995) shows that estimates cluster around 4–16 billion, depending on the standard of living people are expected to maintain. He further shows that studies assuming the current United States standard of living for all nations conclude that our planet could support 2–5 billion people. In sum, it seems that with the current state of technology it is impossible to attain the current United States standard of living for the Earth's population.

The issue of energy is particularly daunting. Assuming there will be 9–10 billion people by mid-century and economic growth will continue at the current rate, world energy consumption will double. Where will this energy come from? As discussed in Trainer (1998), Palfreman (2000), Hoffert (2000), Reuveny (2002), and Harris (2006), there is no magic solution. Oil stocks will decline. Coal could power the world economy for several more centuries, but would likely speed up climate change. Even if methods were found that limit greenhouse gases from burning coal, they would not likely eliminate them. Wind and sun sources are irregularly available and require large areas, and the feasibility of a global hydrogen economy is unclear. Relying on biomass to power a global economy would require areas now allocated to agriculture, and the feasibility of nuclear fusion is debatable at best. Only nuclear energy is a viable option to replace fossil fuels to power a global economy. However, even if we ignore the problems associated with nuclear waste and security, the known amounts of Uranium-235 (a metal used in the generation of nuclear energy) would not sustain the world for long at current consumption rates (Hoffert, 2000).

Can perpetual economic growth be sustained with technological progress? Commercial liberals argue that people will find solutions to existing problems as they have done in the past; there are no limits to economic growth. This argument is supported by using mathematical models assuming that people constantly generate technological progress, and progress continuously promotes total factor productivity, environmentally friendly products, less resource intensive production, and new materials to replace depleted resources. Moreover, it is assumed that all these new methods of production, goods, substitutes, and technologies have no bad side effects, and social institutions and markets work smoothly and perfectly.

These assumptions lead to the commercial liberal conclusion that economic growth can continue forever almost by definition, but they may not hold in the real world. For example, relying on innovation and markets to deliver the solution assumes that actors know all the costs and benefits. When property rights are not well developed, or when innovations exhibit public good characteristics, actors become unsure of costs and benefits. Solving complex global problems requires institutional changes, wealth, and expertise, which are not readily available in LDCs. Innovation to alleviate climate change exhibits these very problems. Moreover, so far many other large-scope projects have been deemed more important than alleviating relatively slow moving environmental problems such as climate change, be it building an international space station, accumulating nuclear weapons, promoting consumerism, or fighting international wars.

The nature of innovation is yet another issue. Commercial liberals assume that progress is always beneficial and has no bounds. However, in reality, technologies can have adverse impacts and may die out. There can also be cognitive limits to understanding the complex dynamic interactions of global ecological, social, political, and economic forces, leading to limits in technological advances. For example, current energy techno-logy causes climate change. Energy efficiency has risen since the mid-1970s in DCs, but this improvement has slowed down. After early successes, the response of crops to synthetic fertilizers is lessening, and agricultural yields have fallen in many Green Revolution regions. Facing these examples, it seems that the effects of perpetual global economic growth may sooner or later lead to a reliance on wide-scope space colonization. However, the feasibility of a project of this magnitude in this century is unclear, to say the least. These examples do not prove that innovation must stop in the future and that solutions will not be found eventually. However, they suggest a need for caution when it comes to formulating public policies that assume perpetual and beneficial technological progress.

W.T.O. AND CLIMATE CHANGE: THE ROAD AHEAD

The gradual removal of trade barriers since 1945 has played a key role in the phenomenal growth in global trade. As long as exports faced significant trade barriers, they remained highly uncompetitive in the importing markets. Once barriers were gradually removed under the auspices of the GATT-WTO regime, national comparative advantages came into effect, pushing countries to specialize in producing what they do most efficiently or least inefficiently, relative to others and exporting these goods, while importing other goods. The growth in trade promoted economic growth, which in turn lead to increased consumption and production, promoting more trade. The effects of these forces on the environment, as we have seen, are debated theoretically. Empirically, the period has seen an increased use of fossil fuels to power the economic growth and larger production, and this has accelerated global warming and climate change (IPCC, 2001a, 2007).

Considering the role of the GATT-WTO trade regime in addressing climate change, many questions come to light. Beyond its direct effect on trade liberalization, what will be the effect of the WTO on climate change? What is the likelihood of conflicts between a Kyoto Protocol-based climate change regime seeking to guard the environment and a GATT-WTO trade regime seeking to promote free trade? Answering these questions is speculative because the bulk of climate change effects are expected in the future, the Kyoto Protocol has not yet produced any substantial results, and the US, so far the chief contributor to climate change, has failed to ratify the protocol.

Nevertheless, analyzing the approach of the GATT-WTO regime to the trade-environment nexus in the past can provide us insights. Is it driven by considerations involving the EKC effect? Is it cognizant of the possibility that the direct effects of trade could harm the environment? Is the WTO aware of studies arguing and demonstrating the impossibility of attaining the DC standard of living for all the people on Earth? Is it cognizant of and condoning a situation in which the planet as a whole produces and consumes beyond its biological capacity, as reflected by its ecological footprint, in effect consuming and producing at the expense of future generations? Is the WTO cognizant of the links moving from trade to climate change through economic growth and the use of fossil fuels? Is the WTO approach motivated by the Precautionary Principle, which calls for avoiding potentially large damages to the environment even if the probability of adverse outcomes is less than 100%? These are all important questions that can and should be addressed in future research.

A related question is whether the WTO slowed or prevented trade-driven environmental degradation in the past. For example, trade in some animals could diminish biodiversity, and trade in some products can damage the environment by intensifying pollution in one place or causing damages in another. Trade in fossil fuels, timber from deforestation, and crops grown in deforested areas may promote climate change by increasing consumption of fossil fuels and by eliminating natural sinks of greenhouse gases. In fact, all trade flows generate greenhouse emissions due to transportation or production. If the WTO has stood by as trade-promoted environmental degradation expanded, or rejected attempts to block it, we would be inclined to conclude that the GATT-WTO trade regime may accelerate trade-related activities that promote climate change, or at least would not be useful in slowing them down and is not a good candidate for monitoring and enforcing trade-related activities of a climate change regime. In this case, we would conclude that we need a new global institution for this purpose, for example, a World Environmental Protection Agency.

In contrast with this bleak possibility, it is also possible that the WTO has been friendly to the environment and has rejected attempts to expand international trade at the expense of reducing environmental quality. If the WTO has been a guardian of the environment, including the environment in its policy considerations, it is possible that they will continue to do so in the future. In this case, we may not need to diminish its ability to monitor and enforce a free trade regime framework and we might even seek to strengthen and expand it. This possibility seems particularly attractive since the Kyoto Protocol climate change regime seeks to slow the rate of climate change by instituting an international system for trading carbon emission permits and clean defense mechanisms. We might conclude that trade mechanisms devised to combat climate change be promoted and implemented by the current WTO.

Any evaluation of the role of the WTO in environmental degradation must begin with the link between the design principles of the GATT-WTO trade regime and the environment. Restating our research question, is this regime good for the environment? Answering this question would further require conducting a set of systematic case studies focusing on the WTO policies in cases that brought environmental issues into the WTO normal deliberations and decision making. Candidates for such studies include the following: (1) Assessing the actions of the WTO Committee on Trade and the Environment (CTE), which was established in 1995 with a mandate to assess trade-environment linkages, and evaluating its effect on WTO policies; (2) Assessing relationships and links between the WTO and Multilateral Environmental Agreements such as those signed by countries to promote biodiversity or reduce the use of certain damaging materials, some of which employ trade measures in enforcing their effects on the countries that signed them; (3) Assessing the WTO case law and jurisprudence pertaining to international trade disputes brought to the WTO court, in which disputants disagree on the legality of certain trade actions that arguably damage the environment; and (4) Assessing the WTO case law and jurisprudence in cases involving use of environmental policy to impose barriers on the entry of traded goods that damage the environment into another country, which exporters argue reflect protectionism, not environmental policy.

The assessment and evaluation of these cases is very important because they could suggest a policy direction for the global community, pointing out the need for either strengthening and expanding the scope of the WTO, or alternatively, scaling down the scope of the WTO and giving priority to the global environment. For example, the global community could decide to create a new World Environmental Protection Agency that would give priority to environmental considerations of trade policy.

The potential impossibility of attaining the DCstandard of living for all people on Earth with the current state of technology suggests that our analysis might conclude that the overall costs, over time, from the WTO promotion of free international trade outweigh the overall benefits. Should that be indeed the outcome of the proposed research agenda, it seems that we would need to reconsider the current global adherence to the idea of free international trade, which was brought to the fore by commercial liberalism. Assuming that the current state of technology would essentially prevail in the coming decades, sooner or later the promotion of free international trade would have to play second to the much more pressing need of mitigating climate change. This global shift in attitudes would bring the era of ever-expanding free international trade volumes and global economic system to a stop, at least until we find a way to completely disentangle the current link between global economic growth and climate change.

REFERENCES

Antweiler, W., Copeland, B. R., and Taylor, M. S. 2001. Is Free Trade Good for the Environment? American Economic Review 91: 877–908.

Barbier, E. B. 2001. The Economics of Tropical Deforestation and Land Use: an Introduction to the Special Issue. Land Economics 77: 155–171.

Cohen, J. E. 1995. How Many People can the Earth Support? New York: W. W. Norton.

Cole, M. A. 2000. Trade Liberalization, Economic Growth and the Environment. Cheltenham: Edward Elgar.

Dean, J. 2002. Does Trade Liberalization Harm the Environment? A New Test. Canadian Journal of Economics 35: 819–842.

Dinda, S. 2004. Environmental Kuznets Curve Hypothesis: a Survey. Ecological Economics 49: 431–455.

GATT 1991. The GATT: What It Is, What It Does, General Agreement on Trade and Tariffs. Geneva: GATT.

Grossman, G., and Krueger, A. B. 1993. Environmental Impacts of a North American Free Trade Agreement. In Garber, P. M. (ed.), The U.S.—Mexico Free Trade Agreement (pp. 13–56). Cambridge, MA: MIT Press.

Harris, J. M. 2006. Environmental Natural Resource Economics: A Contemporary Approach. Boston: Houghton Mifflin.

Hoffert, M. 2000. Beyond Fossil Fuels. What's Up with the Weather. U.S.: Public Broadcast Service. URL: www.pbs.org/wgbh/warming/beyond.

IPCC 2001a. Climate Change 2001: Impacts, Adaptation, and Vulnerability. Intergovernmental Panel on Climate Change. URL: www.ipcc.ch.

IPCC 2001b. Special Report on the Regional Impacts of Climate Change: An Assessment of Vulnerability. Intergovernmental Panel on Climate Change. URL: www.ipcc.ch.

IPCC 2007. Climate Change 2007: The Scientific Basis. Intergovernmental Panel on Climate Change. URL: www.ipcc.ch.

Kuznets, S. 1955. Economic Growth and Income Inequality. American Economic Review 45: 1–28.

Li, Q., and Reuveny, R. 2007. The Effects of Liberalism on the Terrestrial Environment. Conflict Management and Peace Science 24: 219–238.

Lucas, R. E. B., Wheeler, D., and Hettige, H. 1992. Economic Development, Environmental Regulation and the International Migration of Toxic Industrial Pollution: 1960–1988. In Low, P. (ed.), International Trade and the Environment (pp. 67–86). Washington, D.C.: World Bank.

Malthus, T. 1798. An Essay on the Principle of Population. Penguin: New York.

OECD 1994. The Environmental Effects of Trade. Paris: Organization for Economic Development and Cooperation.

Palfreman, J. 2000. Frequently Asked Questions. What's Up with the Weather. U.S.: Public Broadcast Service. URL: www.pbs.org/wgbh/warming/etc/faqs.html

Panayotou, T. 2000. Economic Growth and the Environment. CID Working Paper 56 (pp. 1–118). Cambrdige, MA: Center for International Development and at Harvard University.

Panayotou, T. 2003. Economic Growth and the Environment. Economic Survey of Europe 2: 45–67. Geneva: United Nations Publications.

Perman, R., Ma, Y., McGilvray, J., and Common, M. 2003. Natural Resource and Environmental Economics. London: Pearson; Addison Wesley.

Pugel, T. A. 2007. International Economics. Boston, MA: McGraw-Hill.

Reuveny, R. 2002. Economic Growth, Environmental Scarcity and Conflict. Global Environmental Politics 2: 83–110.

Reuveny, R. 2005. International Trade and Public Policy: The Big Picture. In Robbins, D. (ed.), Handbook for Public Economics (pp. 705–742). New York: Marcel Dekker.

Salvatore, D. 2006. International Economics. New York: John Wiley.

Stern, D. 2004. The Rise and Fall of the Environmental Kuznets Curve. World Development 32: 1419–1439.

Suri, V., and Chapman, D. 1998. Economic Growth, Trade and Energy: Implications for the Environmental Kuznets Curve. Ecological Economics 25: 195–208.

Thompson, P., and Strohm, L. A. 1997. Trade and Environmental Quality: A Review of the Evidence. Journal of Environmental and Development 5: 363–388.

Trainer, T. 1998. Saving the Environment: What it will Take. Sydney: UNSW.

Wackernagel, M., Onisto, L., Bello, P., Callejas Linares, A., Susana Lopez Falfan, I., Mendez Garcia, J., Isabel Suarez Guerrero, A., and Guadalupe Suarez Guerrero, M. 1999. National Natural Capital Accounting with the Ecological Footprint Concept. Ecological Economics 29: 375–390.

Chapter 10

THE E-WASTE STREAM IN THE WORLD-SYSTEM

R. Scott Frey

Globalization and sustainability are contradictory tendencies in the current world-system. Consider the fact that transnational corporations transfer some of the core's wastes to the peripheral zones of the world-system. Such exports reduce sustainability and put humans and the environment in recipient countries at substantial risk. The specific case of e-waste exports to Guiyu, China is discussed. The discussion proceeds in several steps. The nature of the e-waste trade is first examined. Political-economic forces that have increased e-waste trafficking to China are outlined. The extent to which this trade has negative health, environmental, and social consequences is outlined and the neo-liberal contention that such exports are economically beneficial to the core and periphery is critically examined. Policies proposed as solutions to the problem are critically reviewed.

Keywords: e-waste, recycling, hazardous wastes, environmental justice, sustainability, world-systems theory, ecological unequal exchange, capital accumulation.

INTRODUCTION

The editors of The Economist (2007: 14) magazine made several observations about globalization recently that are worth quoting because they raise important questions about environmental justice and sustainability in an increasingly globalized world. Sounding a bit like Marie Antoinette, the editors wrote:

...the best way of recycling waste may well be to sell it, often to emerging markets. That is controversial, because of the suspicion that waste will be dumped, or that workers and the environment will be poorly protected. Yet recycling has economics of scale and the transport can be virtually free-filling up the containers that came to the West full of clothes and electronics and would otherwise return empty to China. What is more, those who are prepared to buy waste are likely to make good use of it.

Despite the internal consistency of their market logic and celebration of the current global system, globalization and environmental justice as well as sustainability can be seen as contradictory tendencies in the current world-system. Consider, for instance, the fact that centrality in the world-system allows some countries to export their environmental harms to other countries (Frey, 1998a, 1998b, 2006a, 2006b, 2012). Such exports increase environmental injustice and reduce sustainability by putting humans and the environment in recipient countries at substantial risk. The specific case of e-waste exports to Guiyu, China is discussed in light of the contradictory tendencies mentioned.

The discussion proceeds in several steps. Environmental justice and sustainability in the world-system are first examined. This is followed by a discussion of the e-waste trade in the world-system. The extent to which this trade has negative health, safety, and environmental consequences in Guiyu, China is outlined and the neo-liberal contention that such exports are economically beneficial to the core and periphery is critically examined. Policies proposed as solutions to the problem of e-waste traffic in Guiyu and the world-system are critically reviewed. The paper concludes with an assessment of the likelihood that existing 'counter-hegemonic' globalization forces will overcome the tensions between globalization and environmental justice and sustainability.

ENVIRONMENTAL JUSTICE AND SUSTAINABILITY IN THE WORLD-SYSTEM

The world-system is a global economic system in which goods and services are produced for profit and the process of capital accumulation must be continuous if the system is to survive (see especially Wallerstein, 2004 for the origins and nature of the world-system perspective and Harvey, 2010, for a recent discussion of continuous capital accumulation under capitalist relations). Proponents of the perspective conceptualize the world-system as a three-tiered system, consisting of a core, semi-periphery, and periphery.

The world-system is an open system that can be understood not only in 'economic' terms but also in 'physical or metabolic' terms: entry of energy and materials and exit of dissipated energy and material waste (Frey, 1998a; Hornborg, 2011; Martinez-Alier, 2009; Rice, 2007, 2009). In fact, the world-system and globalization itself can be described or understood in terms of a process of 'ecological unequal exchange' (e.g., Hornborg, 2011; Rice, 2007, 2009; or a process of 'accumulation by extraction and contamination'). Frey (2006a) has described the process of ecological unequal exchange in the following terms:

- Wealth (in the form of materials, energy, genetic diversity, and food and fiber) flows from the resource rich countries of the periphery to the industrialized countries of the core, resulting often in problems of resource depletion/degradation and pollution in the peripheral zones or the 'resource extraction frontiers'.

- The core displaces anti-wealth (entropy broadly defined) or appropriates carrying capacity or waste assimilation by transporting it to the global sinks or to the sinks of the periphery in the form of hazardous exports. In other words, global sinks and the peripheral zones of the world-system are essentially 'waste-disposal frontiers'.

This paper focuses on the transfer of hazards to the peripheral zones of the world-system, whether hazardous products, production processes, or wastes, with a focus on e-waste. Such hazards damage the environment and adversely affect human health through environmental and occupational exposure. Peripheral countries are particularly vulnerable to the risks posed by such hazards for several reasons: limited public awareness; a young, poorly trained, and unhealthy workforce; politically unresponsive state agencies; and inadequate risk assessment and management capabilities and infrastructure (e.g., Frey, 2006a).

The core (consumers, states, and capital in various countries) benefits from the transfer of hazards to the periphery while the periphery bears the costs associated with such exports. Environmental justice and sustainability are enhanced in the core because environmental harms are displaced to the periphery, while such export practices increase environmental injustice and reduce sustainability in the periphery. Risks associated with hazardous exports or environmental harms are distributed in an unequal fashion within the periphery: some groups (especially the state and capital) are able to capture the benefits while others (those marginalized by gender, age, class, race/ethnicity, and geo-spatial location) bear the costs (see Frey, 2006a).

E-WASTE IN THE WORLD-SYSTEM

Nature and Scope of E-Waste

E-waste consists of discarded computers, cell phones, televisions, and other electrical and electronic products (see Widmer *et al.* [2005] for a review of existing definitions of e-waste). This waste is a byproduct of the information and communication technology-infrastructure underlying the world system's social metabolism. Globalization is dependent (or 'symbiotic', to use Pellow's [2007: 185] term) on information and communication technology, most notably the computer. It is the computer in conjunction with the internet (and the global transport system) that facilitates the transport of wealth to the core and anti-wealth to the periphery, whether it is the movement of bauxite and iron ore from Brazil by large ocean going vessels (Bunker & Ciccantell, 2005) or the recycling, incineration, and/or disposal of e-waste in China.

E-waste is growing more rapidly than other waste streams because the consumption of electronic products is growing at an astonishing pace. Increased consumption of electronic products is due to the constant development of new electronic products, planned obsolescence, and falling prices throughout the developed world. In 1975, for example, there was one computer per 1,000 population in the US, but in 2010 the number was over 800 computers per 1,000 population (United Nations 2010). And it is expected that growth of computers will continue in the core and be overtaken by the developing countries in the next fifteen to twenty years (Yu *et al.*, 2010a).

Anywhere from 30 to 50 million tons of e-waste are discarded each year in the world-system (according to the Basel Action Network and Silicon Valley Toxics Coalition [2002] and the United Nations Environment Programme [2009]). Every year, hundreds of thousands of old computers and mobile phones are dumped in landfills or burned in smelters. Thousands more are exported, often illegally, from Europe, the

US, Japan and other industrialized countries, to countries in Africa and Asia (Bhutta *et al.*, 2010). Recipient countries include Bangladesh, China, India, Malaysia, Pakistan, the Philippines, and Vietnam in Asia and Ghana and Nigeria in West Africa. In fact, it is estimated that upwards of 80 per cent of the US's e-waste is exported to these countries with 90 per cent of the waste going to China (Grossman, 2006: ch. 7).

E-waste follows a path of least resistance: it flows from the highly regulated core countries to low wage countries with limited health and environmental regulation. It is exported to countries for inexpensive, labor intensive recycling, incineration, and/ or disposal (Basel Action Network and Silicon Valley Toxics Coalition, 2002; Nnorom, Osibanjo, & Ogwvegbul, 2011; Tsydenova & Bengtsson, 2011). Valuable materials extracted from computers include copper, lead, plastics, steel, and glass. The state and capital in China (and many peripheral countries) want the 'recycling' industry for economic reasons, including the high demand for used parts and the increasing demand for materials to supply the growing manufacturing sector.

Guiyu Township in Guangdong Province is one of the major destinations for much of the e-waste entering China and it the largest e-waste recycling site in the world-system. (Taizhou city is the second largest site in China and it is located south of Shanghai.) Guiyu is 100 miles northwest of Guanghou and has a population of approximately 200,000 people in seventeen villages (the four main villages are Huamei, Longgang, Xianpeng, and Beilin) (see Appendix Figure 1 for the location of Guiyu). Once a rice producing area, Guiyu became an e-waste recycling center in the early 1990s, though its residents have a long history of waste collection stretching back to the early 20th century when residents would collect duck feathers, scrap metal, and pig bones for sale. Guiyu is now home to an estimated 150,000 e-waste workers (including children, as well as commuters from nearby areas) engaged in e-waste recycling (see Appendix Figure 2). Much of the e-waste recycling that takes place in Guiyu consists of the dismantling of computers and related accessories imported from the US, Japan, Canada, South Korea, Europe, and Taiwan (Basel Action Network and Silicon Valley Toxics Coalition, 2002; Grossman, 2006: ch. 7; Sepulveda *et al.*, 2010).

Steps in Dismantling Computers

Brokers based in Hong Kong and Taiwan sell the e-waste to recyclers through e-waste dealers in China who pay anywhere from 400 to 500 dollars or more per ton for computers (Grossman, 2006: ch. 7). The cost depends on the composition of the e-waste (whether circuit boards, monitors, printers, or the like) and profit margins. Several steps are followed once the e-waste reaches its port of destination in Hong Kong or

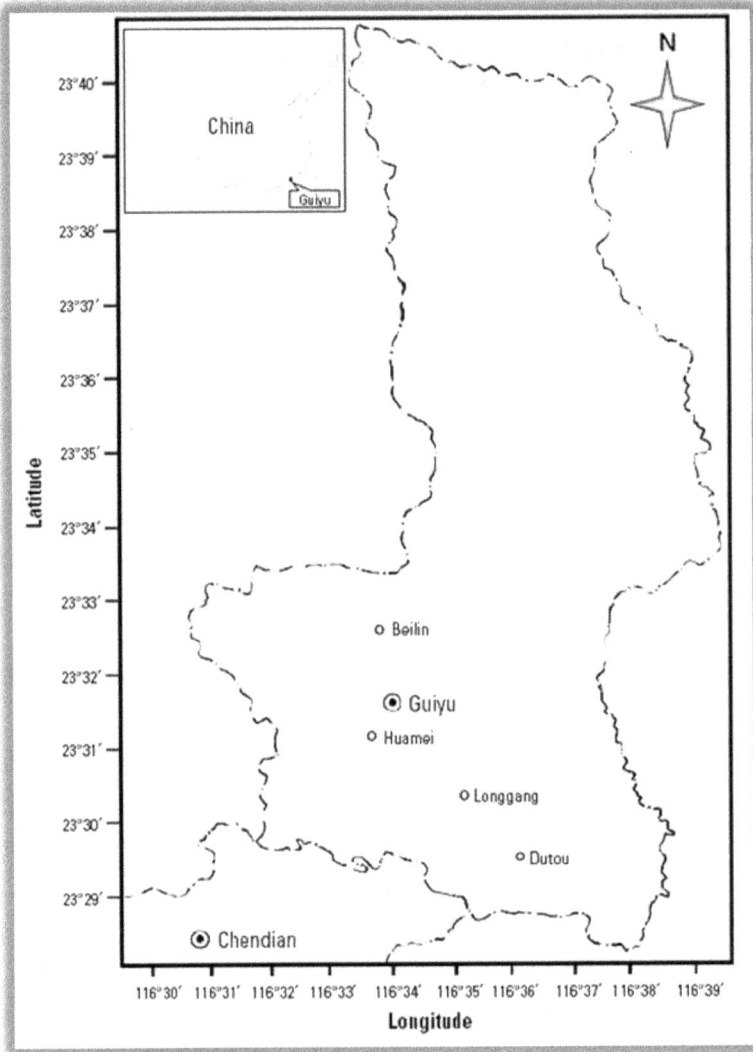

Figure 1 *Location of Guiyu, China*

Shantou, China (see Basel Action Network and Silicon Valley Toxics Coalition, 2002; Grossman, 2006: ch. 7; Tsydenova & Bengtsson, 2011: 51–55).

- Computers are trucked in after they are unloaded from container ships in their port of destination.
- Cathode ray tubes are broken with hammers, exposing the toxic phosphor dust inside. The copper yokes are removed and sold to metal dealers (see Appendix Figure 3).
- Circuit boards are cooked in woks over open charcoal fires to melt the lead solder, releasing toxic lead fumes. The lead solder is collected for metal dealers.

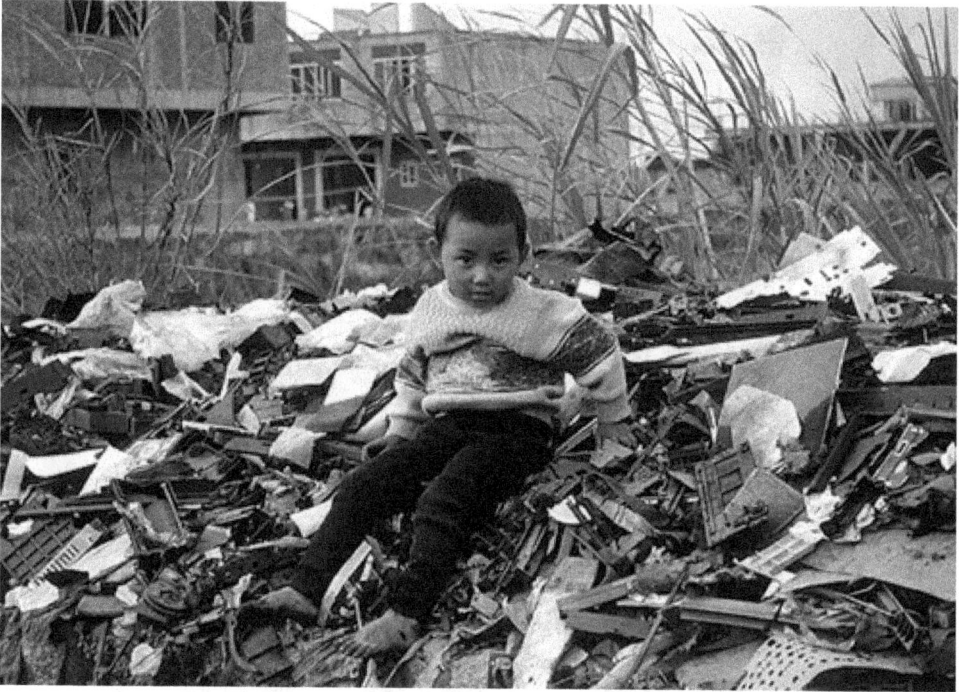

Figure 2 *Child on e-waste in South China*

- Large pieces of plastic are melted into thin rods and cut into small granules and sold to factories that make low quality plastic goods.

- Wires are stripped by hand or burned in open piles to melt the plastics to get at the copper and other metals inside (see Appendix Figure 4).

- Acid baths are used to extract certain materials from microchips such as gold. Nitric and hydrochloric acids are used to release gold from plastic and other commodities and the acids are dumped into the local environment.

- Plastic casings are burned, creating dioxins and furans—which are extremely hazardous to human health.

- Unwanted leaded glass and other materials are dumped in ditches.

- Acids and dissolved heavy metals are dumped directly into local waterways.

Buyers from local factories and outside the area purchase the metals such as copper, gold, aluminum, steel, and other commodities and sell them locally and nationally.

Health and Environmental Risks Associated with Computer Dismantling

The average desktop computer contains valuable recyclable and hazardous materials. Alumnum, copper, gold, steel, and platinum, as well as the more toxic lead are the

Figure 3 *Dismantling computers in South China*

most valuable materials (e.g., Williams *et al.*, 2008: 6447, Table 1). Hazardous materials include heavy metals, brominated flame retardants, and many other toxic materials; lead and cadmium and mercury in circuit boards; lead oxide in CRTs; mercury in switches and flat screen monitors; cadmium in computer batteries; and persistent organic pollutants (dioxins, PVCs, and PAHs) in plastics (Nnorom *et al.*, 2011; Tsydenova & Bengtsson, 2011).

As noted above, materials are extracted in an unsafe fashion. Recycling practices release toxins from hazadous materials and generate new ones. Open air incineration is used to recover copper in wiring and acid baths are used to extract metals such as copper and gold. Waste is dumped in irrigation canals and other waterways, including the nearby Lianjiang River (Sepulveda *et al.*, 2010; Tsydenova & Bengtsson, 2011).

Figure 4 *Stripping Computer Wires in South China*

Working conditions are primitive and unsafe, and workers are exposed to toxic materials but little safety equipment is available. Labor conditions are grave for the workers: they work six day work weeks of twelve hours duration per day for limited pay and they have few rights. Child labor is quite common (Basel Action Network and Silicon Valley Toxics Coalition, 2002; Grossman, 2006: ch. 7).

Available research indicates that human health and the environment are under assualt in Guiyu (see the excellent reviews of the extant literature by Sepulveda *et al.*, 2010 and Tsydenova & Bengtsson, 2011). The air, soil, and water of Guiyu are contaminated with a range of toxic materials, including lead, cadmium, PCBs, benzene, and so on (Bi *et al.*, 2010; Tsydenova & Bengtsson, 2011). A study released in 2007 (Huo *et al.*, 2007) found that a majority of the children sampled in Guiyu had blood levels of lead and cadmium many times higher than limits set by the US Centers for Disease Control and Prevention. A recent study (Yang *et al.*, 2011) undertaken at the other major e-waste recycling site in China, Taizhou, indicates that air samples from the area contain toxic particulate matter that can induce human DNA damage. Exposure to these and related materials are extremely hazardous to human health and represent significant risks to other species, as well as the larger environment and surrounding human communities.

EVALUATING THE COSTS AND BENEFITS

A re the costs associated with the displacement of e-waste recycling to Guiyu off-set by the economic and other benefits as proponents of neoliberalism (Grossman & Krueger, 1993, 1995) and some ecological modernization theorists (Mol, 2001) would suggest? After all, e-waste recycling employs at least 150,000 poor workers desperate for jobs in Guiyu. The materials and parts recovered are recycled and used domestically which reduces dependence on outside sources, reduces pollution associated with mining, provides needed capital for the economy, and reduces energy use and carbon dioxide emissions. In addition, import duties on some of the incoming goods provide a revenue stream for government (see Williams *et al.*, 2008: 6449–6450 for a discussion of the benefits of computer exports).

Answering the question raised above as noted elsewhere (Frey, 2006a) is problematic because it is difficult to identify, estimate, and value the costs and benefits (especially the costs) associated with hazards in monetary terms (see, e.g., Frey, McCormick, & Rosa, 2007; Williams *et al.*, 2008). Despite suggestions and efforts to the contrary (e.g., Logan, 1991), there is no widely accepted factual or methodological basis for identifying, estimating, and valuing the costs and benefits associated with the flow of core hazards to the periphery. Even if the consequences of hazardous exports could be meaningfully identified and estimated, there remains the question of valuing them in monetary terms. Economists typically look to the marketplace for such a valuation, but adverse health, safety, environmental, and socio-economic consequences are not traded in the marketplace. Efforts have been made to deal with this problem by using either expert judgment or public preferences, but such techniques are deeply flawed (see Dietz, Frey, & Rosa, 2002; Foster, 2002a).

When he was Chief Economist of the World Bank, Lawrence Summers (1991) made the argument much like the editors of The Economist mentioned above, that displacing environmental harms to peripheral areas makes economic sense. He wrote in a World Bank memo: 'I think the economic logic behind dumping a load of toxic waste in the lowest-wage country is impeccable and we should face up to that'. Environmental harms should be sent to poor areas because 'measurements of the costs of health impairing pollution depend on the forgone earnings from increased morbidity and mortality. From this point of view a given amount of health impairing pollution should be done in the country with the lowest cost, which will be the country with the lowest wages'.

As noted elsewhere (Foster, 2002b; Frey, 2006a; Puckett, 2006), such reasoning undervalues nature and assumes that human life in the periphery is worth much less than in the core because of wage differentials. Although most costs occur in the periphery and most benefits are captured by the core and elites located in the periphery, the costs to the periphery are deemed acceptable because life is defined as worth so little. In sum, it can be argued convincingly that the costs associated with the transfer of e-waste to Guiyu, China (and elsewhere, for that matter) outweigh the benefits.

What Is To Be Done? And Who Should Do It?

The Chinese government banned imports of toxic e-waste in 2002 and has created additional regulations since then (as recently as January 1, 2011 [Moxley 2011]; see also Yu *et al.* [2010a] for a comprehensive review), but the e-waste continues to flow into the country and thousands of Chinese women continue to cook the core's circuit boards over charcoal burners and the blood lead levels of children remain high. This is a result of lax enforcement of regulations due to bribes and corruption. In turn, local government officials are evaluated by the central government in terms of overall economic growth in their areas, so there is a strong incentive for officials to protect e-waste activities since they contribute to the economic growth of the area. And, of course, China's growth machine requires large material inputs to sustain it and e-waste recycling is an important source for these materials (Grossman, 2006; Yu *et al.*, 2010b: 992–994, 999).

A number of actors have emerged to challenge e-waste recycling in China. These include Chinese government officials and Chinese NGOs. Pan Yue, Vice Minister of the Ministry of Environmental Protection (CMEP) of the People's Republic of China, has been an unwavering supporter of the environment at all levels (Byrnes, 2006). Pan has been very active in promoting partnerships between CMEP and various Chinese environmental NGOs. It is increasingly clear that China (and other developing countries such as India) are going to be confronted with drastic increases in e-waste as domestic computer consumption increases in the next several decades which will further compound the e-waste problem in China and elsewhere (see Yu *et al.*, 2010a for estimates of growth in computer consumption by different regions in the world-system).

What is being done to challenge e-waste exports to China and elsewhere in the world-system? The Basel Convention on the Control of Transboundary Movements of Hazardous Wastes and their Disposal (1989) is a multilateral agreement that was enacted in 1989 under the auspices of the United Nations Environment Programme (and its subsequent amendment to ban e-waste export in 1995 which has yet to be ratified). The Convention, signed by 170 countries, requires that a country can only ship hazardous wastes if it has received written consent from the recipient country.

The Convention has been ineffective in stopping the trade for several interrelated reasons. The US, one of the world's largest e-waste exporters, has not signed the bill and thus the effectiveness of the Convention has been undercut. The Convention has been ineffective because illegal shipments of wastes are pervasive and a general lack of implementation at the border areas in China.

The United Nations Environment Programme (2005, 2009) and international NGOs, including Greenpeace International and Greenpeace China, Silicon Valley Toxics Coalition, the Basel Action Network, and others have been monitoring and studying export flows and actual conditions in and around the recycling centers in China and elsewhere (Basel Action Network and Silicon Valley Toxics Coalition, 2002). The Basel Action Network has been particularly active in pressuring state authorities in the developed countries to enforce higher standards (see Puckett, 2006; www.ban.org). The Basel Action Network, other international organizations, and analysts have made a number of specific recommendations for dealing with the e-waste problem (see, e.g., Grossman, 2006: ch. 7; Nnorom, *et al.* 2011; Pellow, 2007: 203–224; Smith, Sonnenfeld, & Pellow, 2006; United Nations Environment Programme, 2005, 2009; Yu *et al.*, 2010b). A sampling of these recommendations are listed below:

- A fully implemented global regime should be developed to regulate the movement of computer waste.
- The next generation of computers should be constructed to reduce health, safety, and environmental impacts at the time of decommissioning and increase capacity to upgrade computers over time.
- End-of-life electronics and greener design using fewer toxic materials and increased capacity for upgrades.
- Extended Producer Responsibility (EPR) as a new paradigm in waste management.

The European Union's Directive on Waste from Electrical and Electronic Equipment (WEEE), along with the RoHs Directive, were enacted into law in February 2003 and came into force in 2004. The Directives require manufacturers and importers in the European Union countries to take back their products from consumers and ensure safe waste disposal or safe recycling-reuse. Heavy metals (lead, mercury, cadmium, and chromium) and flame retardants (polybrominated biphenyls and polybrominated biphenyl ethers) were to be replaced with safer materials. The directives have not been fully implemented (European Commission, 2002; Geiser & Tickner, 2006).

After years of failure to address the issue of e-waste (see Stephenson, 2008), the US has moved forward in several areas. On November 15, 2010, President Obama issued a presidential proclamation on e-waste recycling and the creation of an Inter-agency Task Force of Agencies within the federal government 'to prepare a national strategy for responsible electronics stewardship, including improvements to Federal procedures for managing electronic products'. He indicated he wanted the Federal Government to lead (Nevison, 2011). The report of the Task Force was released in July 2011. The Task Force identified four major goals, one of which centers on reducing 'harm from US exports of e-waste and improve safe handling of used electronics in developing countries' through five specific actions (Interagency Task Force ..., 2011: 2–3).

- Improve information on trade flows and handling of used electronics, and share data with Federal and international agencies, within the limits of existing legal authorities.

- Provide technical assistance and establish partnerships with developing countries to better manage used electronics.

- Work with exporters to explore how to incentivize and promote the safe handling of remanufactured, recycled, and used electronics at home and abroad.

- Propose regulatory changes to improve compliance with the existing regulation that governs the export of cathode ray tubes from used computer monitors and televisions that are destined for reuse and recycling.

- Support ratification of the Basel Convention on the Control of Transboundary Movements of Hazardous Wastes and their Disposal.

Two Democratic Representatives (Gene Green of Texas and Mike Thompson of California) introduced the Responsible Electronic Recycling Act in the US House of Representatives on June 22, 2011. The bill would ban the export of certain restricted electronic waste exports to developing countries. 'The bill aims to stop U.S. companies from dumping dangerous old electronics on countries where they are broken apart or burned by workers using few safety precautions', said Texas Representative Gene Green. It is reported that several computer manufacturers support the bill (Miclat, 2011).

The Basel Action Network adopted a certification program in April 2010 called the e-Stewards Standard for Reasonable Recycling and Reuse of Electronic Equipment. The program referred to as e-Stewards Certification was established to facilitate responsible disposal of e-waste materials. Basel Action Network announced on July 5,

2011 that Intercon Solutions (a Chicago Heights, Illinois, electronics recycler) would be the first company denied BAN's e-Stewards certification, which aims to recognize e-waste recyclers operating responsibly.[1]

CONCLUDING REMARKS

Counter-hegemonic globalization or 'globalization from below' in the form of transnational networks of NGOs remains one of the most viable means for curbing the adverse consequences associated with the transfer of hazardous processes to the periphery (see Frey, 2006a). Globalization from below may help reduce or mitigate the worst abuses associated with the displacement of computer recycling and other environmental harms to the periphery as suggested by the concrete actions that have occurred in China, the EU, and the US noted above (see Yu *et al.*, 2010a, 2010b and Williams *et al.*, 2008, for a very insightful discussion of why the existing policies noted above are unlikely to solve the e-waste problem in China and elsewhere). Stopping the core's appropriation of the periphery's carrying capacity is another matter, for this process is embedded in the structure of the current world-system. In other words, the process of 'ecological unequal exchange' between the core and periphery is necessary for continued capital accumulation in the core. And it will be some time before environmental justice and sustainability are realized in the peripheral zones of the world-system because of the contradictory tendencies between accumulation in the core and the role of the periphery as resource extraction frontier and waste disposal frontier. To put it another way, the 'metabolic rift' (Foster, Clark, & York, 2010) between the core and periphery is made invisible by globalization and the attendant market ideology espoused by proponents of the neo-liberal perspective, including the editors of The Economist cited at the beginning of the paper.

REFERENCES

Basel Action Network and Silicon Valley Toxics Coalition

2002. Exporting Harm: The High Tech Trashing of Asia. Seattle: Basel Action Network.

Basel Convention on the Control of Transboundary Movements of Hazardous Wastes and Their Disposal. 1989. URL: www.basel.int/text/documents.html.

Bhutta, M. K. S., Omar, A., and Yang, X. 2011. Electronic Waste: A Growing Concern in Today's Environment. Economics Research International 2011: 1–8.

Bi, X., Simoneit, B. R. T., Zhen, W., Wang, X., Sheng, G., and Fu, J. 2010. The Major Components of Particles Emitted during Recycling of Waste Printed Circuit Boards in a Typical E-Waste Workshop in South China. Atmospheric Environment 44: 4440–4445.

---

1. See www.ban.org; www.e-stewards.org/certification-overview.

Bunker, S. G., and Ciccantell, P. S. 2005. Globalization and the Race for Resources. Baltimore, MD: The Johns Hopkns University Press.

Byrnes, S. 2006. Person of the Year: The Man Making China Green. New Statesman, December, 18. URL: www.newstatesman.com.

Dietz, T., Frey, R. S., and Rosa, E. 2002. Technology, Risk, and Society. In Dunlap, R., and Michelson, W. (eds.), Handbook of Environmental Sociology (pp. 329–369). Westport, CT: Greenwood Press.

Editors 2007. Editorial. The Economist, June, 9: 14.

European Commission 2002. Directive 2002/96/EC on Waste Electrical and Electronic Equipment. URL: www.ec.europa.eu/environment/waste/weee/index.

Foster, J. B. 2002a. The Ecological Tyranny of the Bottom Line: The Environmental and Social Consequences of Economic Reductionism. In Foster, J. B. (ed.), Ecology Against Capitalism (pp. 26–43). New York: Monthly Review Press.

2002b. Let Them Eat Pollution: Capitalism in the World Environment. In Foster, J. B. (ed.), Ecology Against Capitalism (pp. 60–68). New York: Monthly Review Press.

Foster, J. B., Clark, B., and York, R. 2010. The Ecological Rift: Capitalism's War on the Earth. New York: Monthly Review Press.

Frey, R. S. 1998a. The Hazardous Waste Stream in the World-System. In Ciccantell, P., and Bunker, St. G. (eds.), Space and Transport in the World-System (pp. 84–103). Westport, CT: Greenwood Press.

1998b. The Export of Hazardous Industries to the Peripheral Zones of the World-System. Journal of Developing Societies 14: 66–81.

2006a. The Flow of Hazardous Exports in the World-System. In Jorgenson, A., and Kick, E. (eds.), Globalization and the Environment (pp. 133–149). Leiden—Boston: Brill Academic Press.

2006b. The International Traffic in Asbestos. Nature, Society, and Thought 19: 173–180.

2012. The Displacement of Hazardous Products, Production Processes, and Wastes in the World-System. In Babones, S., and Chase-Dunn, Ch. (eds.), Handbook of World-Systems Analysis: Theory and Research. New York: Routledge. In press.

Frey, R. S., McCormick, S., and Rosa, E. A. 2007. The Sociology of Risk. In Bryant, C. D., and Peck, D. (eds.), The Handbook of the 21st Century Sociology. Vol. II (pp. 81–87). Thousand Oaks, CA: Sage.

Geiser, K., and Tickner, J. 2006. International Environmental Agreements and the Information Technology Industry. In Smith, T., Sonnenfield, D. A., and Pellow, D. N. (eds.), Challenging the Chip: Labor Rights and Environmental Justice in the Global Electronics Industry (pp. 260–272). Philadelphia, PA: Temple University Press.

Grossman, E. 2006. High Tech Trash: Digital Devices, Hidden Topics, and Human Health. Washington, D.C.: Island Press.

Grossman, G. M., and Krueger, A. B. 1993. Environmental Impacts of a North America Free Trade Agreement. In Garber, P. M. (ed.), The Mexico—U.S. Free Trade Agreement (pp. 13–56). Cambridge, MA: The MIT Press.

1995. Economic Growth and the Environment. The Quarterly Journal of Economics 110: 350–377.

Harvey, D. 2010. The Enigma of Capital and the Crises of Capitalism. New York: Oxford University Press.

Hornborg, A. 2011. Global Ecology and Unequal Exchange: Fetishism in a Zero-Sum World. New York: Routledge.

Huo, X., Peng, L., Xu, X., Zheng, L., Qiu, B., Qi, Z., Zhang, B., Han, D., and Piao, Z.

2007. Elevated Blood Levels of Children in Guiyu, an Electronic Waste Recycling Town in China. Environmental Health Perspectives 115, July, 7: 1113–1117.

Interagency Task Force on Electronics Stewardship 2011. National Strategy for Eletronics Stewardship. Washington, D.C.: Interagency Task Force on Electronics Stewardship.

Logan, B. I. 1991. An Assessment of the Environmental and Economic Implications of Toxic-Waste Disposal in Sub-Saharan Africa. Journal of World Trade 25: 61–76.

Martinez-Alier, J. 2009. Social Metabolism, Ecological Distribution Conflicts, and Languages of Valuation. Capitalism, Nature and Socialism 20: 58–87.

Miclat, M. 2011. Responsible Electronics Recycling Act Introduced in Congress. URL: www.GreenAnswers.com.

Mol, A. P. 2001. Globalization and Environmental Reform: The Ecological Modernization of the Global Economy. Cambridge, MA: MIT Press.

Moxley, M. 2011. E-Waste Hits China. IPS (Inter Press Service). July 21. URL: http://ipsnews.net/news.asp?idnews=56572.

Nevison, G. 2010. Obama Driving E-Waste Reform. Electronics Weekly. November, 15. URL: http://www.electronicsweekly.com/blogs/electronics-legislation/2011/07/obama-driving-e-waste-reform.html

Nnorom, I. C., Osibanjo, O., and Ogwuegbu, M. O. C. 2011. Global Disposal Strategies for Waste Cathode Tubes. Resources, Conservation and Recycling 55: 275–290.

Pellow, D. N. 2007. Resisting Global Toxins: Transnational Movements for Environmental Justice. Cambridge, MA: MIT Press.

Puckett, J. 2006. High Tech's Dirty Little Secret: The Economics and Ethics of the Electronic Waste Trade. In Smith, T., Sonnenfeld, D. A., and Pellow, D. N. (eds.), Challenging the Chip: Labor Rights and Environmental Justice in the Global Electronics Industry (pp. 224–233). Philadelphia, PA: Temple University Press.

Rice, J. 2007. Ecological Unequal Exchange: International Trade and Uneven Utilization of Environmental Space in the World System. Social Forces 85: 1369–1392.

Rice, J. 2009. The Transnational Organization of Production and Uneven Environmental Degradation and Change in the World Economy. Journal of Comparative International Sociology 50: 215–236.

Sepulveda, A., Schluep, M., Renaud, F. G., Streicher, M., Kuehr, R., Hageluken, C., and Gerecke, A. C. 2010. A Review of the Environmental Fate and Effects of Hazardous Substances Released from Electrical and Electronic Equipments During Recycling: Examples from China and India. Environmental Impact Assessment Review 30: 28–41.

Smith, T., Sonnenfeld, D. A., and Pellow, D. N. (eds.) 2007. Challenging the Chip: Labor Rights and Environmental Justice in the Global Electronics Industry. Philadelphia, PA: Temple University Press.

Stephenson, J. B. 2008. Electronic Waste: Harmful U.S. Export Flow Virtually Unrestricted Because of Minimal EPA Enforcement and Narrow Regulation. Testimony before the Subcommittee on Asia, the Pacific, and the Global Environment, Committee on Foreign Affairs, House of Representatives, September, 17. URL: www.gao.gov.

Summers, L. 1991. Selections from a World Bank Memo. The Economist, February, 8: 66.

Tsydenova, O., and Bengtsson, M. 2011. Chemical Hazards Associated with Treatment of Waste Electrical and Electronic Equipment. Waste Management 31: 45–58.

United Nations 2010. World Development Indicators. URL: http://data.un.org/DataMartInfo.aspx#WDI.

United Nations Environment Programme 2005. E-Waste, the Hidden Side of IT Equipment's Manufacturing and Use. New York: United Nations. 2009. Recycling—From E-waste to Resources. New York: United States.

Wallerstein, I: 2004. World-Systems Analysis: An Introduction. Durham, NC: Duke University.

Widmer, R., Oswald-Krapf, H., Sinha-Khetriwal, D., Schnellmann, M., and Boni, H. 2005. Global Perspectives on E-Waste. Environmental Impact Assessment Review 25: 436–458.

Williams, E., Kahhat, R., Allenby, B., Kavazanjian, E., Kim, J., and Yu, M. 2008. Environmental, Social, and Economic Implications of Global Reuse and Recycling of Personal Computers. Environmental Science and Technology 42: 6446–6454.

Yang, F., Jin, S., Xu, Y., and Lu, Y. 2011. Comparisons of IL-8, ROS and p53 Responses in Human Lung Epithelial Cells Exposed to Two Extracts of PM2.5 Collected from an E-Waste Recycling Area, China. Environmental Research Letters 6: 1–6.

Yu, J., Williams, E., Ju, M., and Shao, Ch. 2010a. Forecasting Global Generation of Obsolete Personal Computers. Environmental Science and Technology 44: 3232–3237.

Yu, J., Williams, E., Ju, M., and Shao, Ch. 2010b. Managing E-Waste in China: Policies, Pilot Projects and Alternative Approaches. Resources, Conservation and Recycling 54: 991–999.

Chapter 11

GREAT POWER POLITICS FOR AFRICA'S DEVELOPMENT: AN OVERVIEW ANALYSIS OF IMPACT OF THE EU'S AND CHINA'S COOPERATION WITH THE CONTINENT

Zinsê Mawunou and Chunmei Zhao

After the African countries got independence, the European Union (formerly the EEC) and China had shown a willingness to contribute to the improvement of socio-economic development of Africa with series of measures for its socio-economic prosperity through a partnership that has evolved for a long time. While the long way of the EU countries position has been challenged by China's unprecedented implication on the continent, African countries have been taking advantage of this situation to improve their economic situation during recent years. This article analyzes this statement by exposing firstly a background introduction to the EU–Africa diverse agreements and conventions, on the one hand, and the present Chinese strategy of cooperating with Africa, on the other, in order to highlight the opportunities offered to the continent through this cooperation. Secondly, the article presents economic implications of these actors' involvement in Africa based on the optimistic approach of the EU's and China's partnership with Africa. It finally concludes with some suggestions to those actors as well as to African countries emphasizing that even though the EU's and China's partnership with Africa does move Africa in the right direction it still has some gaps.

Keywords: Africa's development, cooperation, Sino-Africa, EU-Africa.

INTRODUCTION

When African countries got independence, the ambition of their liberation movement leaders was to ensure the development of their countries in every sector. Facing with many challenges in the postcolonial era, this ambition was difficult to achieve due to the scarcity of resources. Therefore, the EU and China decided to help African countries and support their emerging economies by making available ways to achieve their goals. The logic led to the signing of agreements and conventions between the EU and the ACP countries[1] and also to various Chinese development programs that are considered to be a good opportunity for Africa to raise its economic level and launch sustainable development programs. Therefore, what are the approaches from both the EU and China to Africa's development agenda? Theoretically, liberalism appropriate with the logic of the arguments developed in this article will be used to understand and get a meaningful explanation to the issues of the article. So, the first part is the inventory of the different steps from the EU and China towards Africa's development. This inventory is important in the context of this paper because it highlights different strategies of the EU and China for the development of Africa especially the financial ways made available to increase the opportunity for development of the continent. The second part is a discussion aimed at differentiating the nature of the EU and China partnership with Africa by showing how each of them contributes to the improvement of Africa's development. Finally, the article concludes by clearly exposing its position and raises some suggestions to both players for a mutually beneficial economic relation with African countries.

COOPERATION WITH AFRICA: STRATEGIES FROM THE EU AND CHINA

Both sides are willing to boost Africa's economic situation through cooperation. On the EU part, strategies take the form of agreements and conventions, while China also has its own way of dealing with Africa.

Introductory Background of Conventions and Agreements between the EU and Africa

The Yaoundé Conventions

After gaining their independence in the early 1960s, some African countries negotiated with the European Community the continuity of their preferential economic relations, ushering the formula of economic partnership between the EU and Africa

---

1. Africa-Caribbean-Pacific.

(Greenidge, 1997). The European Community (that later became the European Union) and the group of African countries, joined later by the Caribbean and Pacific countries decided to establish a framework for economic, cultural and political cooperation. The first agreements between European and African countries took shape in Yaoundé[2] where two conventions were held, namely Yaoundé I and II.[3] The goals of the association were the diversification and industrialization of these countries' economy to ensure a better stability and strengthen their independence. The means used were granting financial aid and special customs regimes. In all, there had been Yaoundé I and Yaoundé II.

The Convention of Yaoundé I

Under Yaoundé I, signed in July 1963 (Faber & Orbie, 2007), only 18 African countries were signatories. This Convention planned (primarily for the benefit of the AMSA)[4] aid for development (that was called 'technical and financial assistance') and trade preferences. It is interesting to note that the dispositions of Yaoundé related to trade were based on principles of non-reciprocity, thus addressing commercial arrangements of the pre-independence period. To this end, the EDF[5] of Yaoundé I amounted to 730.4 million euro (Morzellec, 2001). There were various bodies forming the association: the Association Council, the Association Committee and a Parliamentary Conference (Steen & Danau, 2006). Six years after this convention, Yaoundé I was followed by Yaoundé II.

---

2. Name of Cameroon capital where these agreements were signed.

3. It is important to note here that the Yaoundé Conventions were preceded by what is called the association regime. Indeed, countries that signed the Treaty of Rome (1957) had expressed their solidarity on this occasion with colonies, countries and overseas territories and undertook to contribute to their prosperity. Thus, the fourth part of this treaty considered the creation of the European Development Fund (EDF) to grant technical and financial assistance to those colonies and overseas territories which had historical links with the European countries. It was called at that time Regime of Association. But this article does not focus on that Treaty for two reasons. First, at that period it was only a treaty between European countries without participation of any African colonies; second, African colonies had not yet attained their sovereignty, at least the major part, while this paper focuses on the independent African countries. In this sense, it is unnecessary to develop the association regime here.

4. African and Malagasy States Associated.

5. EDF: European Development Fund. In general, the EDF funding are divided as follows: First EDF: 1959–1964; Second EDF: 1964–1970 (Yaoundé I Convention); Third EDF: 1970–1975 (Yaoundé II Convention); Fourth EDF: 1975–1980 (Lomé I); Fifth EDF: 1980–1985 (Lomé II); Sixth EDF: 1985–1990 (Lomé III); Seventh EDF: 1990–1995 (Lomé IV); Eighth EDF: 1995–2000 (Lomé IV bis); Ninth EDF: 2000–2007 (Cotonou Agreement); Tenth EDF: 2008–2013(Cotonou Agreement). See Cooperation ACP / European Commission, Financing Instruments European Development Fund, URL: http://www.confedmali.gov.ml/acpue_fed.php, accessed 8 December 2010.

The Convention of Yaoundé II

The Second Yaoundé Convention, signed in July 1969, aimed at increasing the European Development Fund (EDF) resources for development projects, which rose to 887.3 million euro (Morzellec, 2001). It is important to note that it was from Yaoun-dé II that the signatories of agreements between Europe and Africa began to increase, for example, it was under Yaoundé II that countries like Kenya, Tanzania and Uganda decided to join the ACP group (Karingi et al., 2005). However, the peak enlargement of signatories of the agreements between Europe and African countries and the strengthening of partnership between both partners started from 1975 through the various Lomé Conventions.

The Lomé Conventions

Since its first agreement signed in 1975, the Lomé Convention had evolved and the Convention of 1975 was renewed four times. After the expiration of the first convention in 1980, the second Convention took over until 1985. The third one covered the period of 1985–1990 and the last one, the fourth Convention, covered a whole decade, from 1990 to 2000, and has been revised at mid-term. Totally, the successive Lomé Conventions lasted 25 years running from 1975 to 2000. Important amendments have been introduced in the partnership between the EEC and Africa after the expiration of Yaoundé Conventions and the introduction of the first Lomé Convention in February 1975. Since then, the EEC had experienced its first major enlargement including the entrance of the UK in 1973; thus, the adherence of the United Kingdom to the EEC strongly encouraged some Anglophone countries to undertake, too, the privileged partnership with the EEC. The ACP group has since been expanded to 46 members with the participation, for the first time, of the Caribbean and Pacific countries.

The Lomé I Convention

The first Lomé Convention, signed in February 1975, was largely inspired from the Yaoundé II Convention. It was characterized by its contractual nature, its principles of partnership and its various aspects related to aid, trade and politics. Concretely, the EU granted favorable access conditions to its market for ACP countries products, which were not obliged to grant similar concessions to European exporters. The main innovation of this Convention, said Steen and Danau (2006) was to introduce a stabilization system of export income for agricultural products, STABEX[6] and the sugar protocol. The 4th EDF of Lomé I, which was preceded by that of the Association Regime (EDF 1) and the two EDF (EDF 2 and 3) of the Yaoundé Conventions, received a found-

---

6. STABEX: Stabilization System of Export Earnings from Agricultural Products.

ing budget of more than ECU 3 billion.[7]

The second Lomé Convention (Lomé II)

The second Lomé Convention signed in 1980 was the logic extension of the first Lomé Convention. Obviously, Lomé II also introduced innovations compared to the previous agreement; the most important being the establishment of SYSMIN[8] for countries that rely heavily on mine products and record export losses. Lomé II (1980–1985) also focused on strengthening infrastructure and its funding increased to ECU 4.725 billion.

The third Lomé Convention (Lomé III)

In 1985 the third Lomé Convention occurred. It intervened at the time when there were serious questions on the effectiveness of aid to development. One can see an emergence of the political dimension and the introduction of other dimensions. In this convention, there was also the issue about 'human dignity' since a new fund was created and intended to help refugees and the 6th EDF got a funding of ECU 7.4 billion (Karingi et al., 2005) and was directed to development programs especially in the rural sector.

The Lomé IV and Lomé IV bis

The last agreement signed in Lomé under the ACP-EU partnership was the Lomé IV Convention in 1990, which was reviewed in 1995. For the first time, the Lomé Convention took a political aspect and the respect of Human Rights became a fundamental term of ACP-EU cooperation. This Convention strengthened the political dimension and introduced conditions and sanctions. Then, the violation of those principles led to partial or total suspension of development aid. Another innovation of this convention concerned the structural adjustment, debt, role of the private sector, environment, demography and decentralized cooperation. In general, the 7th EDF (under Lomé IV) got 11.583 billion euro and the 8th EDF (under Lomé IV bis) received 13.151 billion euro, which was an envelope of 24.734 billion euro for Lomé IV projects.

The Cotonou Agreement

The signing of the Cotonou Agreement happened within a peculiar context that had significant effects on the content, principles and ideas this agreement contained.

---

7. ECU: European Currency Unit is the predecessor of the Euro. It was equivalent to 3 pounds and (in France) from 1960 1 ECU = 5 silver francs.
8. SYSMIN: System of Stabilization of Export Earnings from Mining Products.

Context of the Signing of the Cotonou Agreement

The Cotonou Agreement was signed when important changes had been taking place on the international arena. The fall of the communist bloc and the end of the Cold war introduced major changes in international relations, because the concept of democracy and market economy suddenly became the dominant ideology and many African countries embraced those concepts. Another effect of the fragmentation of the Communist block was the emergence of new states in Eastern Europe having the ambition to join the EU. This group of new countries faced serious problems during their transition to the market economy and to the democratic system, so the EU felt the responsibility to assist them during the transition. It was in this context that aid programs were created to support the development of Central and Eastern European States (Morzellec, 2001). At the same time, globalization was gaining ground and the world was becoming increasingly interdependent. But African countries in particular appeared to be the big losers in this process with a decline in international trade, investment and production. Parallel to these events, there was growing a concern on human rights.

The Cotonou Agreement and its Innovations

On June 23, 2000 in Cotonou[9] there was signed a new partnership agreement within EU-ACP for a period of 20 years to be reviewed each five years. It was at Cotonou that the free trade agreements called 'Economic Partnership Agreements' (EPA) emerged, replacing the existing non-reciprocal preference system. It was the end of the asymmetric and joint partnership conventions of Yaoundé and Lomé: preferences became reciprocal. The financial protocol of the Cotonou Agreement had a total budget of 13.8 billion euro for the 9th EDF (EU Council, 2000). However, the real resources for the period 2000–2007 amounted to 25.1 billion euro, due to the consolidation of all EDF (residues from previous EDF) and some founds of the European Investment Bank (EIB) that can be used for investment purposes (Gahamanyi et al., 2004). Finally, the 10th EDF (2008–2013) received a funding of 22.8 billion euro and this was for a five-years' period.[10] Overall, the 9th and 10th EDF in the context of Cotonou Agreement have mobilized a total funding of 36.6 billion euro without the leftovers of previous programs.

---

9. Economic capital of Benin Republic.
10. EU Country & Regional Programming to ACP countries. URL: http//:www.acp-progra mming.eu/wcm/; accessed 22 December 2010.

Conventions and Agreements Characteristics	Yaoundé		Lomé				Cotonou
	I	II	I	II	III	IV and IV bis	
Budget	€730.4 millions	€887.3 millions	More than 3 billion Ecu	4.725 billion Ecu	7.4 billion Ecu	€24.734 billion	€36.6 billion (9th and 10th EDF)
Terms	– economic diversification and industrialization of the countries; – operating with various joint bodies; – non-discriminatory trade preferences based on reciprocity	– non-discriminatory reciprocal trade preferences; – membership of three African countries	– massive support from other African countries; – discriminatory non-reciprocal trade agreement; – priority to the construction of roads, bridges, hospitals, and schools; – introduction STABEX; – establishment of the Sugar Protocol; – addition of joint institutions	– the introduction SYSMIN; – expanded areas of intervention	– priority to agriculture and rural development; – issue of 'human dignity'	– political dimension and respect for human rights; – structural adjustment	replaced the existing non-reciprocal preference system

Table 1 *Summary of the different conventions and agreements*

The cooperation between China and Africa has a long history and evolved through many phases. The shifts in China's African policy are closely interlinked with domestic development strategies as well as international events. Accordingly in modern history China's African policy has passed through roughly three phases.

The Evolution of Sino-African Development Cooperation, 1955–1979

In fact, China started its aid to Africa almost at the same time when the Western aid programmes started—in the early 1950s. To be precise, China's aid to Africa and other developing countries started after the Bandung Conference of 1955 and was guided by The Five Guiding Principles of Chinese aid, set out by Premier Zhou Enlai during the India-China bilateral negotiations. But with regard to Africa, these principles included:

- Mutual respect for sovereignty and territorial integrity;
- Mutual non-aggression;
- Non-interference in each other's internal affairs;
- Equality and mutual benefits;
- Peaceful coexistence.

China not only supported African liberation movements (Taylor, 2006) but also provided a great deal of economic assistance on a grant basis despite the fact that China itself was a struggling developing country with a few resources. Between 1973 and 1979, for example, aid to Africa amounted to 6.92 per cent of China's GDP annually, and forty-four African countries had signed economic and technical cooperation protocols with China. It was during that first phase that China constructed the Tanzania—Zambia railway. Despite the increasing allocation of aid to Africa during that period, China avoided the term 'aid' in its cooperation with Africa; instead Chinese officials preferred to use the language of solidarity and friendship—a situation quite different from the often paternalistic Western aid language of poverty reduction and democratization.

The original Five Guiding Principles were later replaced by China's Eight Principles of Economic and Technical Aid, which Premier Zhou Enlai announced on January 15, 1964 during his visit to fourteen African countries. The additional guiding principles emphasized that: Chinese technical assistance should build local capacities, and Chinese experts working in Africa should have the same standard of living as the local experts; economic cooperation should promote self-reliance and not dependency; and respect for the recipient's sovereignty should mean imposing no 'political or economic conditions' on recipient governments. As a result of these diplomatic efforts, the number of African countries recognizing China grew to thirty-seven by the early 1970s. Between 1970 and 1975, some sixteen African heads of states visited China. At the same time, Chinese aid to Africa grew from $428 million in 1966 to nearly $1.9 billion in 1977.

Moreover, with the death of Mao Zedong and the subsequent policy shift towards economic modernization under the leadership of Deng Xiaoping, China entered a new era in world politics, culminated in the establishment of formal diplomatic relations with the United States in 1979.

China-Africa Relations in the Post-1970s Reform Period

In the early 1980s the policy of modernization and economic reform became the centerpiece of China's Communist Party under Premier Deng (Taylor, 2006). This period saw the announcement of the new Four Principles on Sino-African Economic and Technical Cooperation in 1983 by Premier Zhao Ziyang (Davies, 2007) based on equality, mutual benefit, pursuing practical results by adopting a variety of means, and seeking common development. Within the scope of the Eight Principles, these new adjustments were prompted by the weakening of ideological conditions and increasing attention was given to economic relations and the strengthening of humanitarian aid support.

Following these policy adjustments, China supported more than two hundred infrastructure projects in African countries in the 1980s. The overall number of projects in Africa and West Asia exceeded 2,600, amounting to US$5.6 billion.

In fact, between 1970 and 1976, China committed US$1,815 million aid to Africa (Taylor, 2006).

The Post-1990 Reforms

In the 1990s, African countries accelerated the process of multiparty democracy and the liberalization of the economy under the watchful eyes of the IMF and the World Bank. With the trend towards liberalization and privatization in full swing, the Chinese government realized that it would no longer be possible to insist on traditional cooperation between governments, that development aid should be directed towards invigorating private sector development in Africa and that the new policy should also enroll the participation of Chinese enterprise in African markets. The new approach then was consistent with China's broad economic trade strategy of exploiting the opportunities made possible by the process of economic globalization (Lin, 1996: 33–36).

The second important reform was the decision to grant interest-free loans and subsidized export credits to African countries in order to promote Chinese trade and investment in Africa. The results of these reforms are numerous.

According to Weston *et al.* (2011: 7), Chinese government's pledges and announcements of bank loans and deals indicate that China's assistance to Africa is growing rapidly, especially since the 1990s. Accordingly, China's April 2011 Foreign Aid White Paper reported that Africa was the destination for 45.7 per cent of Chinese aid in 2009. The New York University Wagner School (Ibid.) reported Chinese investment projects and aid in Africa to be $10 million in 2002, $838 million in 2003, $2.3 billion in 2004, $4 billion in 2005, $9 billion in 2006, and $18 billion in 2007. Deborah Brautigam, a scholar of China-Africa relations, reported that Export-Import Bank of China pledged $20 billion in loans from 2007–2009 (cited in Weston *et al.*, 2011: 7). She noted that, commercial deals and loans aside, China's ODA to Africa was $1.4 billion in 2007. Since then, Premier Wen Jiabao has pledged an additional $10 billion in low-interest loans to African states in 2009–2012.

Also, some 585 Chinese enterprises received approval by the Chinese authorities to invest in Africa in 2002. South Africa had 98 approvals, amounting to $119 million in value. Other important Chinese FDI destinations in Africa include Tanzania, Ghana, and Senegal (Broadman, 2007). By the end of 2000, the Chinese had established 499 companies in Africa with a total contractual investment of $990 million, of which $680 million was Chinese capital (Taylor, 2006).

Impact of the EU's and China's Cooperation on Africa

The EU's and China's presence in Africa is highly admired both as an opportunity that Africa should grasp to get a place into the international economy and also as a chance to improve its development. Some cases are selected to analyze the effects of the EU partnership with Africa, followed by the Chinese effects on Africa.

The EU Case

First of all, one should mention that for the only year of 2007 (under the 9th EDF), Africa received the funding of 2.694 billion euro from a total sum of 3.4 billion (i.e., 79.24

per cent) (see Figure 1). This investment has funded several projects on the continent. This means that Africa got the bulk of projects to be financially supported by the EU.

Of a total of 184 projects for funding consideration by the EU, Africa had 150 projects (Delcoustal, 2008) corresponding to 81.52 per cent regionally divided as follows: West Africa—70; East Africa—20; Central Africa—37; Southern Africa—23.

One of the considerations of the Cotonou Agreement (under the 10th FED) is to allow greater regional integration through the formation of trading blocs. Supporting the existing sub-regional institutions, the EU has placed special emphasis on strengthening these institutions as a relay and a tool for its policy of assistance to

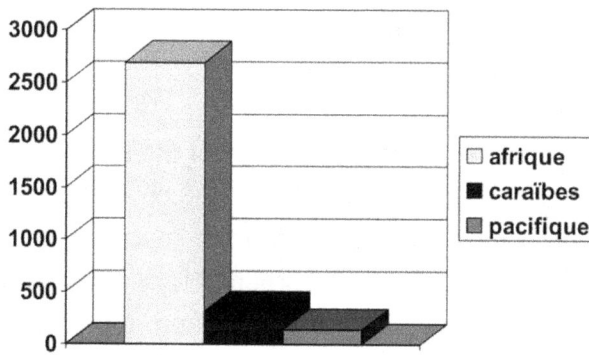

Afrique Sub-Saharienne 2.694 Mds €
Caraïbes 314.08 M€
Pacifique 140.78 M€

Figure 1 *The commitments for 2007, Geographical Distribution*

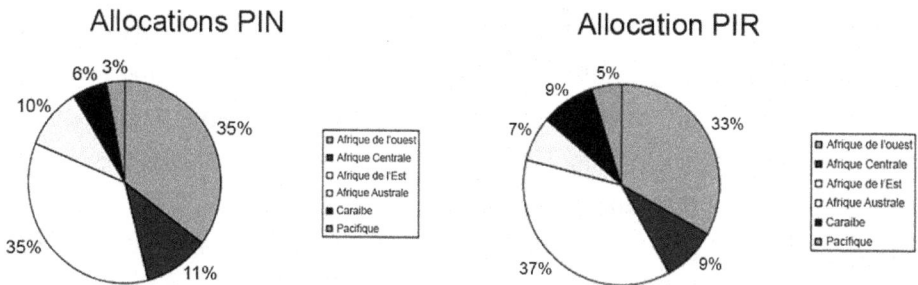

Figure 2 *Budgetary allocations for regional and national programs for ACP countries*

Africa. Indeed, the objectives of strengthening regional blocs are multiple. It could allow many of these countries to overcome the obstacles posed by their relatively small domestic market; then at the regional level, it provides institutional capacity and human resources to adjust their technical and administrative insufficiencies. And finally, the regional approach may allow African countries to pursue their interests with more confidence and strength. Moreover, the terms and obligations of membership of an ambitious program of reforms within a regional organization also facilitate the task for national leaders to implement politically difficult measures, such as reducing rates of protection or the establishment of large-scale reform of regulatory and judicial systems. In addition, monitoring at a regional level a dialogue among partners helps to reduce the risk of slipping to macroeconomic terms.

In this perceptive, these institutions have an ability to negotiate and receive funding for implementation of projects for countries in the region. This article explores the case of ECOWAS and the effect of agreements on the development of this region.

The Case Study of West Africa

Even if ECOWAS is a minor trading partner for the EU (about 2 per cent of its trade), the EU is a privileged trading area of ECOWAS countries. About half of imports and exports of ECOWAS member countries in fact concern the EU. Agricultural and food products represent a significant proportion of total of ECOWAS exports to the EU (28 per cent), and the imports of the ECOWAS from the EU (16 per cent). Among the ECOWAS countries, Côte d'Ivoire, Ghana, and Nigeria produce 83 per cent of food exports from the region to the EU.

The following Table 2 presents the inventory of the EU–ECOWAS trade in 2002 and shows that such a scale of trade increases competitiveness, the flow of European direct investment, intra-regional flows and increased trade volume.

Individually and in some key sectors, the economy of some West African countries has improved through the implementation of the regional agreements within the EU–Africa partnership. For example, in Senegal, fishing became the first sector of the economy from years of drought and the farm crisis (Gahamanyi *et al.*, 2005). Indeed, fishing helps both to reduce the deficit balance of payments and unemployment as well as to supply the population's needs in protein. For a total turnover of 300 billion FCFA,[11] fishing generates an estimated value of 100 billion FCFA, or 11 per cent of total GDP and 2.3 per cent of primary total GDP. Moreover, the authorities pay special attention to the fisheries sector to restore the trade balance, chronically in deficit. Since

---

11. The CFA Franc (FCFA) is the currency of several African countries that constitutes the franc zone.

Category of products	EU exports to ECOWAS ($ million)	In %	EU imports from the ECOWAS ($ million)	In %
Agricultural products	1,864	17.0	2,902	31.3
Raw materials	806	7.3	5,231	56.4
Manufactured products	8,301	75.7	1,147	12.3
Total	**10,971**	**100.0**	**9,280**	**100.0**

Table 2 *Structure of trade between the EU and ECOWAS in 2002
(in millions US $ and percentage)*

1986, the fisheries sector ranks first in exports exceeding peanut products and phosphates handsets by providing nearly a third of the value of overseas sales. Fishery now generates nearly 100,000 direct jobs for nationals and over 90 per cent by artisanal fishing. It also creates many related jobs and occupies nearly 15 per cent of the labor force in Senegal, which is about 600,000 people.

We should note that the whole Africa has had roughly the same pattern in cooperating with the EU as in the case of West Africa. Figures 1 and 2 have shown it. The funds allocated to Africa are spread across the continent according to the level of development of the economic blocs as stipulated by the Cotonou Agreements.

The Chinese Case

Chinese presence in Africa is notable in various areas such as trade, social, infrastructure building, peacemaking and peace building. But here, few cases will be selected, namely trade, infrastructure building and a new developmental model based on Chinese Special Economic Zones experience to illustrate the impact of Chinese involvement on Africa.

Chinese Trade with Africa

With its status as a 'latecomer' in investment in Africa, China does not seem to be handicapped by this. Contrary to all expectations, it becomes Africa's third largest trading partner after the USA and France (Alden 2007). Two-way trade, which stood at less than US$10 billion in 2000, surged to over US$50 billion by the end of 2006 (see Figure 3). Within the same period China's share of Africa's exports jumped from 2.6 to over 9.3 per cent and it has become the leading trading partner for several countries of the continent's commodity-based economies.

As shown in Figure 3, China's trade with Africa has been growing from the mid-1990s to 2006. It also shows that for the same period the trading value increased from approximately US$4.5 billion to US$10 billion in 2006, meaning the Chinese partnership with Africa produced good result in the trading sector.

However, this figure is not equally distributed on the continent. In Africa, there are countries that have deeper trading relations with China than others as the following Figure 4 shows.

This clearly shows the proportion of Chinese trading partners in Africa in 2006, Angola being the first and Benin Republic the tenth. And this top ten accumulated 78 per cent of the whole Chinese trade with Africa. According to Rotberg (2008) another group of seven countries (Angola, Egypt, Congo, Ghana, South Africa, Tanzania, and Uganda), which the Chinese Premier Wen Jiabao visited in the summer of 2006, had a combined trade volume of over $38 billion with China, or 52.2 per cent of total Chinese-African trade in 2007.

Infrastructure Building

The renovation and extension of infrastructure were sorely neglected throughout the continent during the final decades of the twentieth century, when trade and aid to Africa were dominated by Euro-American partners. However, China has committed to participate in the refurbishment, the building and extension of the infrastructure networks throughout the African continent.

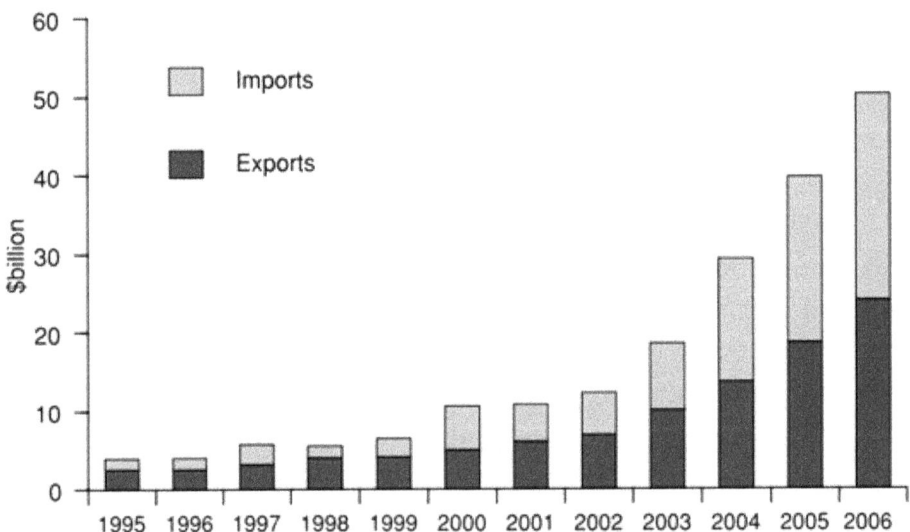

Figure 3 *China's trade with Africa, 1995–2006*

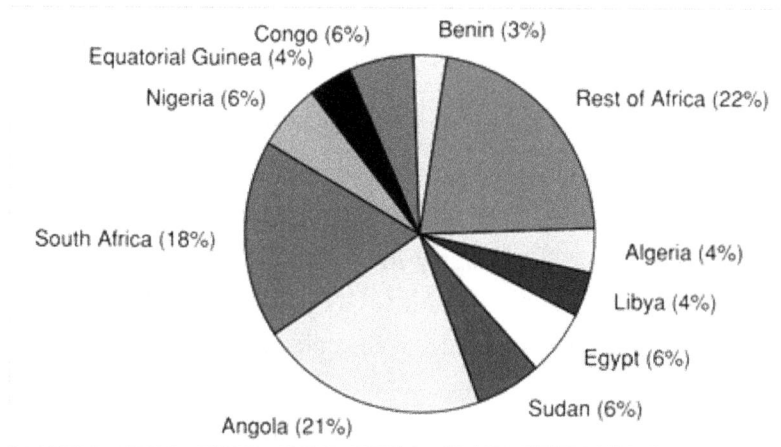

Figure 4 *China's top ten trade partners in Africa, 2006*

In this respect, Chinese (re)development of roads, railroads, ports, and airports are also Chinese priorities by cooperating with Africa. To take but one example, the colonial construction and contemporary Chinese reconstruction of the Benguela rail line in Angola, which runs from the Angolan coast directly eastwards toward the rich mining zones of the Democratic Republic of the Congo and Zambia illustrates such large-scale Chinese projects in Africa (Rotberg, 2008). As part of its massive effort to redevelop Angolan infrastructure that was devastated by the decades-long civil war, in 2004 China extended $2 billion in soft loans to Angola.

Special Economic Zones

The Chinese Special Economic Zones (SEZs) originated at the Forum on China–Africa Cooperation (FOCAC) summit held in Beijing in November 2006, which was attended by over forty African heads of state. By deciding to export its experience of Special Economic Zones to Africa, China believes these zones will provide the liberalized investment environments focused on strategic industries to attract foreign companies. The model of special geographical zones where investing companies enjoy preferential economic policies is by no means unique. Numerous African governments have established or are establishing such zones in their countries in an attempt to attract foreign direct investment (FDI), especially in labor-intensive manufacturing industries. Kenya, Egypt, and Mauritius are the most proactive on the continent with respect to such an activity. These zones are positioned on the continent in order to become Africa's new economic growth nodes. Chinese-initiated SEZs in Africa require large amounts of investment in infrastructure, both within the zones and linking them to ports and regional markets. If completed as planned, the infrastructural corridors will provide the essential linkages between fragmented African markets and will have a positive impact upon regional economic integration.

The terms of these zones are being negotiated between Beijing and targeted African governments that are willing to offer the required policy concessions in order to receive committed Chinese investment. Beijing has strategically selected some key African economies in which it will apply its SEZ model. These designated countries reflect China's commercial priorities in Africa, they are geographically dispersed over the continent, and have had long-term close political relations with the PRC. For a total of five, these zones are: Zambian Mining Hub, Indian Ocean Rim Trading Hub; Tanzanian SEZ: A Logistics Hub, Nigeria as a Gateway to West Africa, Manufacturing Hub in Egypt.

CONCLUSION

From the above-said, it is clear that the interaction between the EU and China with Africa produced satisfactory results on the continent. For proof, Africa's trade with the EU and China takes unprecedented proportions, and increases productivity on the continent. So Africa can be proud of contributing to the international trading system, even if its position in the system is still rather modest. Although the investments of different actors in Africa help the continent to revive its economy after decades of stagnation, the global players' policy with respect to Africa's development still has some gaps.

Regarding the EU, despite the Africans' enthusiasm with respect to the benefits of the EU–Africa partnership, many think this partnership could not reverse the economic decline of Africa. Thus, their argumentation is that the cooperation does nothing more than a financial lifeline to Africa, or worse, a 'pension' or automatic allocations to less credible political regimes, without a clear relation with their performance. So we can say that the EU–Africa partnership is far from being an opportunity for Africa. In addition, the Lomé and Cotonou Conventions focused on the free access of some European products to the African market and vice-versa. Unfortunately, most of the African countries heavily depend on customs revenue. Thus, the application of this principle had serious implications in many African countries' budgets. For example, Côte d'Ivoire lowered 40 per cent of its tariffs in 1986 and this resulted in very significant layoffs in the chemical industries, textiles, footwear, and in automotive assembly plants. In Senegal there was a loss of 1/3 jobs in manufacturing between 1985 and 1990, following a reduction of tariffs from 65 per cent to 90 per cent during the same period. A report of the Institute of International Economics[12] stated that the loss of revenue in 1990–1995 from imported duties could range from 1.6 million euro in Guinea Bissau to 352 million euro in Nigeria; the same report indicated that Cape Verde could expect a drop in revenue of up to 80 per cent.

On the other hand, sometimes critics are also raised about China's efforts to improve socioeconomic conditions in Africa. It would be valuable if China issues some rigorous regulations on how to grant aid and any kind of investment to African countries, especially clauses and guarantees from African countries on the management of these funds, because sometimes such invested funds do not reach their destined goals. Communities that are located in regions where China undertakes its extractive, industrial, or commercial pursuits often do not see direct benefits from the Chinese presence.

Finally, the African countries should know not only what they want and how to obtain it but also how to manage their relationships with these partners in order to maximize profits. It is true that much remains to be done in Africa, but it is also good to start with good management and progress towards effective takeoff of the African economy.

REFERENCES

Alden, Ch. 2007. China in Africa. London: Zed Books.

Broadman, H. 2007. Africa's Silk Road: China and India's New Economic Frontier. Washington, DC: The World Bank.

Conseil de l'Union Européenne. 2000. Accord interne entre les représentants des gouvernements des États membres, réunis au sein du Conseil, relatif au financement et à la gestion des aides de la Communauté dans le cadre du protocole financier de l'accord de partenariat entre les États d'Afrique, des Caraïbes et du Pacifique et la Communauté européenne et ses États membres, signé à Cotonou (Bénin) le 23 juin 2000, et à l'affectation des aides financières destinées aux pays et territoires d'outre-mer auxquels s'appliquent les dispositions de la quatrième partie du traité CE. Bruxelles. URL: www.mci.dj/document/accord_cotonou_fr.pdf

Davies, P. 2007. China: The End of Poverty in Africa—Towards Mutual Benefit? Sweden: Diakonia.

Delcoustal, V. 2008. Le Fonds européen de développement—La mise en œuvre 2007. Bruxelles. URL: http://www.rpfrance.org/ce/doc/pdf08/FED20070812908.PDF

---

12. Institute is situated in Hamburg, Germany.

Faber, G., and Orbie, J. 2007. The EU's Insistence on Reciprocal Trade with the ACP Group. Economic Interests in the Driving Seat? Paper prepared for the EUSA Tenth Biennial International Conference Montreal, Canada, 17–19 May 2007. URL: http://aei.pitt. edu/7991/1/orbie-j-05g.pdf.

Gahamanyi, B. M., Dansokho, M., and Diouf, M. 2005. Explique-Moi, L'Accord de Partenariat ACP-CE. Manuel de facilitation à l'intention des acteurs non étatiques de l'Afrique de l'Ouest. Friedrich Ebert et Enda.

Greenidge, C. B. 1997. Return to Colonialism—The New Orientation of European Development Assistance. DSA European Development Policy Study Group.

Karingi, S., Lang, R., Oulmane, N. 2005. Effets des accords de partenariat économique entre l'UE et l'Afrique sur l'économie et le bien-être. CAPC, Travail en cours No. 22. Addis Ababa: Centre africain pour les politiques commerciales, Commission économique pour l'Afrique.

Lin, T.-C. 1996. Beijing's Foreign Aid Policy in the 1990s: Continuity and Change. Issues and Studies 32(1): 32–56.

Morzellec, J. 2001. Les Pays ACP. In Ami, B., and Christian, P. (dir.), Dictionnaire juridique de l'Union européenne. URL: http://www. cremoc.org/articles/paysacp.pdf.

Rotberg, I. R. 2008. China into Africa: Trade, Aid, and Influence. Washington, D.C.: Brookings Institution.

Steen, D. V. D., and Danau, A. 2006. L'Accord de Partenariat Economique (APE) entre l'Afrique de l'Ouest et l'Union européenne. Quels enjeux pour les exploitations paysannes et familiales? Bruxelles: Collectif Stratégies Alimentaires.

Taylor, I. 2006. China and Africa: Engagement and Compromise. New York: Routledge.

Weston, J., Campbell, C., and Koleski, K. 2011. China's Foreign Assistance in Review: Implications for the United States. Washington, D.C.: U.S.—China Economic and Security Review Commission, Staff Research Backgrounder.

Chapter 12

CONNECTING LOGISTICS NETWORKS GLOBALLY VIA THE U.N. SINGLE WINDOW CONCEPT

Michael Linke

The UN Single Window concept is a proven approach to facilitate cross border business including transport, customs and other government-related regulations by enabling seamless trade with a central IT platform, in a hub and spoke like system. Several approaches and implementations already exist, although one needs a proper planning for a further penetration worldwide. Enterprise Architecture Management (EAM) as a specialized IT strategy discipline can help to manage this complex challenge of integrating application landscapes into different existing UN integration frameworks.

Keywords: logistics networks, (UN) Single Window concept, cross border business, transport regulations, customs regulations, government regulations, seamless trade, EAM (Enterprise Architecture Management), Information Technology (IT) strategy, UN integration frameworks, international trade.

In recent decades, the world has become increasingly globalized. Countries today are more connected than ever and rely heavily upon international trade of goods and services in order to function in an appropriate manner. Advanced communication networks seem to play an important role in this acceleration (Ardalan, 2010). The results of international trade seem to be more positive than negative and evidence suggests that it has led to economic prosperity in many countries, subsequently resulting in an improved quality of life for its citizens which is stated also in recent relevant studies in that field (Sharmin & Rayhan, 2011). Nonetheless other voices point out that potentially backsides, especially with respect to inequality could still exist (Baumann, 2011). The United States, China and the European Union's 27 member states together account for billions of dollars in trade each year. According to the World Trade Organization, the EU exported 1.8 trillion USD worth of merchandise and imported nearly 2.0 trillion USD in 2010.

As trade between countries becomes a more integral part of the world economy, the need for fast and efficient methods of customs and security processes grows ever more crucial. While no one country operates under the exact same processes or policies, very few have streamlined practices when it comes to trade and customs regulations. It should be mentioned that globalization seems to have an additional impact, especially in the sovereignty domain of certain countries or economic unions, which might have derived from that, also an impact of customs regulations as such (Grinin, 2012). In fact, in many countries, businesses who wish to involve themselves in international trade are required to submit documentation (manual, automated or a mix) to several regulatory bodies in order to legally conduct trade. This range of documentation, paperwork and procedures depends on a number of factors including the type of goods or merchandise involved, their value and destination country. The result is what many businesses view as an inhibitive and stifling system that is overly complicated and slows the process of trade.

A recent survey conducted by the World Customs Organization (WCO) Compendium revealed that of 56 countries who participated in the survey, customs procedures involved on the average 15 separate agencies, with 96 % of reporting countries requiring at least 5 different regulatory bodies. From these findings, it is quite clear how the combination of paperwork and customs regulations can slow or even stifle overall economic prosperity at the macro level. At a more individual level, the regulations and documentation surrounding the import and export of goods can prevent companies from participating in international trade on the whole.

TRANSPORT
ELECTRONIC (EDI)
CUSTOMS
PAPER
AGRICULTURE
HEALTH

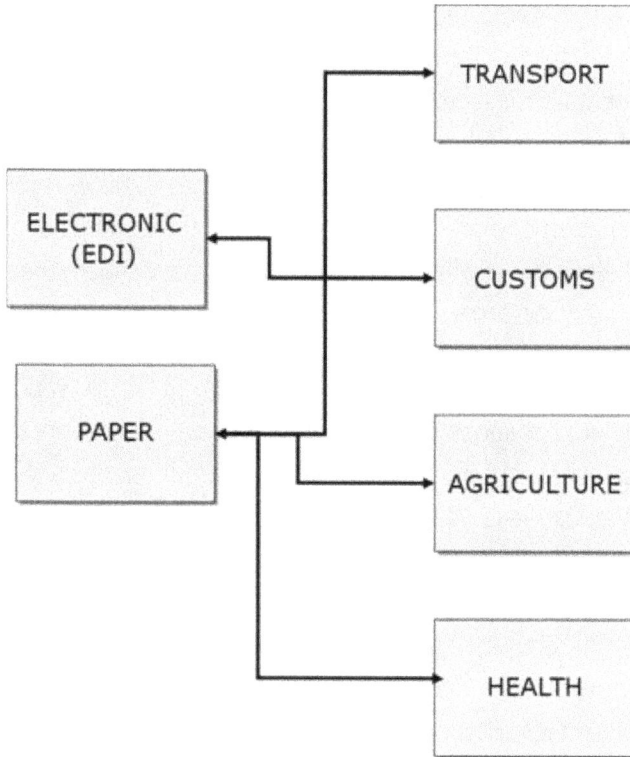

Figure 1 *Current trade entities in various trading situations*

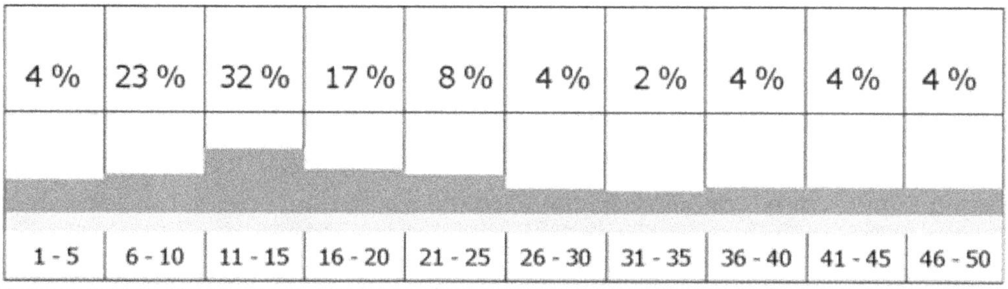

4 %
23 %
32 %
17 %
8 %
4 %
2 %
4 %
4 %
4 %
1 - 5
6 - 10
11 - 15
16 - 20
21 - 25
26 - 30
31 - 35
36 - 40
41 - 45
46 - 50

Figure 2 *Number of government agencies involved in cross border transactions*

The UN Single Window concept was introduced in order to reduce the aforementioned issues and inefficiencies involved with the importing and exporting of goods. Backed by a number of international organizations including the United Nations Economic Commission for Europe (UNECE), the World Customs Organization (WCO) and the Association of Southeast Asian Nations (ASEAN), it is defined by the UNECE as:

> *A facility that allows parties involved in trade and transport to lodge standardized information and documents with a single entry point to fulfill all import, export, and transit-related regulatory requirements. If information is electronic then individual data elements should only be submitted once* (GFP, 2012; ASEAN, 2007; APEC, 2007).

As an organization dedicated to the facilitation of both trade and electronic business, the UNECE Centre for Trade Facilitation and Electronic Business (UNECE, 2012) has been instrumental in researching the Single Window concept in depth as well as creating a set of recommendations for future implementation based on their findings (Centre for Trade Facilitation and Electronic Business, 2005). Through a careful research and examination of existing implementations, the benefits of adopting a single window are clear. Both the public and private sectors have much to gain by this methodized streamlining of customs and trade regulations.

Governments stand to benefit from implementing a single window at a number of different levels. Any changes put in place would be in an effort to streamline and regulate processes across a number of agencies simultaneously. This consolidation and standardization can lead to reduced overheads and a reduction in process errors resulting in reduced risk. In addition, having all customs and trade information flowing through a single window will allow governments to monitor more easily what is coming in and out of the country. This is important not only for economic and statistics agencies who report on such matters, but also to the nation's security. Security agencies will be able to access all pertinent information about goods entering the country through the single window instead of being forced to collect information from a number of different departments, allowing security to move more swiftly and effectively.

Overall, the effect of a single window on a government is far reaching. Customs agencies, permit departments and trade monitoring agencies can work together under a standardized umbrella and works towards making customs procedures faster, safer and more efficient. These changes will propagate through the economy and allow business to engage in international trade more easily.

As to the government, the private sector would benefit enormously from the use of a single window. Many scholars predict that it will eliminate or reduce the existing non-tariff related barriers to trade (Dobson, 2010), thus lowering the costs of international trade. In addition, due to the standardization of required documentation as a part of the single window, businesses would no longer be required to expend as much energy and resources submitting and keeping up to date with paperwork from more than one agency. This process would be made much more efficient and all required documentation would go to one source, or window. The positive effects of a single window propagate even further to shipping and delivery companies, the banking and accounting industries, and eventually to consumers.

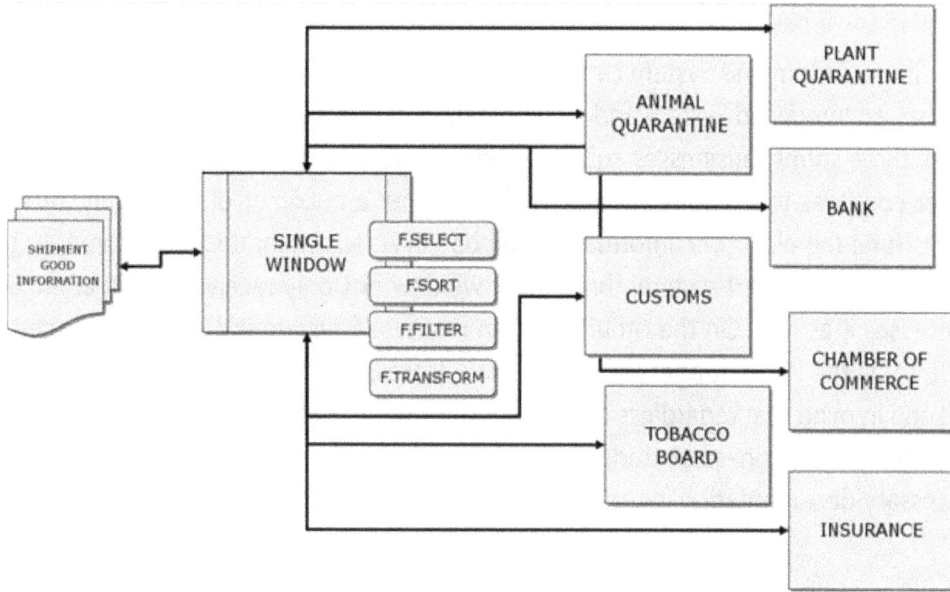

Figure 3 *Single window meta concept*

TECHNICAL DIMENSIONS OF A SINGLE WINDOW

There does not exist single stringent set of specifications which outline the definition and a scope of a Single Window. Every country has developed its own processes regarding customs and trade and consequently, will require different solutions in order to achieve a single window. The UN/CEFACT has created a number of guidelines for single window created with a view to 'enhance the efficient exchange of information between trade and government' (Centre for Trade Facilitation and Electronic Business, 2005). In these guidelines there are suggested three models for a single window—Single Authority, Single Automated System and Automated Information Transaction System.

Single Authority

In the Single Authority model, there is one body or agency that acts as a singular retainer for electronic or paper documentation related to a defined unique business function or service. Upon receiving documentation, this single retainer acts as an authority and either manually or automatically files necessary paperwork and disseminates the required documentation to the respective agencies or authorities. As a part of its functionality, this singular authority should only disperse documentation once overviewed and formatted to the recipient's specifications. See Figure 4 for an illustrated diagram of the Single Authority model.

Single Automated System

The Single Automated System can take three forms—an Integrated System (illustrated below), an Interfaced System and a combination of integrated and interfaced systems. In all three forms, businesses submit all electronic information related to trade with other countries to a singular source which is either a public or private entity, or window. Here, the electronic information will be collected, integrated and stored. In the case of an integrated system, this central window not only receives information, but processes it as well. On the other hand, an interfaced system will send the formatted data to all the relevant agencies rather than process it itself. However, it is important to keep in mind that regardless of the type of Single Automated system, a user experience from the business or trader perspective does not change as in any scenario all necessary documentation is submitted electronically to one authority.

Automated Information Transaction System

The Automated Information Transaction System is the most complex single window model, but it also the most advantageous for businesses. In this model, entities involved in international trade are only required to submit electronic information through a singular application. This application contains in its application backend the

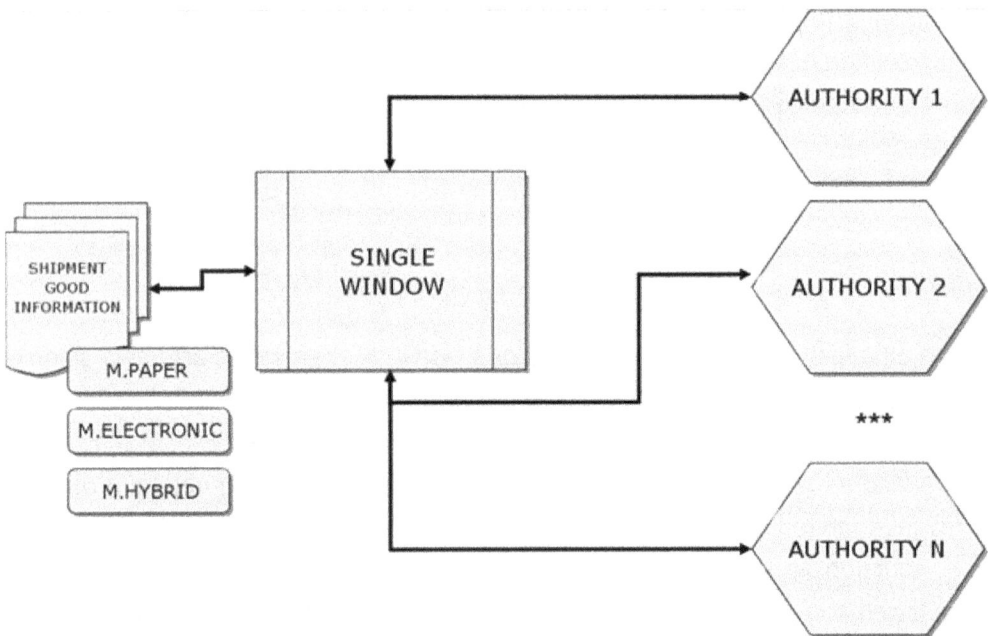

Figure 4 *Single Authority conception*

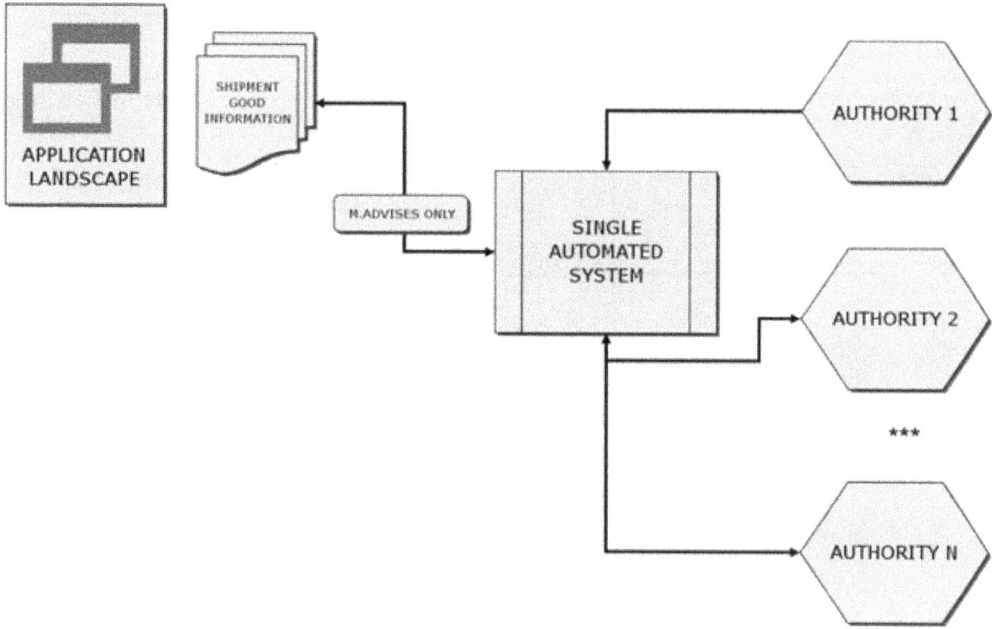

Figure 5 *Single Automated System conception*

integration with all concerning agencies and regulatory authorities. In many cases, custom fees, tariffs and taxes can also be calculated and integrated within this application allowing businesses not only to submit their information, but also make the necessary payments for their trades.

According to a survey conducted by the WCO Compendium, in 2011 only 33 per cent of participating customs administrations operated on a single window model, while the rest were still in the process of developing one. A breakdown of the survey results is as follows (Choi, 2011):

- 4 % operate Single Window—Integrated Model;
- 7 % operate Single Window—Interfaced Model;
- 22 % operate Single Window—Hybrid Model;
- 13 % operate One-stop Service;
- 44 % operate Stand-alone system;
- 9 % operate other systems.

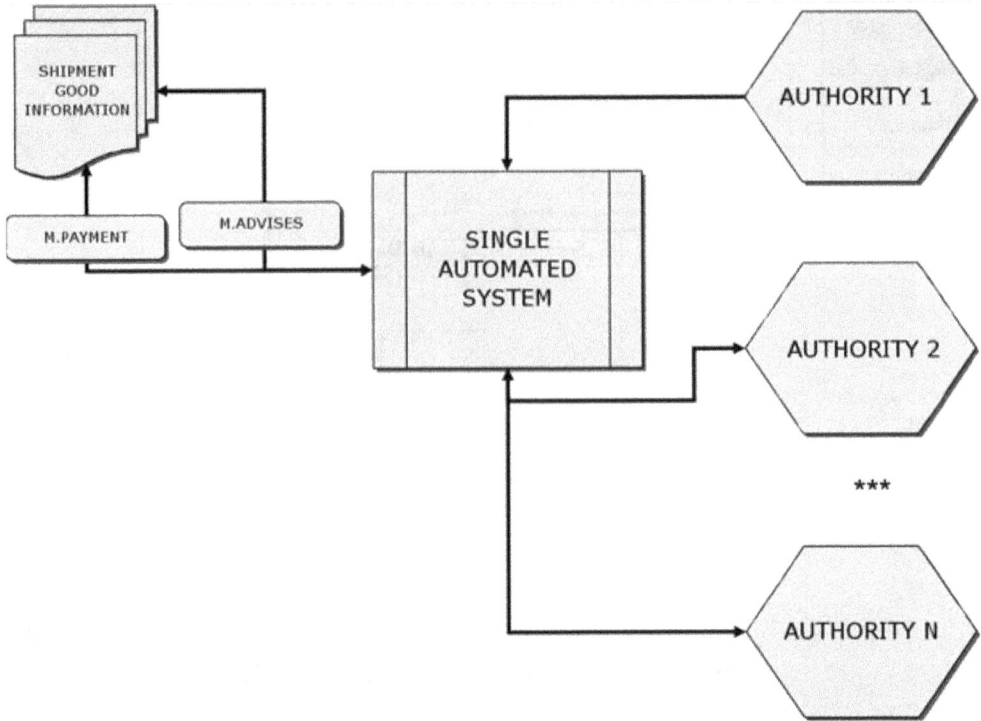

Figure 6 *Conception Automated Information Transaction System*

SINGLE WINDOW CASE STUDIES

Several organizations began making strides towards achieving the Single Window concept in the 1990s, while many more started work in the last decade. Every country faced unique situations and goals which in turn led to various implementations of methods. It was clear through each case study, however, that there were certain common steps that each organization took in order to achieve their goals.

The first was clearly a preparation. Due to the scale of a single window project encompassing departments that involve customs, imports and exports, an enormous preparation process was required in order to achieve a proper foundation upon which to build a single window. The UNECE emphasizes the importance of choosing a suitable agency to lead all others in the single window effort. While this agency can be public or private, it is important to ensure that is has enough legal power and government funding to act as an effective leader.

After a leading agency is chosen, it becomes crucial to set the requirements. It is important to identify early which processes, organization units and cross-organizational counterparts in related agencies should be integrated into the system's initial release. Each of these bodies should define its own requirements and only after this phase is completed there should begin the feasibility assessment and initial design work.

Variations in Single Window Implementation

The UNECE has revealed through study and survey of the countries that have or are in the process of introducing a single window, that there is a wide range of options in reference to the methods in which these systems are set up. For instance, while nations such as Finland and the United States fund the development of a single window through their government, other single window systems, including the one in Germany, are paid by the private sector (Butterly, n.d.). The country of Mauritius, which will be further discussed, received funding from both the public and private sectors.

Countries also vary in the way in which the single window is used. It is mandatory in some countries (Finland, Senegal, Mauritius), while intended to be voluntary in others (Germany, Sweden). Additionally, Germany, Senegal, Malaysia and a few other nations charge for use of their single window system. Each nation inevitably runs into challenges when implementing new systems, though these can vary as well. The countries such as the United States that have a complex and long-existing infrastructure to handle the trade, find it difficult to make the transition from older legal systems to a single window. Other nations find it difficult to get support from all participating agencies, while others run into problems finding initial funding and development power for the project (Butterly, n.d.). The UNECE emphasizes the importance, but not the necessity of technology in single window development. Though it is really advantageous to incorporate computerized and automated processes within the single window, the overall methodology can be executed manually in cases where funds for technology cannot be secured (Centre for Trade Facilitation and Electronic Business, 2005).

Mauritius

Mauritius, despite having an economy ranked 128th in the world (CIA 2011), was one of the first countries to have an information transaction system, proving that existing economic wealth is not a prerequisite to implementing a Single Window system. Through a corporation called Mauritius Network Services Ltd., consisting of public and private sector representatives combined with outsourced technical assistance, the TradeNet application was developed.

TradeNet is a completely proprietary electronic data interchange (EDI) application designed to receive information from IT applications including content related to customs, imports, exports, duties and tariffs. Since the launch of the program in 1994, businesses are able to submit all their information electronically through TradeNet as well as to make bank payments in the system in order to pay for any necessary duties, taxes or tariffs.

According to UN/CEFACT, 'it is estimated that TradeNet has decreased the average clearance time of goods from about 4 hours to around 15 minutes for non-litigious declarations, with estimated savings of around 1 % of GDP' (Centre for Trade Facilitation and Electronic Business, 2005). Thus, it is no surprise then, that Mauritius is bucking the trend in Africa and has been on the receiving end of consistent economic growth over the last 15 years. Global investment and exports seem to have increased and the countries could experience therefore a possible healthier distribution of wealth. The success of the single window in Mauritius has attracted attention of other African countries such as Uganda and Rwanda, both of which are in the development stages of a Single Window system (Hitimana, 2012; TradeMark East Africa, 2012).

Sweden

As a part of an ongoing initiative to provide more government transparency to its citizens, Sweden created a single window system known as the Virtual Customs Office (VCO), which is aimed at electronic processing of customs declaration as well as import and export licenses. The single window incorporates a large number of national organizations including the Swedish Customs Authority, Swedish Board of Agriculture, National Board of Trade, The National Inspectorate of Strategic Products, Swedish Police and The National Tax Administration and Statistics of Sweden (Centre for Trade Facilitation and Electronic Business, 2005).

The VCO aims at providing a user-friendly service to traders in an attempt to make the filling customs declarations and import/export licenses as simple as possible. Integrated into the virtual office are real-time updates of taxes, tariff codes and duties, which traders can receive either via email or SMS. Fully financed by the Swedish government, a survey of VCO users revealed that 80 % of traders saved time, 54 % saved money as a direct result of using the system, 72 % believed it provided increased flexibility, and 65 % thought the quality and speed of served had improved (Ibid.).

The Netherlands

Air cargo handled through Schiphol Airport in the Netherlands is processed through a single window system headed by the customs department. Called VIPPROG, the Netherlands' single window is integrated with a private documentation system called Cargonaut, which handles cargo manifest paperwork. The government pays Cargonaut in order to maintain and have access to relevant records (Centre for Trade Facilitation and Electronic Business, 2005).

In this situation, the customs department acts as the leading agency in the single window initiative and they process all paperwork in the form of a single automated (integrated) system. Here all cargo is given a risk assessment and depending on the results, any risk factors are sent to one or more of the corresponding agencies. The system is designed to integrate with the customs department and ten other agencies including immigration, and various health and agriculture offices (Ibid.). If any of these agencies wish to further inspect the cargo, the customs department arranges an inspection appointment, where any and all interested agencies can examine the cargo during a certain scheduled time. This ensures that all goods can be checked at once, accelerating the time in which cargo is generally processed through the airport while at the same time mitigating risks.

The United States of America

In the United States of America, there is a large concerted effort to implement and utilize a single window integrated with many of the country's government agencies in order to improve the trade process. The United States is one of the largest importers in the world and exports quite a bit as well with a combined total of three trillion USD worth of merchandise coming in and out of the country in 2011 (GFP, 2012). As a result, the country has set up a group known as the International Trade Data System (ITDS), aimed at establishing a 'single window through which the data required by government agencies for international trade transactions may be submitted' (ITDS, 2012). By implementing a secure government-wide system to collect, store, integrate and disseminate information related to trade, the ITDS hopes to reduce public and private sector overhead, comply more easily with a number of government requirements, and improve national security allowing multiple agencies to have access to pertinent information (Ibid.).

In the United States there seem to be more than a hundred agencies who require access to trade documentation (Ibid.). The ITDS has the monumental task of setting up a system whereby members of the trade and transportation communities are required to submit relevant documentation through a secure EDI only once, leaving the single window to take care of the rest. Much like VIPPROG in Schiphol Airport, the goal is to have the single window perform a security and risk assessment, then forward on the findings to any government agencies who are qualified to further assess compliance or security risk.

Still in the process of development, it is hoped that the country will benefit from this new system in a number of ways. Though providing efficient means of transporting goods across the country's boundaries will no doubt reduce overheads for both the government and private businesses, one of the most important goals is to increase government compliance and security. Storing and handling all data under a central hub or single window will facilitate the sharing of information between government agencies, allowing them to collaborate on security and compliance efforts.

ENTERPRISE ARCHITECTURE MANAGEMENT (EAM) AS AN I.T. DISCIPLINE

The more or less new discipline of Enterprise Architecture Management (EAM) as partially a discipline of the organization studies and IT can be described on the basis of two partial entities, which are already indicated by the combination of words.

Conceptual and Historical Dimension of the Term EAM

An enterprise is an activity that contains a well-defined target. Currently, this can mean a large number of organizations and sub-organizations, which pursue a common target or produce a common result. An enterprise can thus mean anything—from a big group to a state or public institution—in practice, also summarized into holdings, trusts, and other divisionally separated legal forms. Thus, they also have several Enterprise Architectures. An enterprise in this context can also be an Extended Enterprise, which includes all the partners, suppliers, and clients of the actual business in its value-added or administration chain into its own IT-based value added. The business architecture within the framework of Information Technology (IT) describes the interaction of the elements of information technology and the business activities within the business. It distinguishes above all due to the sub-elements, for example, the information architecture or the software architecture with a global view on the role of information technology within an organization. The official definition of the term architecture according to the ANSI/IEEE standard 1471–2000 in the IT environment is:

*An architecture is the fundamental organization of a system, embodied in its
components, their relationships to each other and the environment, and the principles
governing its design and evolution.*

The definition used, however, is narrower: an architecture is a formal description
of a system, a detailed plan of the system and its components, the structure of the
components, their mutual effects, their principles and guidelines, which control their
draft, their development, and their implementation. In larger groups, several different
Enterprise Architectures can exist at the same time. However, in all cases, an Enterprise
Architecture includes several technical systems. The enlarged concept of Enterprise
Architecture dates from the 1980s. One of the leaders of the architecture movement,
John Zachman, saw the value of the use of an abstract architecture for the integration of systems and their components. Zachman developed the analogies in the field
of traditional construction architecture and later used concepts from the airplane industry in order to cover the business process aspects in his framework. Since then, a
number of frameworks have been published, which all aim at describing a business in
a structural way (Zachmann, 2008).

Architecture Frameworks as an Auxiliary

An Architecture Framework divides a complex task of the IT architecture management into several partial layers, which can be described separately to partially reduce
complexity. Each partial layer (Layers) should be specified in the Meta model of the
Framework. An approach is the ISO standard 15704, which defines general demands
towards the company architecture. In this standard, the architecture is considered as
a description of the fundamental structure of the system parts and the links between
the individual subsystems.

THE RELATIONSHIP BETWEEN THE SINGLE WINDOW
CONCEPT AND ENTERPRISE ARCHITECTURE

The goal of a single window system for trade is to consolidate existing processes
and simplify existing procedures. When incorporating technology, this concept
aligns itself well to enterprise architecture, which is defined as 'the process of
translating business vision and strategy into effective enterprise change by creating,
communicating and improving the key requirements, principles and models that describe the enterprise's future state and enable its evolution' (Gartner, 2012).

In fact, the UNECE has created a framework meant to assist in the creation of a single window called the Single Window Implementation Framework (SWIF). It is based very heavily upon an existing standardized enterprise architecture framework known as The Open Group Architecture Framework (TOGAF) and its Architecture Development Method (ADM), which has evolved from initial work performed by the US Department of Defense. The figure above (Zachmann, 2008) illustrates the core entities of the TOGAF model.

Overview of SWIF Methodology

SWIF is divided into a preliminary stage and additional eight phases, each consisting of a defined set of objectives, activities and outputs/results. Designed to be a dynamic and iterative process, these phases together are intended to outline the general steps necessary to establishing and maintaining an EA-based Single Window System.

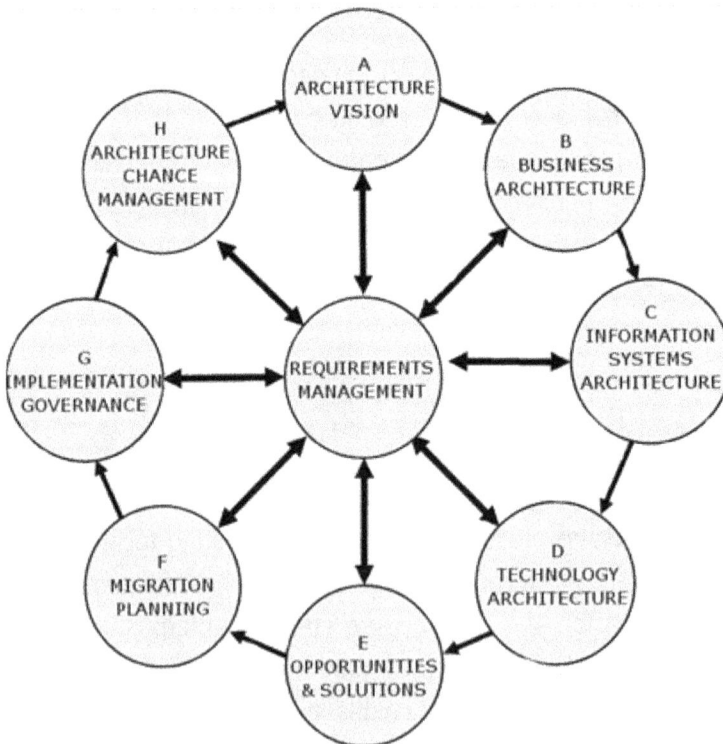

Figure 7 *TOGAF model*

Phase A: Architecture Vision

This is the highest-level phase, where the project can identified through broad definitions. The keys in this phase are to identify stakeholders. As the Single Window concept pertains to customs and trade, there are four categories of stakeholders—Authority, Supplier, Customer, and Intermediary. These refer to relevant government agencies, exporters, importers, and auxiliary parties such as financial and shipping institutions respectively (van Stijn *et al.*, 2011).

Once all the stakeholders are identified, the goals of this phase are to create a very broad overview of the requirements of the stakeholders, and establish key performance indicators for the project.

Phase B: Business Architecture

Much of the work that goes into streamlining processes occurs in Phase B. In this phase, existing business processes are examined and weaknesses identified. Functions which can be automated or consolidated should also be identified.

Phase C: Information Systems Architecture

This is the first phase which heavily involves IT. One of its main goals is to harmonize data, which will be a key component to facilitate future modifications and scalability. Standardized data allows for increased interoperability between business processes, allowing for more transparency and ease of use. A data model should be designed in this phase, incorporating all consolidated and streamlined business processes (Phase B), along with any data which will be utilized.

Phases D–H

Phase D deals with obtaining, designing and/or modifying any hardware or software required to implement the new business processes. In the next phase, a plan should be put in place for 'implementing, deploying and operating the Single Window' (van Stijn *et al.*, 2011). Phase F involves the final preparations required to ensure that all the sub-systems in place fulfill the requirements of the original high-level plan. The last two phases entail the implementation of a monitoring system and identifying ways to improve the system.

Throughout each of the phases, the management of requirements should be always kept in mind. It is important to ensure that all the work going towards the implementation of the enterprise structure does not ever stray from the business requirement established during the preliminary phase and Phase A.

The trade industry is very dynamic and experiences constant changes in regulations, duties and tariffs as a result of a various factors. Therefore, a single window system must be designed to be alterable, dynamic and growth scalable. The TOGAF ADM and SWIF account for the dynamic nature business in the design of their methodology. The previously described phases are intended to work in a cyclical format on several different levels. The framework is flexible enough to support the cycling of a single phase, between phases and around the entire ADM itself, allowing for changes, new initiatives and sub projects to be implemented during the life of the single window. This dynamic capability combined with the properties of enterprise architecture will ensure fewer faults associated with updates to regulations or tariffs, resulting in a more secure trade industry and lower overheads.

Integrating Security into a Single Window System

Security is an important factor in international trade and any new system put in place to facilitate the processing of goods across national boundaries should not compromise a country's security. Due to the structure of SWIF, in which harmonized data and business processes allow for an easier propagation of necessary modifications and alterations throughout the system, adding security features to a single window more easily executed.

Single windows designed through the SWIF benefit from having data, business processes and documentation in a standardized format. This clear, hierarchical structure reduces the possibilities of data security issues that can arise from having a group of separate legacy systems. Furthermore, the adoption of security standards such as ISO 28000 and BS 7799 (as recommended by UNECE) are made easier due to the simplified architecture created through SWIF.

On the user-end, countries which already employ a Single Window system use a number of techniques to secure their application(s). Amongst the most common security implementations are a PIN/Password system, Public Key Infrastructure (PKI), Authentication Tokens, Biometrics and Smartcards (Choi, 2011). Additionally, in most countries they protect raw data through an additional level of security incorporated during the implementation of the architecture. As a result, other government agencies are not able to access any raw data. Instead, they rely upon the single window to provide them with the processed information, thus reducing the exposure of raw information.

Of course, these security features are all additional to the more secure borders created by the single window itself. The window is responsible for receiving all the data and information associated with customs and trades, which should provide it will all the tools necessary to assess risk (ITDS, Centre for Trade Facilitation and Electronic Business). This is in stark opposition to many existing border control methods when several departments are responsible for different pieces of information, making risk assessments more difficult.

CONCLUSION

The Single Window concept as proposed by the UNECE is beneficial to the world economy and security in a number of ways. Its structure, mainly drawn from an existing approach to enterprise architecture, is aimed at simplifying customs procedures while at the same time improving security techniques. No doubt, a streamlined method of international trade requires less time to bring imported goods to the market, which will be very much appreciated by business.

Case studies performed on the countries already benefiting from a single window system have received overwhelmingly positive reviews from the private sector. The single windows not only decrease the amount of time needed to clear goods, but also saves business money by reducing overheads. From a public sector perspective, the ability to update duties and tariffs through a single window has been shown in countries such as Mauritius to increase revenue from foreign trade. Governments are also on the receiving end of steep overhead reductions, which is typical for organizations converting from legacy to enterprise architecture systems. Citizens, too, have much to gain from the single window as the ripple effects of this implementation are widespread. Mauritius is a prime example of a country which has transformed itself from a localized agricultural economy into a significant member of the world trading community.

The UN Single Window concept could therefore be a next logical step in trade globalization. It will help facilitate international trade, enable governments to give businesses an opportunity to reduce their overheads and simplify international shipments while at the same time keeping borders secure and documentation updated and in check.

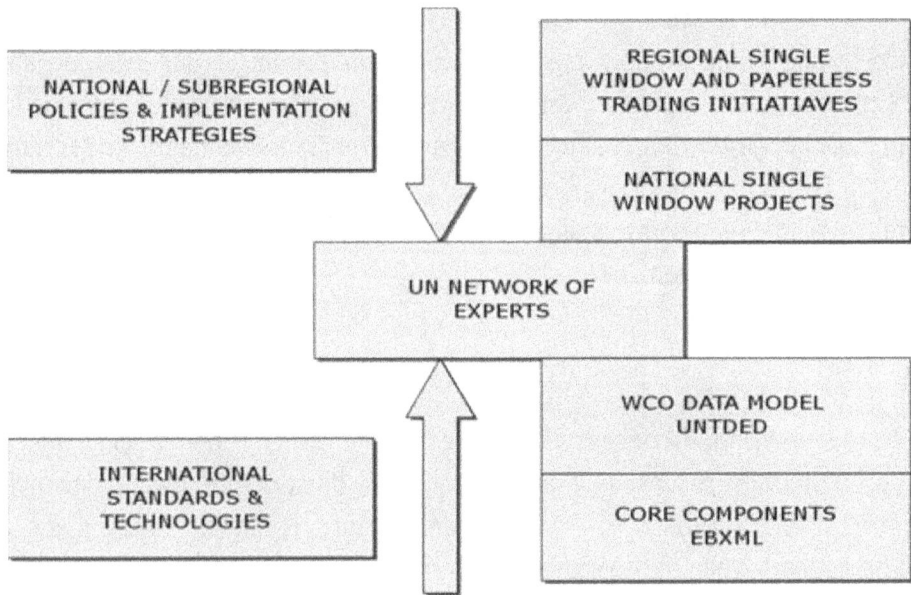

Figure 8 *UN Single Window Implementation Roadmap*

REFERENCES

Ardalan, K. 2010. Globalization and Finance: Four Paradigmatic Views. Journal of Globalization Studies 1(2): 41–67.

APEC—Asia-Pacific Economic Cooperation 2007. Single Window Strategic Plan. URL: http://www.apec.org/About-Us/About-APEC/~/media/DE912B04B0AC485A803823C32EB8750C.ashx.

ASEAN—Association of Southeast Asian Nations 2011. Trade and Facilitation. URL: http://www.aseansec.org/Fact%20Sheet/AEC/ AEC-01.pdf.

Bauman, Z. 2011. From Agora to the Marketplace, and Whereto from Here? Journal of Globalization Studies 2(1): 3–14.

Butterly, T. n.d. Single Window Implementation Framework. URL: http://css.escwa.org.lb/edgd/1476/d2s4-3.pdf

Centre for Trade Facilitation and Electronic Business 2005. Recommendation and Guidelines on Establishing a Single Window to Enhance the Efficient Exchange of Information between Trade and Agreement. New York: United Nations Centre for Trade Facilitation and Electronic Business. URL: http://www.unece.org/ fileadmin/DAM/cefact/ recommendations/ rec33/rec33_trd352e.pdf.

Choi, Jae Young 2011. A Survey of Single Window Implementation. URL: http://www.wcoomd.org/files/1.%20Public%20files/PDFandDocuments/research/17_SW_Survey%20Analysis_Choi_EN.pdf

Dobson, W. 2010. Gravity Shift: How Asia's New Economic Powerhouses will Shape the 21st Century. Toronto: University of Toronto Press. URL: http://www.lob.de/cgi-bin/work/ such e2?titnr=259464463&flag=citavi.

Gartner 2012. IT Glossary: Enterprise Architecture (EA). URL: http://www.gartner.com/it-glossary/enterprise-architecture-ea/.

GFP—Global Facilitation Partnership for Transportation and Trade 2012. Single Window Environment. URL: http://www.gfptt.org/entities/TopicProfile. aspx?name=single-window.

Hitimana, B. 2012. Uganda: Clearing Goods. URL: http://allafrica.com/stories/201203271181. html.

Grinin, L. 2012. New Foundations of International System or Why do States Lose Their Sovereignty in the Age of Globalization? Journal of Globalization Studies 3(1): 3–38.

ITDS—International Trade Data System 2012. What is ITDS? URL: http://www.itds.gov/xp/itds/ toolbox/background/background.xml.

Sharmin, S., and Rayhan, Md I. 2011. Does Globalization Always Increase Inequality? An Econometric Analysis in Bangladesh Perspective. Journal of Globalization Studies 2(2): 160–172. UNECE—United Nations Economic Commission for Europe 2012. UN/CEFACT: About Us. URL: http://www.unece.org/cefact/ about.html.

TradeMark East Africa 2012. Electronic Single Window Pilot Launch. URL: http://www. trademarkea.com/ site/?000=1&001=23&003=news&004=771. van Stijn, E., Phuaphanthong, T., Kerotho, S., Pikart, M., Hofman, W., and Tan, Y. 2011. Single Window Implementation Framework: D5.0:4b. URL: http://www.unece. org/fileadmin/DAM/cefact/ SingleWindowImplementationFramework.pdf.

Zachman, J. 2008. John Zachman's Concise Definition of the Zachman Framework. URL: http://www.zachman.com/about-the-zachman-framework.

Chapter 13

THE RECENT GLOBAL CRISIS UNDER THE LIGHT OF THE LONG WAVE THEORY

Tessaleno C. Devezas

In this paper it is presented the secular unfolding of four economics-related agents, which when considered as a whole allow comprehending what happened in the past in the global economy and shed some light about possible future trajectories. The four agents considered are: world population, its global output (GDP), gold price and the Dow Jones index. The joint action of these actors, in despite of being only a part of the whole, might be seen as a good depiction of the great piece representing the world economic realm. The application of analytical tools such as spectral analysis, moving averages, and logistic curves on time series data about the historical unfolding of these actors allows the demonstration that the recent global crisis seems to be a mix of a self-correction mechanism that brought the global output back to its original learning natural growth pattern, and that it carries also signals of an imminent transition to a new world economic order. Moreover, it is pointed out that fingerprints of Kondratieff long waves are ubiquitous in all observed time-series used in this research and it is demonstrated that the present decade will be probably one of worldwide economic expansion, corresponding to the second half of the expansion phase of the fifth K-wave.

Keywords: economics, long waves, world GDP, economic recessions, gold price, Dow Jones Industrial Average.

1. INTRODUCTION

Since the onset of the present global financial crisis started in the fourth quarter of 2007 that at least two 'faqs' are omnipresent in the technical or amateur discussions on the unfolding of world economic affairs: Why was not it foreseen? And where are we presently in the framework of the long wave theory?

It became very complex to speak about causation of this crisis; there is no consensus about an economic theory that could explain its genesis, and much less about the hypothesis of a timely forecasting. On the other side, there has been some consensus that the crisis has a pure financial and monetary policy nature and is not the consequence of any kind of overproduction as observed in previous economic shocks. Some strange names have been given to this financial turbulence: subprime crisis, real state crisis, super bubble, and more recently it was even coined as the Great Recession to differentiate from less severe 'normal' recessions of the last 80 years and from the Great Depression of the 1930s.

As usual in times of big economic recession comparisons with previous crises abounded in the technical literature. Most commonly we have seen the obvious comparisons with the Great Depression of the 1930s, but also comparisons with the world-wide panics of 1873 and of 1907 have been pointed out. But the fact is that none of these comparisons passed the necessary stringent tests. Its general character, as we will try to demonstrate in this work, seems to be unique, carrying in its structure clear symptoms either of a self-correcting mechanism or even an anomaly of the current socioeconomic system.

Strange still economists and financial analysts insist on looking at this crisis with the very narrow lenses of the current economic and financial theories and models, neglecting the potential of the overwhelming evolutionary world system approach when trying to understand the unfolding of human affairs on this planet. Economics has taken a far too narrow view not only of its modeling and assumptions, but on its reliance on definitions. Models and definitions are maintained even when they are obsolete and no more suitable.

This piece does not intend to offer an exhaustive analysis of the causes of the present crisis. Our goal relies mainly in presenting a new vision about the evolution of some economics-related agents during the last century (more exactly since 1870), which when considered as a whole allow a better comprehension on what is happening and shed some light about possible future trajectories.

2. THE FOUR AGENTS

Economics is above all the surface manifestation of all human activities related to the exchange of goods and services that as any other system in the universe has to follow some iron rules of nature. Humans, human activities, organizations, Earth's material resources, are all parts of the natural order. Following this line of thought we have to describe the behavior of large populations, for which statistical regularities should emerge, just as the law of ideal gases emerge from the incredibly chaotic motion of individual molecules, as recently stated by Bouchaud in a short paper published in Nature with the suggestive title 'Economics Needs a Scientific Revolution' (Bouchaud, 2008). The present author in a paper written in 1996 has already pointed out the same observation (Devezas, 1997). The fact is that during the last twenty years we have witnessed the birth of the new science of Econophysics (a term coined by Gene Stanley in 1995 [Bouchaud, 2009]) which applies the conceptual framework of physics to economics and has been very successful in explaining the endogenous behavior of financial markets, demoting accepted axioms and debunking myths of mainstream economics like the rationality of agents, the invisible hand, market efficiency, etc. We will turn to this point in a later section of this article.

Socioeconomic systems are complex systems and free markets are wild markets. No framework in classical economics is able to describe wild markets. Physics' modern branch of Chaos Theory, on the other hand, has developed models that allow understanding how small perturbations can lead to wild (very big) effects. Devezas and Modelski (2003) have shown that the world system evolution consists in a cascade of multilevel, nested, and self-similar (fractal) processes, exhibiting power law behavior, which is also known in physics as self-organized criticality. Wild oscillations are part of the far from equilibrium chaotic behavior. In a more recent complement of this research Devezas (2009) has demonstrated that the world system is prompt to a very important transition in the near future. The results described in the present paper, using other sets of data and different mathematical tools, come to reinforce this result.

It is very important to keep in mind that complex systems is perhaps a misnomer, because their manifestation and their subjacent laws are not really complex—their imperatives are very simple and usually translated in beautiful patterns like that of fractals, power laws and logistic growth curves. All that we need is to choose the suitable sets of data and apply to them simple mathematical tools. Consider that Einstein demonstrated the time dilation phenomenon using only high-school mathematics.

Let us be simple and call to the stage only four actors (agents) that, in despite of being only a part of the whole, might be seen as a good depiction of the great piece representing the world economic realm. Their historical unfolding translated by time series data represents the result of collective actions involving people, organizations, networks, nations, etc., whose interactions unfold in space and time and manifest some simple patterns that ease us to grasp recent and past economic events.

The considered agents are: the world population, the world aggregate output known as Gross Domestic Product (GDP), the historical leader of all commodities—Gold, and the still most important financial index, the DJIA (Dow Jones Industrial Average). In this paper we will examine the interplay among these agents using historical time series regarding their quantitative evolution, as well as the patterns emerging from their secular behavior when subjected to some simple analytical tools.

3. NOTES ON THE USED SETS OF DATA

The figures for world population and GDP were taken from Maddison's historical series (Maddison, 2007b, n.d.), which are considered to be one of the most reliable sources for economical and population data for the past 2000 years.

The macroeconomic variable—GDP—is undoubtedly a very good measure of global and region-wide economic activity, for it works as an aggregator covering the whole economy. In the technical literature there has been a hectic discussion about the validity of GDP statistics as a good measure for living standards and nation's productivity (see for instance the recent short comment on this theme by the Nobel Prize winner Joseph Stiglitz [2009]). But regarding this controversial point we wish to clarify that the approach followed in the present analysis is one of comparisons between countries and/or regions, and moreover we compare the historical rates of growth, and not the absolute values of GDP estimates.

Add to that the fact that Maddison uses in his figures the purchasing power parity (PPP) converters, which eliminates the inter-country differences in price levels, so that differences in the volume of economic activity can be compared across countries, allowing a coherent set of space-time comparisons. In order to normalize the temporal variations of the used currency Maddison takes constant 1990 US dollars converted at international 'Geary-Khamis' purchasing power parities (see for details Maddison 2007b: ch. 6).

Still regarding the GDP data series it is important to point out that Maddison's figures are not complete along with the entire time span (since 1870) we want to focus in the present analysis. Maddison's tables present complete data between 1870 and

2006 only for the USA, 12 Western European countries, Japan, Brazil and Indonesia. For India the numbers are complete since 1884, for Russia/USSR there are numbers for 1870, 1890, 1900, 1913, and is complete after 1928, and finally for China there are numbers for 1870, 1890, 1900, 1929–1938, and is complete since 1950. For all other countries the figures are complete since 1950. For this reason when designing the graphs for the historical unfolding of the world GDP only a given set of countries was chosen for some given periods, as will be discussed later. Data for the most recent years of 2007 and 2008, as well as the projections for 2009 and 2010, were taken from a recent report of the International Monetary Fund (IMF, 2009), converted using Maddison's criteria.

The time series for the weekly Gold price since 1900 were taken from Kitco historical charts (Kitco, n.d.) and for the Dow Jones index also since 1900 from the webpage of Analize Indices (Analize Indices…, n.d.).

4. SPIKE-LIKE GROWTHS

Graphed on a time-line of two millennia both the Earth population as well as its economic output (world GDP) presents a spike-like growth, as depicted in Figures 1 and 2. Both these mega-phenomena began sweeping the planet in the past century conducing nowadays to very serious concerns about materials/energy consumption, carbon dioxide concentration in the atmosphere, shortage of water, and extinction of species. These mega-phenomena account for the proliferation of afflictions swamping mankind at this very onset of the 21st century. It is not exaggerated to

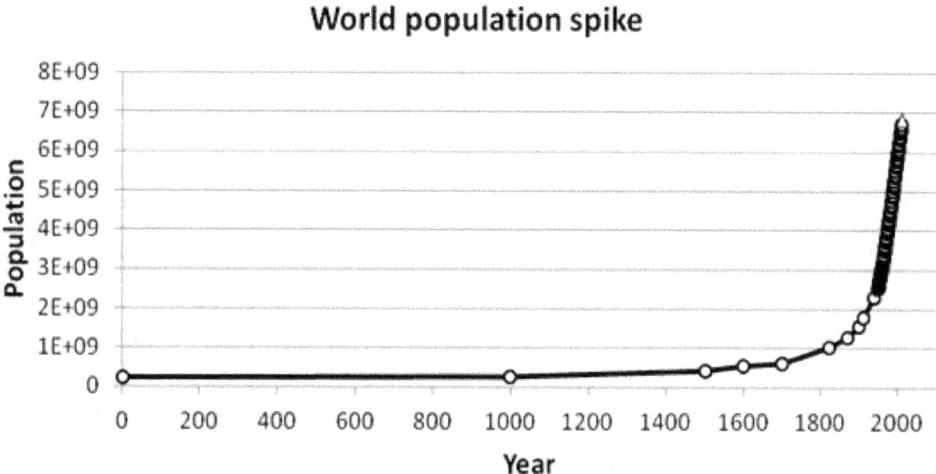

World population spike

Figure 1 *Spike-like growth of the world population in the last two millennia*

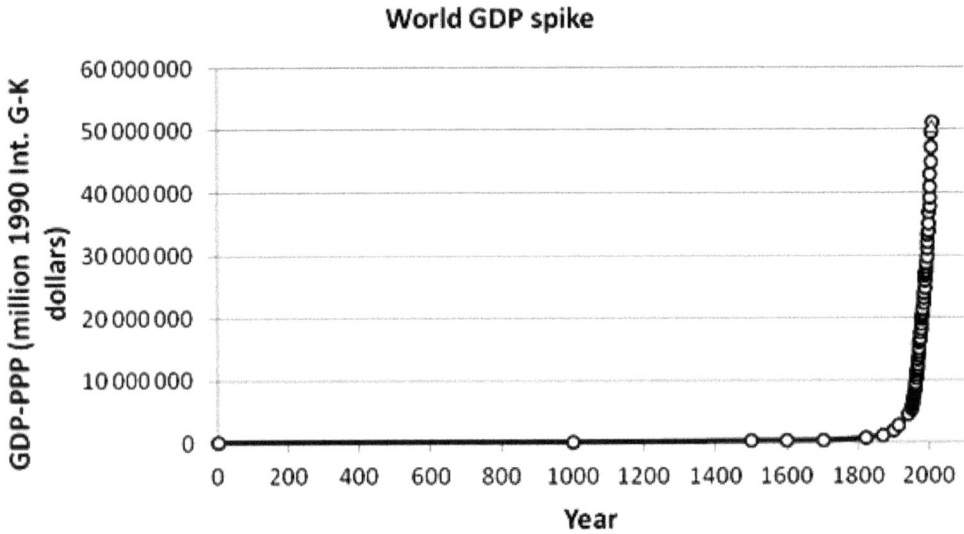

World GDP spike

Figure 2 *Spike-like growth of the GDP in the last two millennia*

say that humanity is presently in a very World War (or World Revolution) whose main goal is its own surviving, spending large amounts of its own GDP trying to win this war. There is already a growing planetary consciousness that some extreme measures have to be undertaken immediately if the human race intends to endure as a species.

On the other hand, one can ask: Is that really so? Is there a real menace pointing to a possible worldwide catastrophe that could definitively jeopardize human life on Earth? Another question then naturally emerges: could not Gaia as a resilient system find its own way out of this apparently imminent disaster? As will be seen in our analysis ahead in this paper, this kind of graphs evincing explosive growths is always misleading and used frequently for apocalyptic propaganda. In order to get the correct conclusions about the real trends we should look for the details hidden behind the considered growth phenomenon and this is usually done expanding the x-axis and narrowing the focus on its unfolding in shorter time spans.

We know that this is a very controversial theme of debates and equally know that there are many scientists voicing against the exaggeration of simple extrapolations of the observed trends. Our objective in this work is not properly to deliver answers about this scientific puzzle, but the fact is that the approach we are pursuing in last years and the results of our ongoing research, as well as the results of other recent investigations, point to this very concrete possibility—the World System is approaching an Era of Transition that will conduce naturally to a new order within which these troubles will be overcome. What we do not know yet is if this transition will be a smooth one or much on the contrary, a very turbulent one as already happened in the past. We hope that the present results may help in shedding some light on the road ahead.

We have already pointed out that Devezas and Modelski (2003) have demonstrated that the World System is prompt to a very important transition and demonstrated that the dominating order has already reached 80% of its millennial learning path (see Devezas & Modelski, 2003: Figure 9). In another recent work Devezas *et al.* (2008) have shown that the increasing efficiency of energy systems is following an irreversible path toward the usage of carbon free energy sources, a process that will be completed before the end of the present century (see Devezas *et al.*, 2008: figures 10 and 11).

Very recently econophysicists Johansen and Dornette (2001) have given an important contribution in this direction. They have shown that, contrary to common belief, both the Earth's human population and its economic output have grown faster than exponential (i.e. in a super-Malthusian mode). These growth rates are compatible with a spontaneous singularity occurring at the same critical time around 2050 signaling an abrupt transition to a new regime. But the abruptness of this transition might be smoothed, a fact that can be inferred from the fact that the maximum of population growth was already reached in the 1960s, in other words, a rounding-off of the finite-time singularity probably due to a combination of well-known finite-size effects and friction, suggesting that we have already entered the transition region into a new regime.

Closing this section it is shown in Figure 3 the spike-like growth of the Dow Jones Industrial Average (DJIA) considered weekly from 1900 until September 2009, and in Figure 4 the historical growth of gold price for the same time span, also considered weekly. In the case of gold, which will be subject of a detailed analysis ahead in this paper, we do not have what can be coined as a spike-like growth, but anyway it can be observed a spectacular growth with wild oscillations, exhibiting two very strong peaks separated by approximately 30 years.

DJIA weekly price

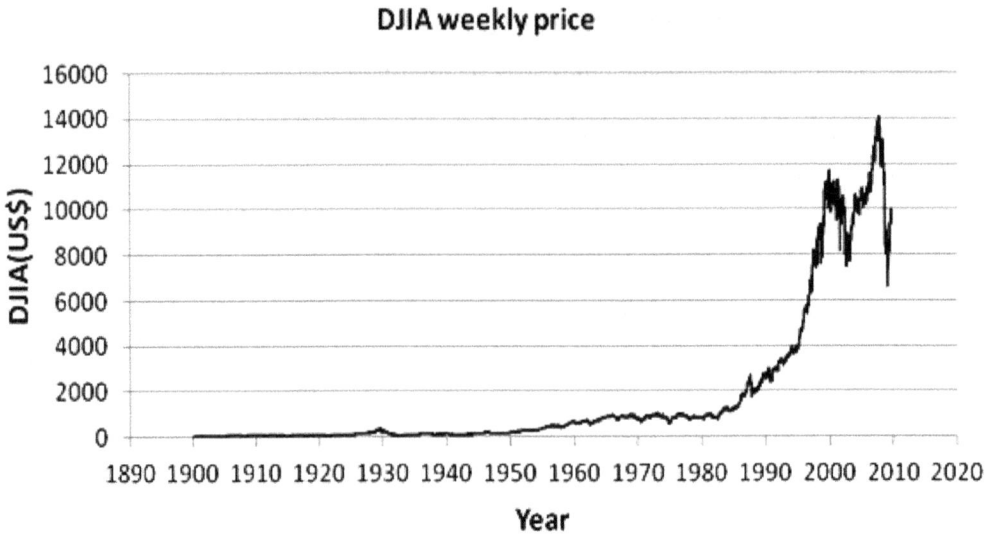

Figure 3 *Dow Jones Industrial Average weekly price since 1900 until September 2009.*

Gold weekly price

Figure 4 *Gold weekly price per troy ounce since 1900 until September 2009.*

5. SIGNALS OF SATURATION

Let us begin looking at the evolution of the two most important agents, Earth's population and its aggregate output, but initially narrowing our observation to their recent unfolding after 1950, a period for which the most reliable data are available.

In the previous section we have already pointed out the fact that human population growth rate has already reached its maximum, as depicted in Figure 5. A peak of 2.2% was reached in 1962–1963, and after this date has decreased steadily be-ing nowadays of the order of about 1.13%. Looking another way around, the annual change in the world population peaked in the late 1980s when the world population experienced a net addition of about 88×10^6 individuals (obviously because the popu-lation in the 1980s was much bigger than in the 1960s). These figures were taken from the International Data Base of the U.S. Bureau of Census (n.d.), whose estimate for the world population in 2050 is of about 9.316×10^9 people.

An important point to refer about the Figure 5 is the pronounced dip appearing in 1958–1960 that was due to the so-called Great Leap Forward that occurred in China in this period, amidst with natural disasters, widespread famine and in the wake of a massive social reorganization that resulted in a toll of tens of millions of deaths. As we will observe in the next section, this dip is also very visible in the historical evolution of the world GDP and warns us about the weight of China and its very important role in economics-related world affairs.

Curiously, and in despite of the data (calculations!) of the U.S. Bureau of Census, the recent evolution (since 1950) of the world population can be finely fitted by a logistic curve, which delivers a slightly different result regarding both—the extrapola-tion to the year 2050 and the turning point corresponding to the maximum growth rate. This fitting is shown in Figure 6a (the logistic curve) and 6b (the same in the form of a Fisher-Pry plot), which were obtained using the IIASA's LSM II program (IIASA, 2011). As can be observed the fitting is absolutely perfect ($R^2 = 1$), what implies that we are amidst a natural growth process, with a characteristic time Dt of about 160

Annual growth rate (%)

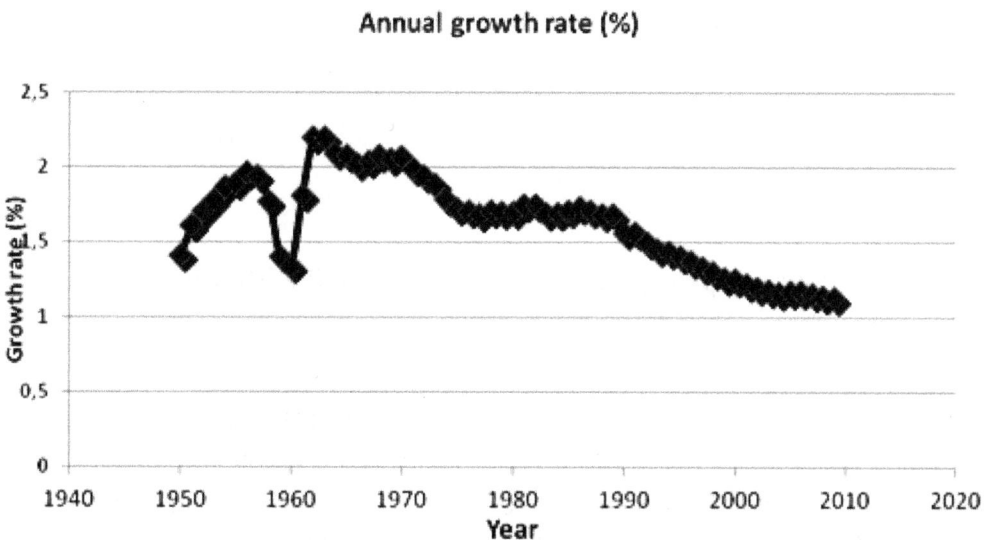

Figure 5 *Annual rate of growth of the world population 1950–2009*

Earth's population

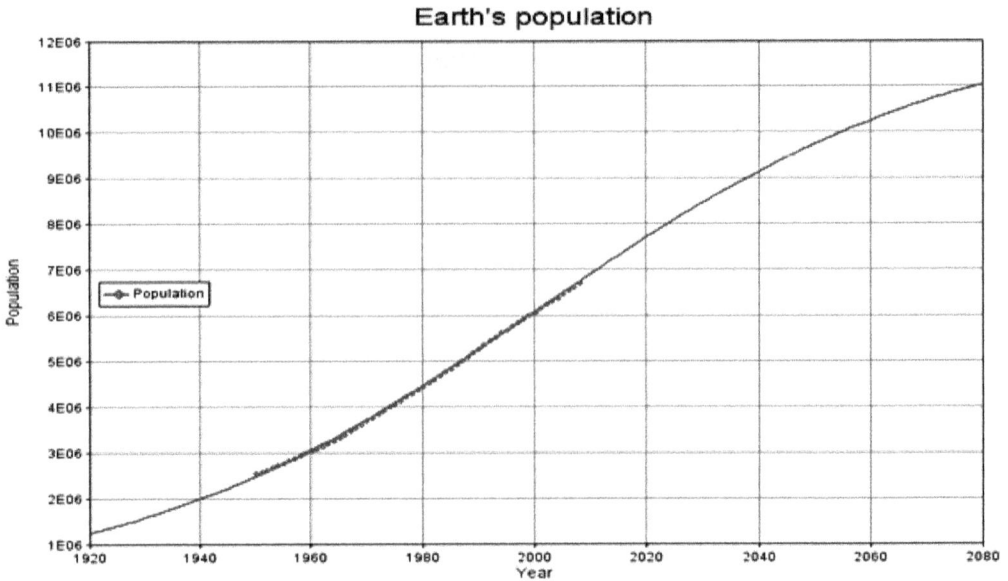

Figure 6a *Logistic growth of the world population 1950–2009 using IIASA's LSM2 software*

Earth's population

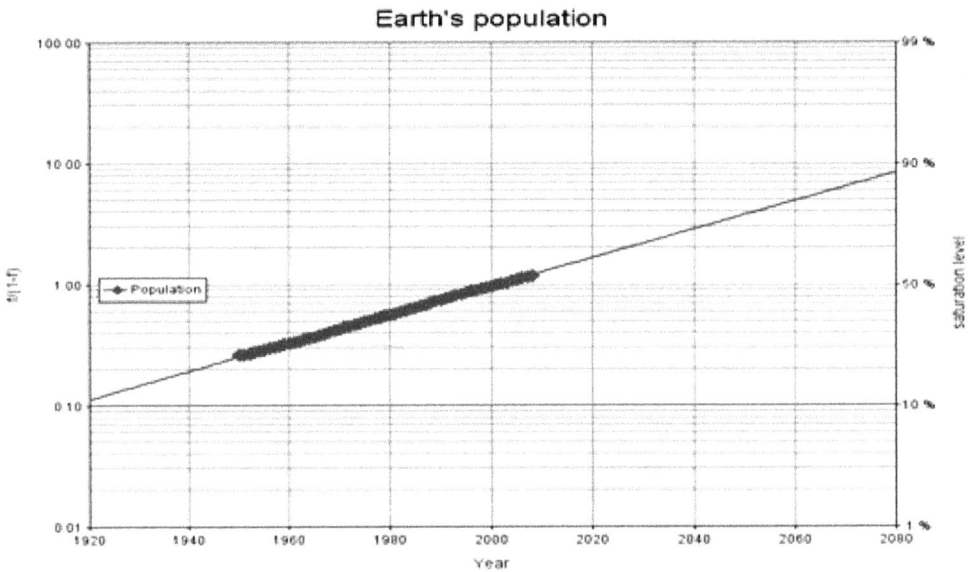

Figure 6b *Fisher-Pry plot of the world population 1950–2009 using IIASA's LSM2 software*

years (1920–2080), with an inflexion point in 2000–2001 (maximum growth rate). The maximum carrying capacity of this process points to a population of about 12×106 people to be reached by the end of the century, but that can stabilize before this maximum (say by about 10×109 people, considering that the end of a logistic growth process implies the transition into a new regime). Our curve points to a population in 2050 of about 9.7×109 people.

In recent paper Boretos (2009) performed the same fitting using the U.S. Bureau of Census' data set until 2005 and has found a somehow moderate result, with a characteristic time Dt = 117 years and a turning point in 1995. Accordingly to the set of data used by this author the extrapolation to the year 2050 matches the projection of the U.S. Bureau of Census.

Let us now call our second agent, the aggregate world output, or in other words, the world GDP. Using Maddison's data since 1950 we have also fitted a logistic curve and the result is depicted in Figure 7a (logistic curve) and 7b (Fisher-Pry plot). The fitting is not so perfect (R^2 = 0.996) as in the previous case of Earth's population, but works equally well.

The resulting logistic corresponds to a natural growth process with a characteristic time Dt ~ 110 years that will saturate about 2080 with a turning point (peak of the growth rate) around 2030. Boretos (2009) has tried the same fitting using a different dataset and numbers only until 2005 and has found a similar result with a characteristic time of about one century and a turning point in 2015. Unnecessary to stress that these differences are absolutely irrelevant considering that we are using different datasets and in our fittings we have used more recent data (until 2008), which has

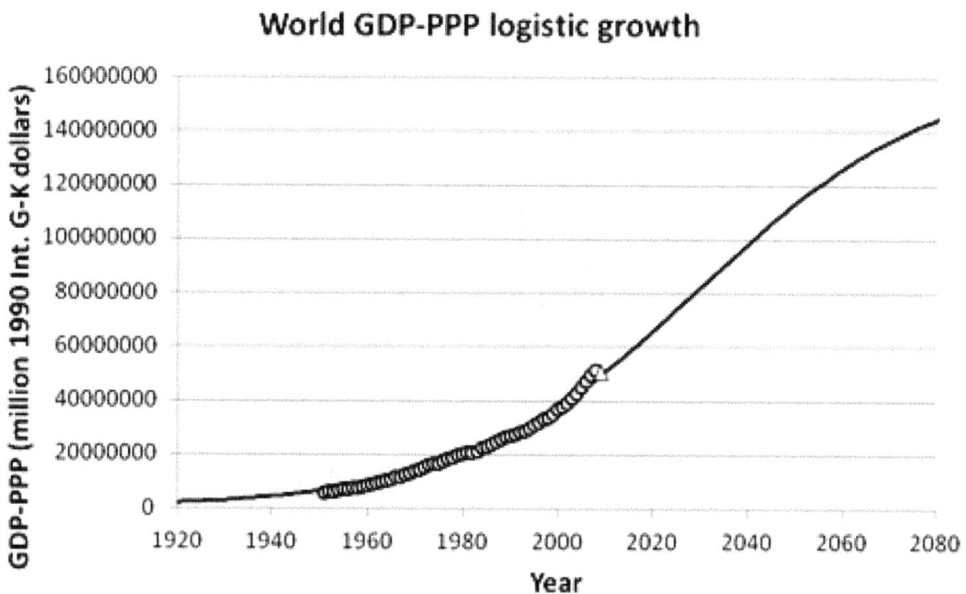

Figure 7a *Logistic growth of the world GDP-PPP 1950–2008*

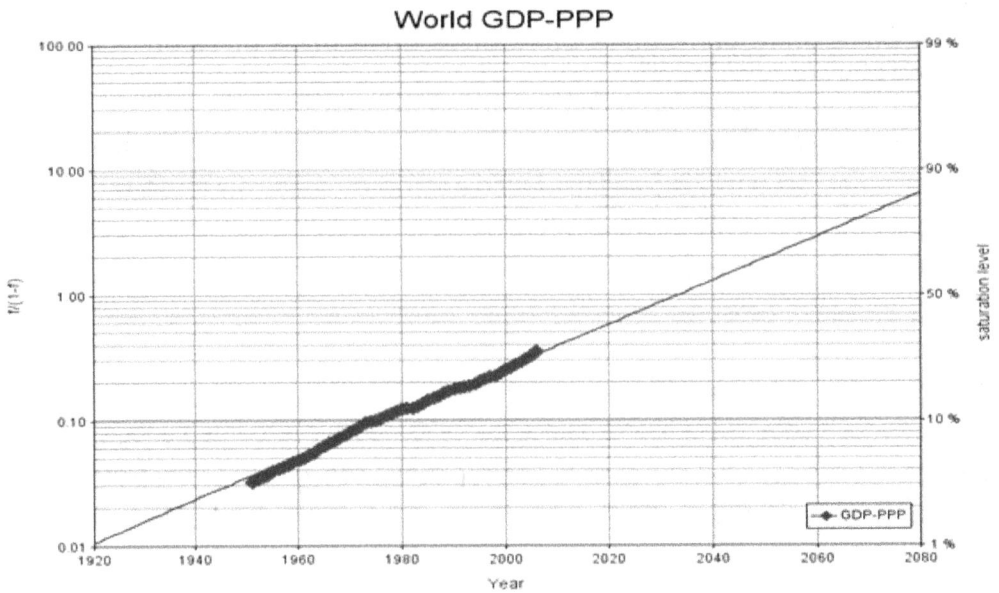

Figure 7b *Fisher-Pry plot of the world GDP-PPP 1950–2008*

naturally contributed to a slightly higher carrying capacity and pushed the turning point ahead in time. The main reason for this difference lies in the higher world GDP growth rates observed in the period 2006–2008, which we will further analyze in the next section.

These results require some further thought. What is the meaning of these natural growth processes? Why the GDP has grown faster than population? There are no simple answers to these questions and their deep analysis deviates from the purpose of this piece. But a few words about their meaning are worth putting.

Human population and its output are growing since the onset of civilization some five millennia ago. But contrary to a widespread impression, the story of world population of the last 5000 years is not one of continuous exponential growth. Rather, it can be best described as a series of three major surges, each more substantial than its predecessor, but both of the first two surges also followed by a long period of population stability (Devezas & Modelski, 2003). The graph depicted in Figure 1 shows only the last stable period and the last spike-like surge respectively. As already shown by Devezas and Modelski (Ibid.), this 2000-year process corresponds to the formation of the global system, one of the global-institutional processes that monitor the progress of agents, and program their developments. Nested within this longer process there are other shorter global-institutional processes like the global economy process

(~250 years; see Devezas & Modelski, 2003: table 2) that corresponds to the process being analyzed in this paper.

At this point we wish to make stand out the first important result of the present investigation, which can be easily discerned through the comparison between the actual points and the path of the logistic growth process shown in Figures 7a and 7b. In these graphs we have also included the estimated projection for 2009 (the triangle in both graphs, using data from IMF [2009]). As can be seen the actual points, mainly between 2005–2008, evidence a slight deviation upwards, and the point corresponding the estimate for 2009 seems to pull the curve downwards in order to match the original path. In order words, the present crisis seems to work as a kind of self-correction mechanism of the system.

The next step was to look at the behavior of the unfolding of the global output per capita. Using the recent data the fitting of a logistic curve does not work well, a result that diverges from those got by Boretos. In Figure 4 of his paper this author shows a logistic fit, but the substitution curve is clearly right skewed and the author does not present the error estimates. Boretos (2009) states that world GDP has increased faster than population at all times, but this is not true as alias we can infer from the linear fitting of the GDP/capita exhibited in Figure 8 below.

As can be observed the overall linear fitting is not bad (R2 = 0,975), but most important the linear trend is perfect until 1981, deviating downwards after this date and until at least 2001, what implies a growth rate of GDP below the population's growth

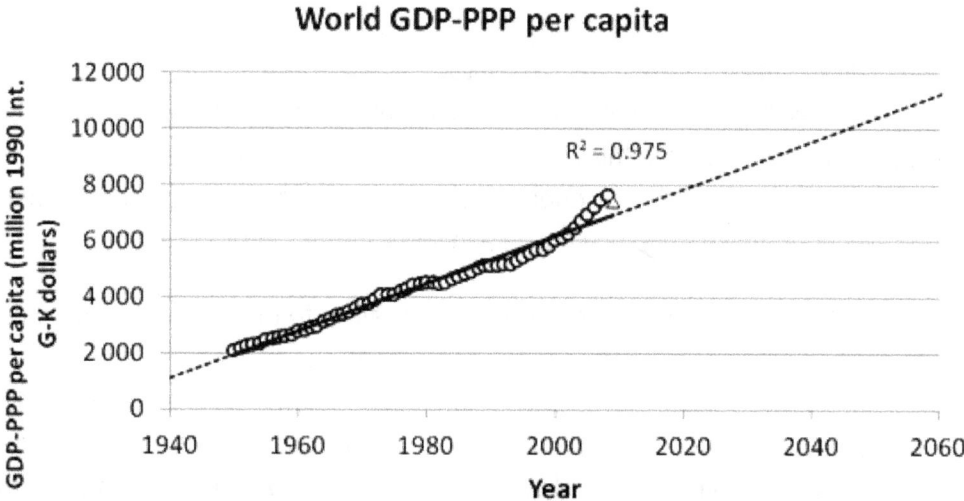

World GDP-PPP per capita

$R^2 = 0.975$

Figure 8 *Linear fitting of the world GDP-PPP per capita 1950–2008*

during approximately a time span of 20 years. After 2004 and until 2008 the actual data exhibits an inverted behavior, that is, the world GDP has grown faster than population—but this trend stopped abruptly in 2009. Again the extrapolated point that contains the outcome of the actual crisis seems to pull the trend downwards. It is clear that if we use the extrapolation for 2010 the corresponding point will be located still closer to the straight line.

Resuming the results of this section we have:

1. The present crisis seems to be a kind of self-correction mechanism that brought the global output back to its original logistic growth pattern.

2. This pattern corresponds to a final phase of the ongoing global economy process, which will saturate before the end of this century, signaling that we are entering into a new regime (a new learning process) of the socioeconomic world system.

In the next sections we will see how results from other analysis and approaches reinforce these preliminary conclusions.

6. COMPARATIVE ANALYSIS OF THE GLOBAL OUTPUT UNDER A LARGER TIMEFRAME

Figure 9 shows the timely evolution of the GDP-PPP for a set of 18 selected countries for which the most complete data are available since 1870. These countries together contribute today for ~70% of the global GDP (74% in 1950, and 73% in 1970). This result is well known; everyone is acquainted with the fact that China is the country exhibiting the most dramatic GDP growth during the last decades, and certainly will surpass the USA in the next decade or so. India and Brazil are also growing at fast paces, but still far below China, while Europe and Japan demonstrate that they are losing momentum in this race. It is very evident that the former USSR was hit at the late 1980s by its political-economical transformation and disaggregation, but is also recovering momentum led by Russia and some of their former members.

This kind of graphical representation does not allow to discern details and much less to perform reliable forecasts. On the other hand, the picture is completely different if we look at annual movements in aggregate activity, or in other words, the annual growth rate of GDP. As will be seen in this section, such visualization allows discerning changes that have appeared systematically across countries, due to catastrophes, political and/or social upheavals, wars, recessions, etc. Moreover, it permits also

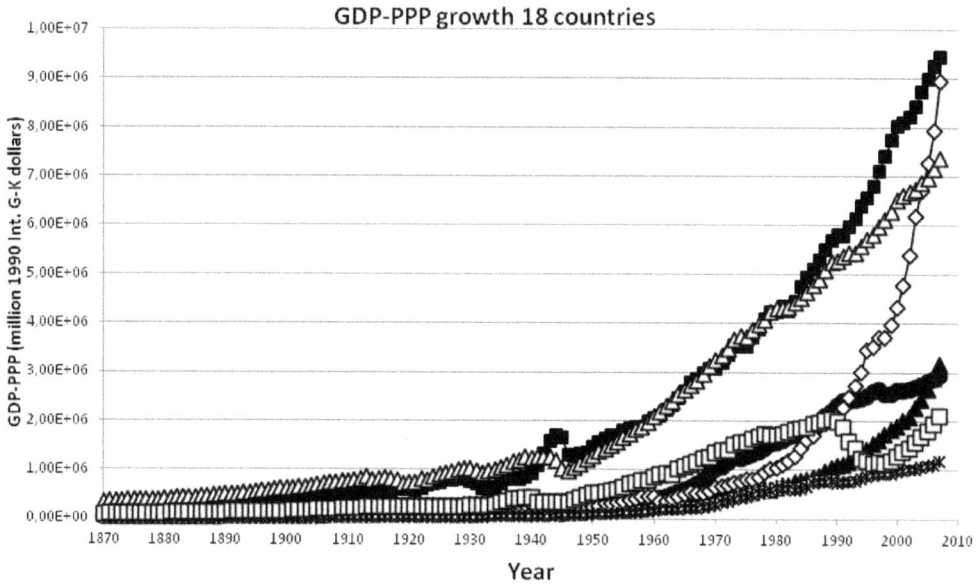

Figure 9 *GDP-PPP growth 1870–2008 for 18 countries—USA (■), China (à), 12 Western European countries (Δ), India (▲), Japan (•), former USSR (□), Brazil (x)*

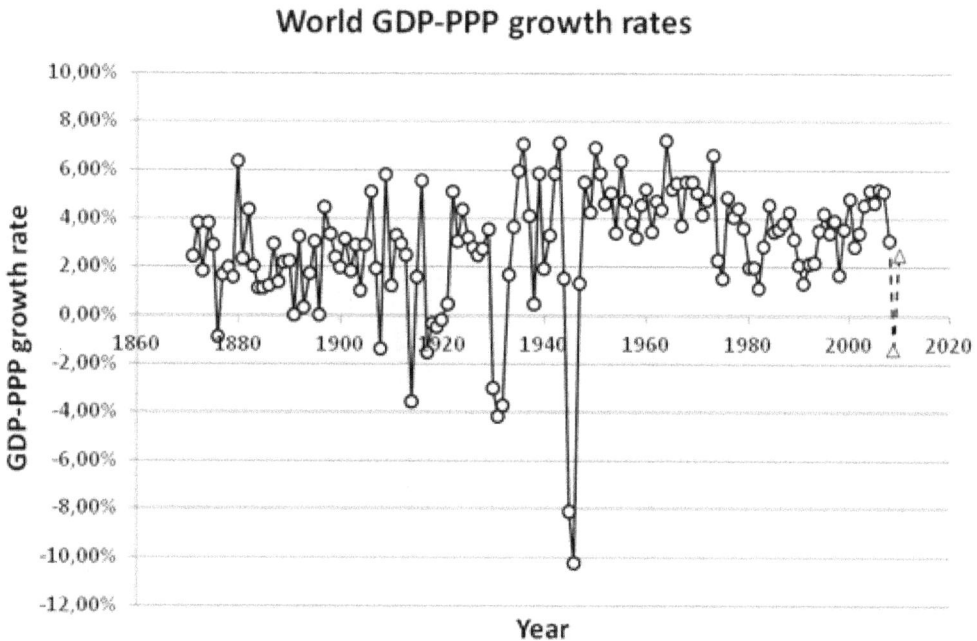

Figure 10. *World GDP-PPP growth rates 1870–2008 (data from Maddison 2007b). Δ—Estimates for 2009 and 2010 from IMF (2009)*

to distinguish some patterns, as for instance the different phases of K-waves observed since 1870.

In Figure 10 that follows it is depicted the historical record since 1870 of the annual growth rate of the world GDP-PPP using Maddison's data. Before advancing commenting on some important details of this picture, it is important to clarify some aspects considered in the construction of this graph.

As already explained in the third section of this work, Maddison's data set is not complete for the entire time span since 1870. For the construction of the graph shown in Figure 10 the data corresponding for the interval 1870–1884 are the numbers for USA and 12 WE countries, where undoubtedly at this time the leading economies in the world (in 1880 corresponding to 55% of the world GDP). Between 1885 and 1927 the numbers include also India, Japan, Indonesia, and Brazil (in 1900 corresponding to ~61% of the world GDP), and between 1928 and 1949 the USSR was added to this group (corresponding in 1940 to ~71% of the world GDP). From 1950 onwards the numbers include all countries.

The validity of this approach can be inferred from the behavior of the two superposed graphs shown in Figure 11, showing the unfolding of the GDP growth rates for the world and for the USA plus 12 WE countries. As can be observed, the move-

GDP-PPP growth rates
World vs. USA & 12WE

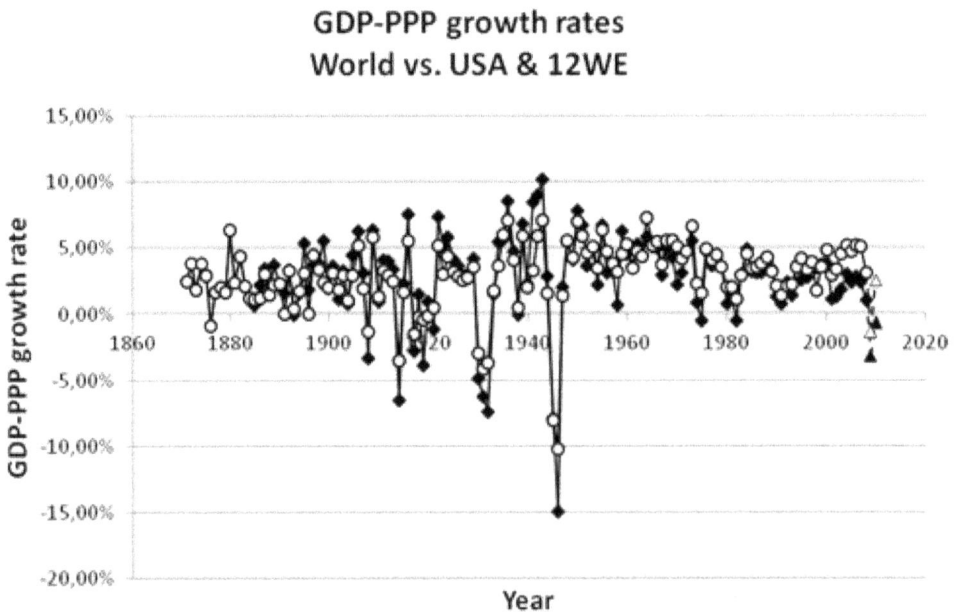

Figure 11 *Comparison between the GDP-PPP growth rates for the world (○) and USA plus 12 WE (♦). The last points Δ (world) and ▲ (USA + 12 WE) are the estimates for 2009 and 2010 from IMF (2009)*

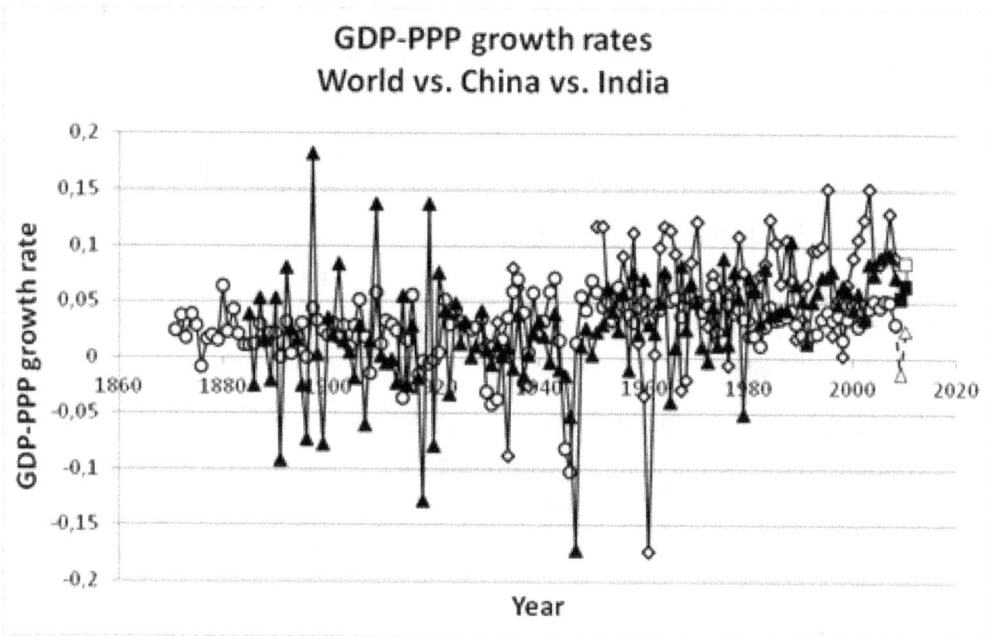

GDP-PPP growth rates
World vs. China vs. India

Figure 12 *Comparison between the GDP-PPP growth rates for the world (O), India (▲), and China (◊).*

ments—ups and downs—are perfectly 'in phase', the only clearly observable difference is that the peaks (maximum growth rates) and dips (minimum growth rates) for the world are damped, due to the fact that the performance for some individual countries are not exactly synchronized with the leading countries. This 'damping effect' seems to work well until at least the year 2000, when an opposite effect seems to enter in action. But the general aspect of the graphs suggests clearly that the USA plus the 12 WE countries leaded the world economy for the most of the time. The picture is completely different when we compare the behavior of individual countries, like India and China (data for China are shown only after 1950), both with very troubled history, as shown in Figure 12 in comparison with the same world graph. It is very clear that the fluctuations are much more radical for the individual countries and not synchronized with the rest of the world.

Note that estimates for 2009 and 2010 (from IMF) were included in all these previous graphs. It is also important to point out that we have not used weighted averages in these graphs; weighted averages contribute to a biased picture of the whole. What we have in all three graphs represent the very fluctuations of the aggregate output.

Now let us try to present in a resumed form, point by point, the main aspects unveiled when looking at these graphs, or in other words, when observing the secular unfolding of the aggregate world output.

1. The most striking aspect exhibited by the graph in Figure 10 is the very turbulent time during the first half of the 20th century, which carried within with the effect of two world wars and the most painful economic crisis already experienced by the world economy; note that the 'dip' corresponding to this Great Depression is placed exactly in the middle of the 'double dip' corresponding to the two world wars, roughly equidistant by ~15 years.

2. This turbulent time is confined between two periods of 'peace times', the first one from 1870 until 1913 (then 43 years), and the second one from 1950 until 2008 (then 58 years).

3. The first period of relative stability and 'peace times' is marked by two pronounced dips with negative growth rates, a first one in 1876 and a second one in 1908. The first dip corresponds to the panic of 1873, which gave place to a strong recession of the world economy, but that was especially severe in the USA. The NBER (National Bureau of Economic Research) statistics (NBER n.d.) consider it as the longest recorded contraction cycle in the USA (65 months, 1873–1879), and some authors (Nelson, 2008) have compared it with the current financial crisis due to many common characteristics. The second dip appears in 1908 and was a consequence of the panic of 1907 (Bruner & Carr, 2007), with also dramatic global consequences, but shorter in duration (in the NBER statistics [Ibid.] for the USA a contraction cycle of only 13 months). Despite its short duration it can be considered as a Great Recession comparable in numbers (GDP contraction) to the present crisis.

4. Still regarding this first 'peace times' period, we can distinguish two subperiods: one with a downward trend (decreasing growth rates, considering the mean values) that extended until at least 1896, soon followed by one with an upward trend (increasing growth rates, considering the mean values), that extended until the middle 1920s but was disturbed by the onset of the WWI. We have then two subperiods: ~1870 to 1896 and 1896 to 1922, each with ~26 years, that as suggested by many adepts of Kondratieff long waves correspond respectively to the downwave phase (or phase B) of the 2nd K-wave and to the upwave (or phase A) of the 3d K-wave.

5. Regarding the second 'peace times' period that followed WWII and started after 1948, we can more easily distinguish different subperiods—more exactly three. A first one located between 1948 and 1973, when the global output averaged a growth rate of about 5%, a second one between 1974 and 1992, when the global output averaged circa 3.5%, and a third one after 1993 when an upward trend is observable, reaching and surpassing the mark of 4% (with a brief interruption

in 2001—the dot.com bubble). The reader should note that there is a dip in the world-series corresponding to 1998, but comparing with the graphs shown in Figs 11 and 12 we can see that it was not a crisis in the USA or Europe, but the consequence of the famous Asia Crisis (Kaufman *et al.*, 1999), which started in July 1997 in Thailand and spread quickly to many other Asian countries, including China and India. Again we have subperiods with time spans averaging two decades—in this case now 25 years and 18 years respectively. K-waves' adopters usually associate these subperiods with the up and downwave phases of the 4th K-wave. Following this schema it seems that after 1992 the 5th K-wave might already be started. We will turn to this point in the next section.

6. Regarding now the actual crisis, translated by the extrapolated points for 2009 and 2010 (small triangles in Figure 10), we cannot draw so easily the same conclusion expressed in the previous section of a self-correction mechanism that is pulling the general trend towards its original path. The points for 2009 and 2010 resemble much more a pathological symptom signaling that something is wrong with the existing economic system, or perhaps more exactly expressed, with the existing global financial system. We use here 'a pathological symptom' because we are facing neither a world war, nor a worldwide social upheaval. Something else seems to be hidden behind the facts.

7. A closer look again to our graphs of Figures 10, 11 and 12 may help to shed some light upon the facts. A very important detail to stress is that we have historically a very important precedent that happened in 1907, that is, exactly one century ago (or, in other words, two K-waves ago!). The phenomenon, known as the '1907 Bankers Panic' (Bruner & Carr, 2007), was very similar to the actual crisis under at least two important aspects: it occurred during an upward trend of the global economy (i.e. during the A-phase of a K-wave) and was a pure financial crisis involving market liquidity that led to bankruptcy of many important agents of the banking system, which quickly spread from New York to Europe and to some Asian countries (see for comparison graph of Figure 12). The remedy at that time was the same as nowadays: the injection of large sums of money to shore up the banking system, soon followed by a profound reform of the U.S. financial system, which included the creation of the Federal Reserve System (FED, created in 1913). The reader should observe in the graph of Figs 10 and 11 that the dip in 2009 mirrors the one in 1908!

8. As already referred to in the paragraph preceding Figure 11, it is very evident from the graphs comparing the unfolding of the world GDP and the sum of the USA plus 12 WE GDPs, that after 2000 a different trend emerged: the growth rates of the world GDP from this date onwards are higher than for the USA and

the European countries together, an inverse behavior of the GDP evolution to this date. This push upwards is clearly motivated by the rocketing GDP growth rates observed for India and China, as can be inferred from Figure 12.

9. Such an inverted trend seems to be a clear signal that we are already witnessing a transition to a new global socioeconomic system, which will carry with a profound restructuration of world economic affairs. In a few words it means that real growth rates of low-income countries have been growing increasingly apart from those of high-income countries. See more details in the Conclusions.

10. Ajar with the times, the present crisis seems to sum up a mix of self-correction mechanism (or at least the urgent necessity of finding the necessary measures for correction) and signals of an imminent transition to a new world order.

Before closing this section it is worth bringing to the reader's attention the fact that negative fluctuations of the world GDP are not sufficient condition to characterize a great depression. There are more things at stake when we wish to speak of economic recessions with a worldwide impact and severe consequences across countries. In a very recent book the economists Carmen Reinhart and Kenneth Rogoff (2009) have shown that in order to characterize a real great depression it is necessary to observe not only a considerable contraction of the GDP, but also a significant retraction of the worldwide commercial exchange. For these authors this phenomenon has only occurred three times in the recorded history: in 1907/1908, 1929/1933 and now in 2007/2009. Many other crises, like those of 1873/1879, 1945/1946, 1987, 1998 or 2000/2001, have not had the same global impact like these three mentioned, because they have not hit equally both measures (GDP and commerce) or have had only regional effects (like the 1998 Asian crisis). This aspect is a very important one regarding our previous conclusions and the parallel between the actual crisis and the 1907 Panic.

7. SCRUTINIZING THE RECENT RECORD OF THE GLOBAL OUTPUT

Keeping in mind the fact already mentioned in our fifth section (Signals of Saturation) that the most reliable data for the global output are those that followed WWII, it is worth scrutinizing further this recent period, which we coined as the second 'peace times' period.

Figure 13a shows the result of applying an 11-year moving average to the data of Figure 10 (world GDP-PPP growth rates) in the period 1947–2008. As can be seen it is evident a wave-like behavior suggesting the fingerprint of a complete long wave. Fig-

World GDP-PPP growth rates
11 years moving average

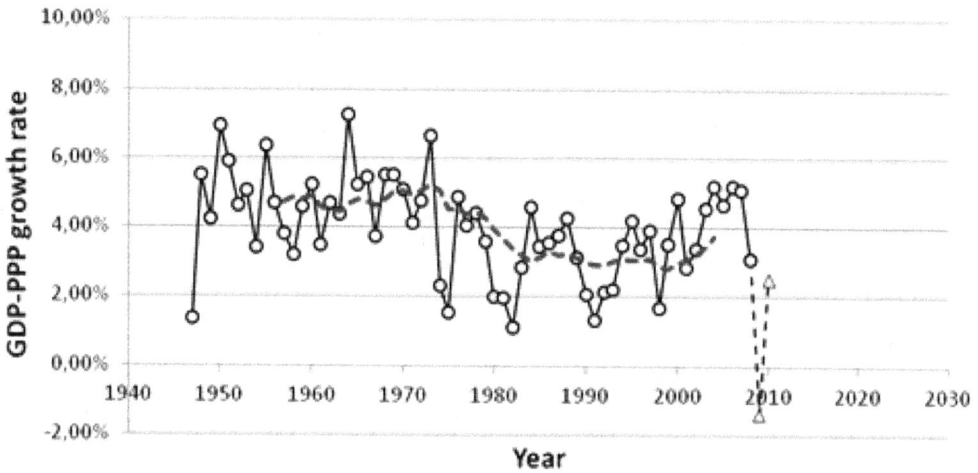

Figure 13a *11-year moving average applied to the world GDP-PPP growth rates in the period 1947–2008. The estimates for 2009 and 2010 (Δ) were not included in the MA.*

World GDP-PPP growth rates
11 years moving average period 1947-2008

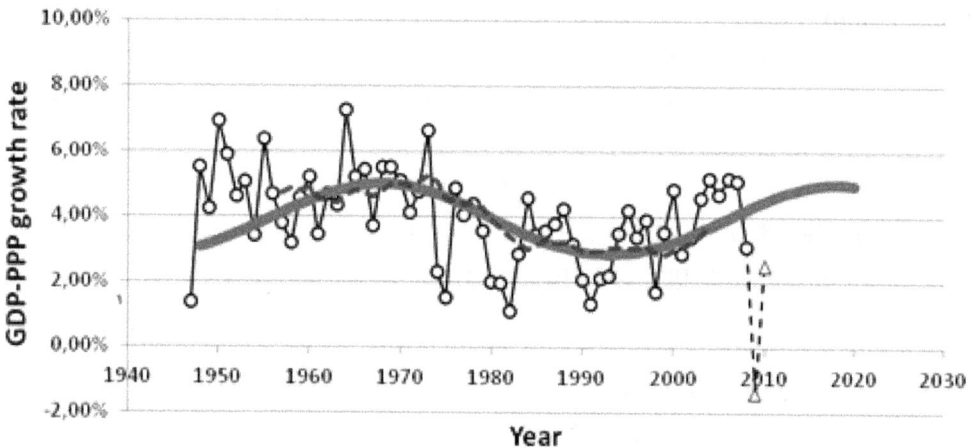

Figure 13b *Result of fitting a simple sinus series P(t) = 4 + 1.03 sin (2p t/50.14) + 0.03 sin2 (2p t/50.14), evincing a periodical movement with a period of about 50 years (the points for 2009 and 2010 were not included in the fitting).*

ure 13b presents the result of fitting a simple sinus series of the type P(t) = P0 + A sin (2p t/T) + B sin2 (2p t/T) + ..., whose solution is P(t) = 4 + 1.03 sin (2p t/50.14) + 0.03 sin2 (2p t/50.14), evincing then a periodical movement with a period of about 50 years (the points for 2009 and 2010 were not included in the fitting).

This result comes to reinforce our conclusion resumed in point 5 of the previous section that we can divide this recent period in three subperiods—the first and second corresponding to an entire K-wave and the third corresponding to the upward movement of the following K-wave. The entire K-wave in this curve matches very well the dates that many different authors have presented for the 4th K-wave, which started about 1947/48, reached a maximum about the 1970s, and was completed in the first half of the 1990s.

The extrapolation for the fifth K-wave points to a maximum to be reached shortly before 2020, or in other words, the present expansion movement, although disturbed by the recent crisis, may well continue for more one decade. The much discussed apparent recovery still on course (crisis 2007/2009) seems to hint that the system is indeed resilient.

8. SHRINKING RECESSIONS AND CONTRACTIONS

In a recent paper the Italian economist Mario Coccia (2010) brings to attention the fact that the duration of business cycles' contraction phases are far shorter than the duration of expansion phases. The author observes also that the duration of the recessions corresponding to the downwave phase of a long wave is in average shorter than the upwave phase. In the case of business cycles the author uses statistics from NBER (n.d.) and from the U.S. Bureau of Economic Analysis (BEA),[1] comparing data for the USA, the UK and Italy. In the case of long waves the author uses an extensive comparison of the dates proposed to this phenomenon by many different long-wave theorists.

His results point to a mean duration of business cycles' contractions in the USA, between 1854 and 2001, of about 17 months, and a mean duration of expansions of about 39 months, or in other words, an average of 31% of the time experiencing economic contraction and 69% experiencing economic expansion. Regarding the K-waves the author points to an average of about 29 years for upwaves (53% of the total time) and 26 years for downwaves (47% of total).

---

1. URL: http://www.bea.gov.

In this research we decided to explore also this phenomenon using the NBER statistics for the USA and were confronted with two very interesting and unexpected results: first, there exists an increasing trend towards shorter contractions and longer expansions, and second, the fingerprint of K-waves is clearly visible also in the history of the U.S. Business Cycles.

Figure 14a shows the graph resulting from the distribution in time of the succession of economic expansions and contractions in the history of business cycles in the USA since 1850. In despite of the star field-like aspect of the distribution of the points, one can clearly distinguish the enduring trend towards longer expansions and shorter contractions translated by the straight trend line. The last point in this graph corresponds to the expansion period that lasted from the end of 2001 to the end of 2007 (73 months) and ended with the onset of the actual crisis.

Figure 14b presents the resulting 20-year moving average applied to the same historical statistics. The trend line reveals a wave-like behavior that coincides with the dating schema used by many long-wave authors and matches very well our conclusions in the previous sections. In this graph we have added a point to the actual cri-

NBER business cycles statistics
Succession of contractions - expansions since 1870

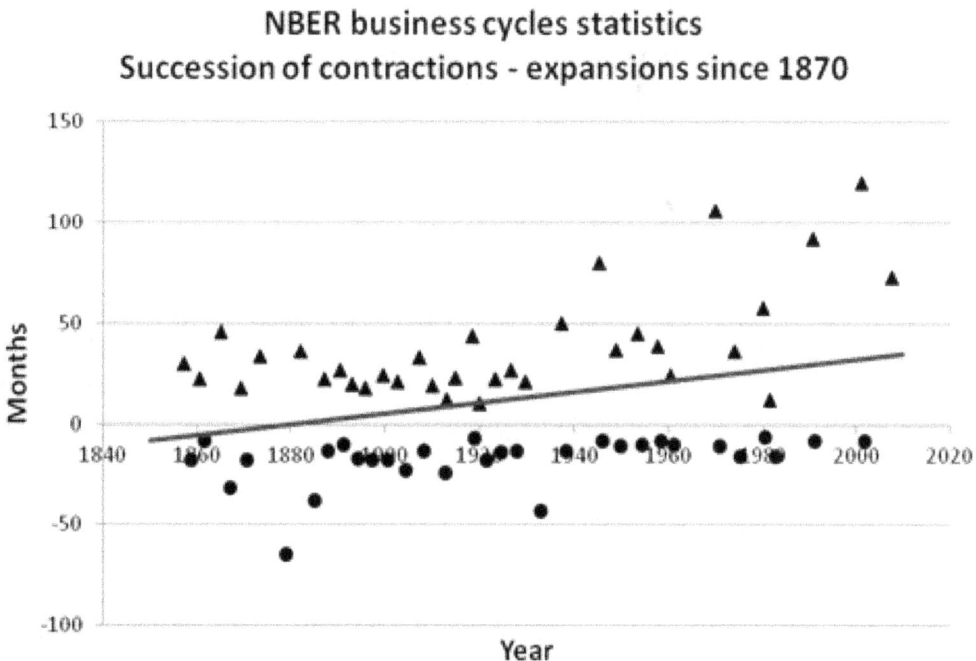

Figure 14a. *Star field-like aspect of the distribution of the succession of economic expansions (▲) and contractions (●) in the history of business cycles in the USA since 1870 (NBER n.d.). The straight trend line translates the trend towards longer expansions and shorter contractions in business cycles.*

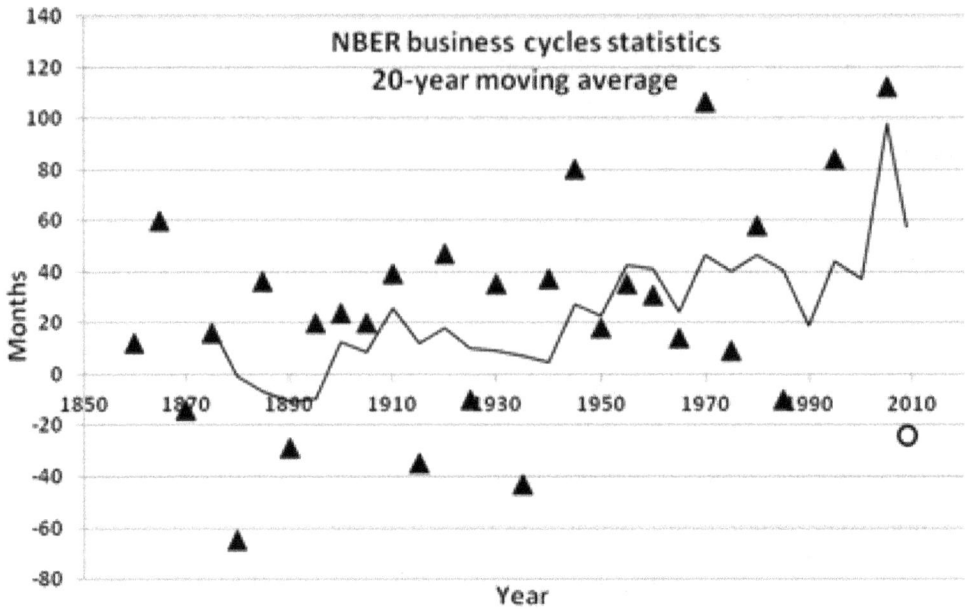

Figure 14b *20-year moving average applied to the points of Figure 14a. The trend line reveals a wave-like behavior that coincides with the dating schema used by many long-wave authors corresponding to the 2nd, 3rd, 4th and 5th K-waves. In this graph we have added a point (○) corresponding to the actual crisis considering it with a supposed duration of 24 months.*

sis considering it with a supposed duration of 24 months. This point was considered in the moving average in order to observe the path of the trend line. Again we are induced to the same conclusion drawn in point 1 at the end of the fifth section (Signals of Saturation)—this last point suggests the action of a self-correction mechanism bringing down a period of excessive growth!

Coccia (2010) suggests that these shrinking contraction periods may be due to a learning process during which government(s) have developed functioning methods to undermine the effects of economic recessions. This suggestion comes to reinforce our second conclusion in the fifth section about a secular learning process of the socio-economic world system.

9. MADDISON'S PHASES OF ECONOMIC GROWTH

n a publication from 2007 Maddison (2007a) performs a balance of his impressive and massive historical research about the evolution of the world GDP and GDP/capita since the beginning of the 19th century, as well as a detailed analysis of the works of some long-waves theorists (Kondratieff, Kuznets, Abramovitz, Schumpeter, and long-wave revivalists like Rostow, Mandel and Mensch). Maddison concludes that the existence of a regular long-term rhythm in economic activity is not proven and states further that there is no convincing evidence to support the notion of regular or systematic long waves in economic life.

Based mainly on his own data on aggregate performance Maddison however concedes that there have been major changes in growth momentum of capitalist development since 1820, which he coins as phases of economic growth. He recognizes five phases: 1820–1870 (transition from merchant capitalism to industrial accelerated growth), 1870–1913 (liberal phase), 1913–1950 (beggar-your-neighbor phase), 1950–1973 (golden age), and a last one from 1973 onwards (neo-liberal phase). Curiously there is some coincidence between these dates and some very important dates used by long-wave adopters either to characterize the duration of a full wave or to mark the transition between phases (up and down) of long waves.

But there are some oddities to point out in Maddison's whole analysis. In first place his review of authors contributing to bring empirical evidence on the existence of long waves is far from complete and does not include very important vast research work of authors that have brought robust empirical evidence using most effective mathematical tools. Maddison reviews basically only classical authors that have tried either to advance economic models to explain the long-wave phenomenon or to present evidence based only on economic statistics (with the exception of Mensch).

As robust empirical and mathematical evidence one must considers at least two authors that have carried during decades (1980s and 1990s) extensive work on long waves: the American economist Brian Berry and the Italian physicist Cesare Marchetti, whose works were published in the pages of TF&SC and elsewhere. Berry (1991) used convincingly chaos theory and spectral analysis to prove the existence of long waves and Marchetti,[2] leading a research team at the International Institute for Applied Systems Analysis (IIASA), produced some hundred graphical analyses applying the logistic substitution model on physical measures of human aggregate activities. In our point of view there is a touch of nonsense and exaggeration in simply refusing all the massive evidence brought by both authors.

Indeed it is very difficult to prove the existence of long waves using only economic statistics. There are many variables that must be considered simultaneously and this consists in an almost impossible task. But we must recognize that in despite of this inherent difficulty there is the register of at least two bold forecasts in recorded

2. See Cesare Marchetti's publications at Cesare Marchetti Web Archive (URL: http://cesaremarchetti. org/publist.php).

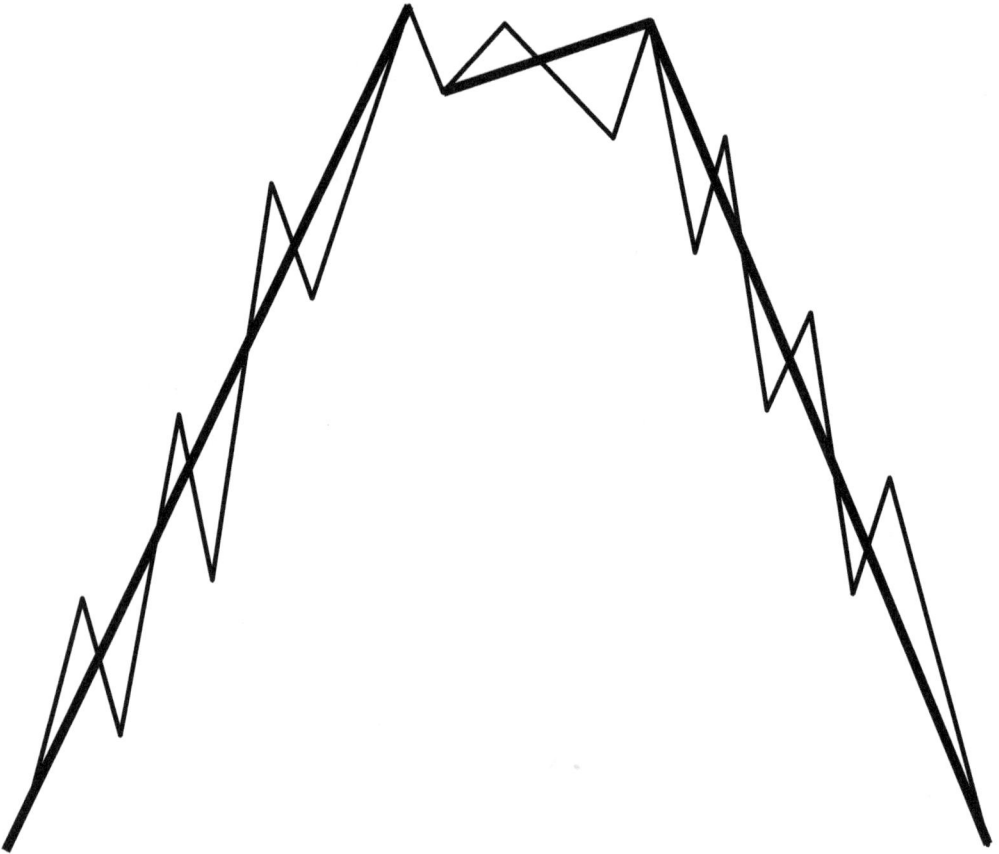

Figure 15 *Schematic depiction of a hypothetic long wave with nested shorter business cycles. As explained in the text this is just a schematic portrait of a very complex phenomenon and does not intend to render a real depiction of a single variable.*

history: Kondratieff himself, writing between 1922 and 1926, predicted accurately the Great Depression of the 1930s and there is the famous graph published in 1974 by Media General Financial Services that had been widely reproduced by dozens publications on long waves since then (the graph was also reproduced in one of our previous publications in the pages of TF&SC [Devezas & Corredine, 2001]). This graph, a schematic depiction portraying the cresting unfolding of K-waves since the 1790s, predicted also very accurately the behavior of the world economy in the following decade (the 1980s), when was observed a global reduction of economic growth and retraction of the world commerce, as alias evinced too through the timely evolution of the world GDP-PPP growth rates shown in Figures 13a and 13b.

This kind of schematic depiction of K-waves has been the preferred target of many criticizers of Kondratieff waves, who insist on the fact that such regular long-term oscillations do not exist. It is clear that such a monotonic upward movement

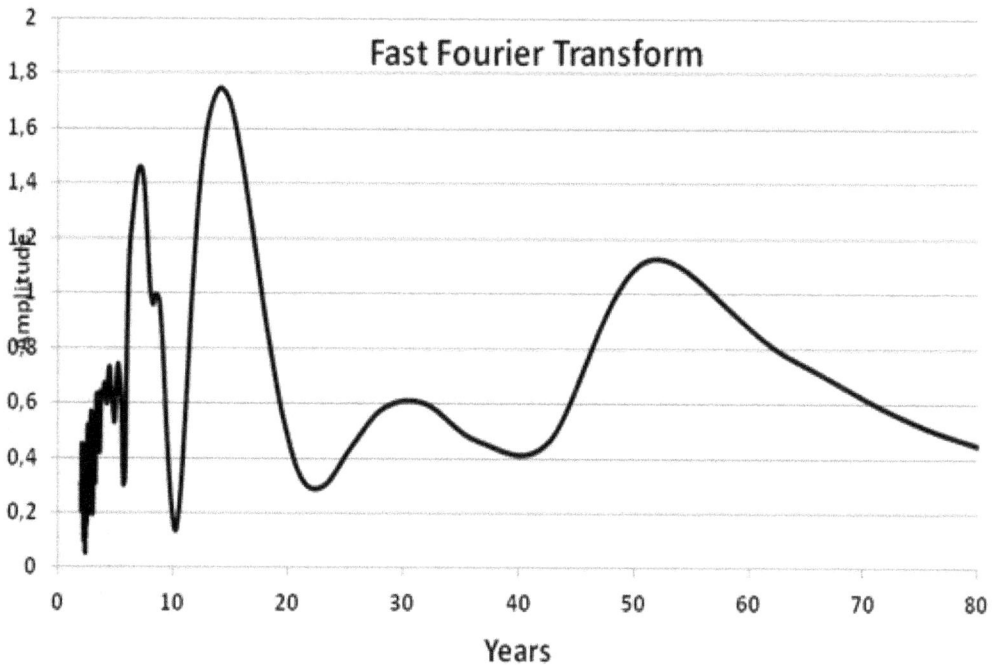

Figure 16 *Fast Fourier Transform using the Sigview software (Kondratieff 2004) applied to the historical unfolding of the GDP-PPP growth rates presented in Figure 10. We can clearly discern the existence of four frequency peaks: 7.5 years, 15 years, 32 years, and 52 years.*

during about two decades, followed by a subsequent two decades-long downward movement does not exist indeed—what is necessary to comprehend is that such a representation is just a schematic portrait of very complex behavior that includes the timely unfolding of several variables and does not try to translate the evolution of a unique variable. Perhaps, a bit more realistic representation should include in the upward and downward movements the within nested shorter business cycles, as we try to express through Figure 15. But again it is very important to stress that this is just a schematic depiction of a much complex phenomenon and does not intend under no circumstances to render a real depiction.

As a second oddity in Maddison's whole analysis we wish to point out the lack of graphical analysis. One really wonders why Maddison does not use graphs in his publications. In his famous and very frequently referred book (Maddison, 2007b), for instance, among 124 tables, Maddison presents only seven graphs, and just for comparisons of GDP cumulated growth (or comparative levels of GDP/capita) for pairs of countries, like UK/Japan, UK/India, USA/China, etc. In his own words he says to use 'inductive analysis and iterative inspection of empirically measured characteristics' (Idem 2007a: 147), but the most of his analysis and conclusions are drawn only based on tabular constructions, which do not allow perceiving long-term trends and details of an evolutionary process. As can be seen in this work, a simple glance at some graphs allows the perception of fingerprints of K-waves, as well as the observation of details related to the temporal behavior of a given economic-related quantity.

It is hard to understand why Maddison is so adamant in his statements about the lack of evidence on K-waves if he has never applied mathematical analysis on his monumental set of data, as for instance spectral analysis. We have already mentioned above the contribution of Brian Berry. This author, in his 2001 paper (Berry *et al.*, 2001) has demonstrated the existence of low frequency waves of inflation and economic growth using digital spectra analysis. He and his collaborators have found ~9 and ~18-year oscillations linked to business and building cycles, and additional ~28 and 56-year rhythms linked to inflation alone.

Very recently Korotayev and Tsirel (2010) have examined minutely the entire data set of Maddison's GDP-PPP growth rates under the optic of modern spectral analysis and have found very similar results, or in other words, two strong frequency peaks corresponding to the shorter business cycles (in this case ~8 years and ~15 years), and two long-term frequency peaks (~30 and ~52 years) related to long waves—the shorter probably corresponding to upwaves and downwaves movements and the longer corresponding probably to complete K-waves oscillations.

In our research we decided then to verify these results and have applied a simple Fast Fourier Transform using the Sigview software.[3] The result is shown in Figure 16 where we can clearly discern the existence of four frequency peaks, in this case 7.5 years, 15 years, 32 years (very weak) and 52 years—again practically the same result as those of Berry and Korotayev-Tsirel. It is important to stress that ours and Korotayev-Tsirels' results were found in the same data set where Maddison says that there is no convincing evidence to support the existence of systematic long waves in economic life.

Closing this section we wish to briefly discuss a statement of Maddison, where he wrote:

> *The government regulatory role in the economy has greatly increased. One result of the latter is that the stability of financial institutions has improved. Before the Second World War, depressions were often reinforced by major bank failures, but these are now rarer and their impact is cushioned (Maddison, 2007a: 161).*

---

3. http://www.sigview.com/download.htm.

What is curious in this statement is that it is partially true—in fact, there have been a learning process during which governments have learned a lot how to reduce the impact of economic shocks, as we have already stressed previously, and that explains the phenomenon of shrinking recessions and contractions portrayed in the Figure 14a. But on the other hand, it is completely false regarding the stability of financial institutions. Let us give a discount to Maddison—he has written these lines shortly before the big financial crash of the end of 2007.

10. GOLD—THE MASTER OF COMMODITIES

At the end of the closing chapter of his 1922 book Kondratieff (2004) has made a very important observation about the behavior of gold during the unfolding of K-waves, which has been bypassed by most of long-waves analysts up to present days. In this chapter, with the suggestive title 'The crisis of 1920–1921 in the system of general movement of conjunctures' Kondratieff paved the way for his dangerous idea of an incoming (temporary) collapse of the world economy and used gold to reinforce his damned prophecy. The inclusion of the word 'temporary' here is very important, because Kondratieff's dangerous idea was not the forecast of a final collapse (as wished by his Bolshevik opponents), but the anticipation of a new downward wave, which should be followed by another upward wave—or, in other words, a general picture of a wave-like movement of the capitalist system.

Kondratieff wrote:

Gold output, on the other hand, showed a remarkable movement, too. Since mid-1890s its output was surging to come to a maximum in 1915 and a subsequent continuous decline... The output of gold is quite likely to plunge into a long depression, which is the most remarkable feature of the current epoch (Kondratieff 2004: 121).

He follows referring to a study of Joseph Kitchin and presents a table from a publication of this author with data about the annual average growth of gold output, in which can be seen a minimum in 1810, a maximum in 1847, again a minimum in 1868, followed by a new maximum in 1891, and declining again after this date. In the following paragraphs he wrote:

It can be readily seen that the dates and periods displayed match closely the turnarounds and periods of upward and downward waves of the long cycles. It is also quite obvious that the upward waves are coincident with periods of a high annual growth of gold, and vice versa. In this case, we enter upon the area of relatively low annual growth of gold, which is going to affect the downward conjunctures of the long cycle... Again this process promises to follow the line of the 1870 (Kondratieff 2004: 141).

Indeed a bold forecast and what happened in the following years is the history everyone knows very well. But after these lines Kondratieff has made also another very important point:

We can therefore relate to the world economy as being quite likely to enter upon a downward phase of the long cycle. This by no means goes to say that this phase will be clear of its own ups or downs or depressions in terms of minor capitalist cycles. They have always been present in such phases in whatever long cycles of conjunctures of the past. They will surely be present in a downward phase of the long cycle. In a general frame of their variation, however, the conjunctures are most likely to keep downwards. Consequently, elevations in minor cycles of the oncoming period will lack the intensity they would display while on an upward wave of a long cycle. By contrast, crises of this period promise to be sharper, and depressions of minor cycles lengthier (Kondratieff 2004 [1922]: 142).

Again a bold forecast, and the reader must keep in mind that these lines were written in 1922. What Kondratieff voiced in this last paragraph is exactly what we have tried to express through Figure 15 in Section 8.

The question now is: What does our important actor in world economic affairs, gold, allow us to say about the present trend and what may be forecasted regarding the forthcoming years? As we will see some forecast is indeed possible, but we have first to consider that the behavior of gold has changed dramatically along with the last century, after Kondratieff inspired vision.

The graph depicted in Figure 4 (Gold weekly price) cannot tell us much about the future of gold price, except perhaps the fact that we are presently witnessing a strong momentum upwards. Such growth however cannot continue indefinitely, nothing in the universe growths forever. But on the other hand, this graph tells us a lot about the gold's past and recent history. As can be seen, since 1900 gold experienced a long period subjected to two levels of constant prices until 1971, when suddenly began to raise, reaching a first modest peak in 1975, soon followed by a strong peak by the end of 1980, outreaching the level of US$ 800. This record was immediately followed by a continuous trend of decreasing prices that endured for 20 years, reaching a minimum of about US$ 270 by the end of 2000, when gold entered a new phase of an apparently unstoppable trend towards ever increasing prices.

The long period of constant prices belongs to the old times of the 'gold standard', which started in Britain after the Napoleonic wars. In the second half of the 19th century, a number of nations in Europe and elsewhere followed suit, and the United States adopted the gold standard de facto in 1879, by making the 'greenbacks' that the Government had issued during the Civil War period convertible into gold; it then formally adopted the gold standard by legislation in 1900, when our graph begins. By 1914, the gold standard had been accepted by a large number of countries, although it was certainly not universal.

During the 1880–1914 period, the 'mint parity' between the U.S. dollar and sterling was approximately $4.87, based on a U.S. official gold price of US$ 20.67 per troy ounce (31.1035 gr) and a U.K. official gold price of £4.24 per troy ounce. This system worked well during almost forty years when the world economy entered the turbulent phase already referred to when commenting on the graph of Figure 10.

We can state that this first period of relative peace corresponded to the real entrenching stage of a successful international capitalist system, when there were no changes in the exchange rates of the United States, UK, Germany, and France (though the same did not hold for a number of other countries). There were few barriers to gold shipments and few capital controls in the major countries. Capital flows generally seem to have played a stabilizing, rather than destabilizing, role. After the outbreak of the First World War, one combatant country after another suspended gold convertibility, and floating exchange rates prevailed. The United States, which entered the war late, maintained gold convertibility, but the dollar effectively floated against the other currencies, which were no longer convertible into dollars. After the war, and in the early and mid-twenties, many exchange rates fluctuated sharply. Most currencies experienced substantial devaluations against the dollar; the U.S. currency had greatly improved its competitive strength over European currencies during the war, in line with the strengthening of the relative position of the U.S. economy.

But in the very beginning of the turbulent phase that followed WWI (and when Kondratieff issued his first publications!), there was a widespread desire in Europe, especially in the UK, to return to the stability of the gold standard, and a worry about the growing attractiveness of the dollar—which was convertible into gold—and of dollar-denominated assets. Following a disastrous five years back on the gold standard, the UK abandoned it in 1931, and others followed over the next few years.

Things began to worse and after the onset of the Great Depression in 1929 Keynesian economics was the evident remedy found to recover the agonizing patient.

In April 1933, U.S. President Franklin Roosevelt through the Gold Reserve Act imposed a ban on U.S. citizens' buying, selling, or owning gold. While the U.S. Government continued to sell gold to foreign central banks and government institutions, the ban prevented hoarders from profiting after Congress devalued the dollar (in terms of gold) in January 1934. This action raised the official price of gold by more than 65 percent (from $20.67 to $35 per troy ounce) and this fact is translated by the first jump to a new level observed in our graph of Figure 4.

In 1971, when the Bretton Woods system broke down, President Richard Nixon ended U.S. dollar convertibility to gold and the central role of gold in world currency systems ended, giving birth to a new era of complete liberalization of capital flows. The consequences are very clear in the graph of Figure 4: the dollar and gold floated and in January 1980 the gold price hit a record of US$ 850 per ounce, soon followed by a decrease that endured for almost 20 years. Only after 2000 gold started to escalate reaching new levels again that make look overt the 1980's record. What can be learned from this picture?

The first quite obvious lesson is that the remedy found to fight the system's illness does not hold for a long time. It is as if the doctors (economists) were combating only the symptoms and not really fighting the true intrinsic system's sickness. The relief measures insistently applied until now by mainstream economics consists in failed contra-cyclical policies that systematically overlook some strong forces underlying the global economy. These strong forces are mainly the inexorable human propensity to hoard and the physical-biological imperatives acting upon the complex socioeconomic system. The latter was already analyzed in deep in some of our previous publications (Devezas & Corredine, 2001, 2002) and we do not intend to discuss in this paper. It is looking at the former that we can discern some important hints that can help us to correctly read the historical unfolding of the role played by this important actor—gold—in the whole piece of economic capitalist development.

The reason for our title—the master of commodities—lies in the fact that gold is the most hoardable commodity. Gold does not tarnish or fade; it resists the entropic laws of decay, and its high specific gravity contributes to the fact that the opportunity cost of hoarding gold is far lower than that of hoarding any other commodity. Gold is essentially money of last resort and has been the most effective hedge against turbulent times, be they caused by wars or economic depressions. For all over the recorded history humans have shown an inexorable trend to hoard gold bullions and all the sudden changes observed in the unfolding of the graph depicted in Figure 4 were due to governments measures trying to oppose this strong economic force. Unnecessary

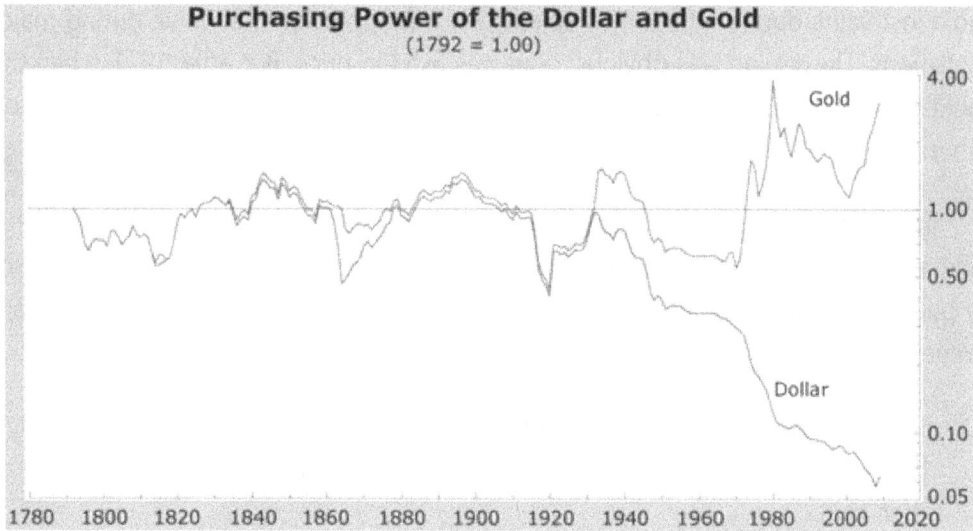

Figure 17 *Purchasing Power of Gold (PPG) compared to the Purchasing Power of US Dollar (PPD) since the 1790s*

to point out that such measures have never worked (in the long range) in favor of the health of the socioeconomic realm. The increasing price trend evinced since 2001 is the clearest proof of the action of hoarding per se.

But in order to draw effective conclusions about the future path of the world economic system is necessary to look at gold's history other way around. In 1977 Berkeley's Professor Roy Jastram in his seminal work The Golden Constant—The English and American Experience, 1560–1976 (Jastram, 1977) demonstrated, for the first time, how gold's purchasing power had been maintained over the centuries. Dividing the gold price index by the wholesale price index he found that the Purchasing Power of Gold (PPG) has fluctuated around a broadly mean value. However, Jastram's research ended in 1976, and therefore he barely foresaw the impact of the new era of floating gold price, still at its genesis.

Very recently Jastram's original work was updated by Leyland (Jastram & Leyland, 2009) in a research supported by the World Gold Council. The new edition contains two additional chapters (and the relevant statistics) examining the period from 1971 to 2007. The conclusions about the behavior of the Purchasing Power of Gold differ somewhat between the periods before 1971, when the gold price was controlled, and after, when it was free. Nevertheless, one conclusion remains unchanged—that gold maintains its purchasing power over long periods of time even though, over shorter periods, it has fluctuated significantly. But more importantly, this new research demonstrates that now gold moves just the opposite of what it used to do. Before 1971 gold lost value during inflationary spirals, while it appreciated in value during major deflations. The reason was obvious: gold was fixed in price. But after 1971, when the gold was delinked and set free to fluctuate, the price of gold goes up when inflation goes up, and falls when deflation hits.

In Figure 17 we present a graph portraying the Purchasing Power of Gold (PPG) and as comparison the Purchasing Power of US Dollar (PPD) since the 1790s recently published in the Web by the American Institute for Economic Research. There are some very important points to infer from this graph that we try to resume below:

1. Both purchasing powers have unfolded perfectly in phase until at least the early 1930s, when they began to diverge and this diversion aggravated substantially after 1971.

2. There is evidently a wave-like behavior and the maxima and minima of the fluctuations before the early 1920s match closely the dates for the turnarounds of long waves pointed out by Kondratieff that we referred to at the beginning of this section; the dip in the early twenties also matches Kondratieff's forecast.

3. The 50-year beat of the maxima of these long fluctuations is absolutely evident—1840s, 1890s, and late 1930s. Even the peak reached in 1981 falls within the long wave timeframe. It is indeed hard to understand the intestine refuse by mainstream economics in believing in the existence of long waves.

4. In 1971 for the first time in history PPG jumped suddenly from a value below to a value above its historical average, and no more returned to the field below <1.00. After a brief hesitation in the mid-1970s, PPG rocketed again in 1980–81, when gold price reached the first maximum shown in Figure 4. This was a decade (1970–1980) not just of high inflation but it also included the two oil price 'shocks' and what appeared at the time to be the end of the post-war 'miracle' growth of the 1950s and 1960s.

5. After the maximum reached in 1980–1981 PPG entered a 20-year long declining period, during which a self-correction mechanism seemed to act in order to bring it down to its original path along with its historical average. That was the time of the 'great moderation' of the decades 1980s and 1990s, a period of disinflation, generally improving economic circumstances, mostly strong stock markets and marked politically by the fall of communism.

6. In contrast, since 2001 the PPG has risen again due to the well-known concern over global imbalances and rising debt, which culminated in the current economic and financial crisis.

7. Comparing the last decreasing period of PPG (1980–2001 = 21 years) with the preceding ones (1842–1870 = 28 years, 1895–1920 = 25 years, 1940–1971 = 31 years) we can say that it was relatively shorter, but not very far away. Associating this fact with the observation that PPG is presently going away from its historical average we can suspect that we are facing an anomaly, or at least we are experiencing a transition phase as we have already pointed out when analyzing other economic indicators.

8. Such an anomaly, or if we prefer, the imminence of a transition phase, is evident from the 'bifurcation' (perhaps better, divergence) presented in the graph of Figure 17. It is quite possible (in fact it is the case since 1971) that a portion of the increase of PPG is really just the outcome of the decrease of PPD, considering that the change in gold price is simply a mathematical recalculation of an ever-changing US Dollar value.

9. The history of fiat currencies is that they lose their purchasing power over time. Because a limited amount of gold exists in the world and paper money can be created without limits, gold has been an ultimate protection against the debasement of currencies. If we look at the historical charts of the purchasing power of major currencies as well as the amount of these currencies in circulation (see, e.g., the graphs presented by Financial Sense University [Hewitt and Petrov 2009]) what we see is that all major currencies have lost steadily purchasing power since 1971—US Dollar is now at 20% of its level in 1971, GB Pound at 18%, Canadian Dollar at 18%, Australian Dollar at 10%, Japanese Yen at 70% and

Figure 18 Ratio DJIA/Gold price considered weekly since 1900.

Swiss Franc at 70%. Opposed to this decrease the amount of circulating paper money of the same currencies grew by a factor 8 (USD), 5 (GBP), 10 (CAD), 20 (AUD), 10 (JPY), and 3 (CHF) respectively. On the other hand, the amount of mined gold has grown slowly and almost linearly, from about 95×103 metric tonnes in 1971 to about 160×103 metric tonnes in 2008 (a factor of only 1.6 in more than three decades; [Hewitt and Petrov 2009]). Resuming this point, the amount of available gold (or gold output) is not the cause of the movements of PPG after 1971; the subjacent cause lies in the combination of two other linked factors—an ever-increasing debasement of currencies and declining (1971–1981) or improving (1981–2001) economic circumstances.

10. But supposing that in despite of the changed circumstances the system is resilient and that the PPG will not deviate very much from its historical average (considering also that hoarding has its natural limits), we might conjecture that the actual increasing trend of PPG (and naturally also of gold price) can continue until 2010–2011 (a decade after 2001), but will return to its historical mean value, a process that may involve one or two decades of economic growth that will coincide with the upward phase of the 5th K-wave peaking about 2020. This forecast matches well our previous considerations when discussing the world GDP.

There is also another way to look at the historical unfolding of gold price calling to playing other of our important agents—the Dow Jones Industrial Average (DJIA). We can calculate a ratio dividing the DJIA weekly index by the weekly price of gold, or in other words, to determine the historical record of the answer to the question: how many ounces of gold does the Dow Jones Industrial index buy? The Canadian financial analyst Ian Gordon originally developed this method, which he uses as an economic forecasting tool. The resulting graph is shown in Figure 18, and as we can see there is also a clear regular wave-like pattern.

The pattern, however, is quite different from that presented in Figure 17—it seems inverted with relation to PPG, some of the PPG peaks are now pronounced dips and the waves have now a skewed aspect, evidencing two or three decades of growth followed by sudden falls. The first quick movement downward was soon after the stock market crash of 1929, and lasted only until 1933, recovering after Roosevelt's Gold Reserve Act. It followed an upward movement during almost three decades, which stopped around 1965–1966 in consequence of a hesitating stock market. In 1971 again a sudden drop after the end of the US dollar convertibility, which extended until

1974 and was followed by the profound dip in 1981 that was due to the combination of a bearish stock market and an accentuated gold rally in prices.

The last wave begins then in 1981 and one can read in this curve the timid stock market crash of 1987, which was followed by a rapid increase of the ratio DJIA/Gold (mainly after 1992), not only due to a worldwide bullish stock market, but also due to the healthy economic growth (and consequently to the cheaper gold) of the 1990s, which the Nobelist Joseph Stiglitz (2003) coined as 'The Roaring Nineties'. A peak in the ratio happened in 2000, and after the dot-com bubble burst it has followed a steadily downward trend.

The actual situation is one of a hesitating stock market, mainly due to fears of an imminent inflation, and of a gold rally that many financial analysts (Wiegand 2009) want to believe that will continue for a while with gold prices escalating until over US$ 2900! May be such a so high price level will never be reached, but a simple extrapolation of our curve of Figure 18 induce us to hope that a minimum of the ratio might be reached very soon, which may be soon followed by an upward movement, implying in a recovering economy. Considering also the regular beat of the peaks—1929–1965–2000, or in other words, a period of about 30–35 years (or a half K-wave), we can speculate that the next peak might be reached by about 2030 or earlier.

Concluding this section we can state that the historical evolution of gold allow us to foresee that the present circumstances of a weakening dollar, a bearish stock-market, and increasing gold prices will reach the end very soon and a renewed economic upsurge may well take place lasting at least until the decade 2020–2030.

CONCLUSIONS

In this paper we have investigated the global secular evolution of four important economic-related actors, whose interplay when scrutinized with the suitable analytical tools evince some historical patterns that shed some light on what is going on with the world economic system. These actors are: the world population, the world aggregate output known as Gross Domestic Product (GDP), the historical leader of all commodities—Gold, and the still most important financial index, the DJIA (Dow Jones Industrial Average). Also the succession of economic depressions and expansion periods in the US was examined.

The main conclusions of this research are summarized below:

1. Fingerprints of Kondratieff long waves are ubiquitous in all observed time-series used in this research—world GDP growth rates, succession of economic expansions-contractions in the USA, purchasing power of gold and the historical ratio DJIA/gold price.

2. Regarding the present crisis we can state that it has some unique characteristics, which distinguish it from all previous economic depressions. But in despite of its unique characteristics a parallel with the panic of 1907 may be drawn—both have occurred amidst a strong international growth period and are perfectly symmetric in the observed space-time pattern.

3. The most important conclusion concerning this crisis is that it seems to sum up a mix of a self-correction mechanism that brought the global output back to its original logistic growth pattern, and signals an imminent transition to a new world economic order.

4. The next decade will be probably one of worldwide economic growth, corresponding to the second half of the expansion phase of the fifth K-wave, but that will saturate soon after the 2020s.

5. There are strong signals that we are already witnessing a transition to a new global socioeconomic system, which will carry within it a profound restructuration of world economic affairs, with a multipolar world leadership and a new world currency. The trend analysis applied in this research using logistic curves, spectral analysis and the singularity approach converge to the same general result of an evolutionary trajectory leading the world system toward a true age of transition.

Regarding this last conclusion it is important to make stand out the fact translated by our results shown in Figure 11 (and commented on in point 9 of Section 6) that real growth rates of low-income countries have been growing increasingly apart from those of high-income countries. Since the onset of the Industrial Age high-income countries have contributed with at least about 70% for the global output measured as world GDP growth rate. Recent numbers of the United Nations Development Programme presented by Marone (2009) show that this historical trend was maintained up to the mid-1990s, with the contribution of all income categories being roughly constant. But after this point and up to 2007 growth contribution from low-income countries surged by more than threefold, from around 10% (mid-1990s) to almost 35% (2007). In the mid-1990s high-income countries contributed with 77% for the global output growth, and low/middle-income countries contributed with 23%. Presently these numbers have radically changed to 95% from low/middle-income and only 5% from high-income countries. Indeed, we are amidst a great transformation.

In this work we have applied a broad perspective approach with the main goal of exploring past events encompassing the action of the four actors/variables/agents together in order to find patterns of behavior that can concede us to comprehend what is going on. We just tried to construct a 'timescape' using these variables that allow us to discern for instance that an incoming transition seems to be on marsh and that the present crisis exhibits symptoms of a saturating world economic system. We avoided bold forecasts and have speculated only about the very near future, within a time horizon of about two decades, a future that somehow is already determined by today's actions (and non-actions) and circumstances.

But as we all know very well, contingency exists and there are much more variables that must be considered in order to construct the most probable scenarios. We hope that our present results may contribute for more embracing studies that applying the multiple perspectives approach may lead to the enhancement of our ability to think constructively about the future of economics on a global scale.

REFERENCES

American Institute for Economic Research 2009. The Value of Gold, October 5th. URL: http://www.aier.org/research/briefs/2099-the-value-of-gold.

Analyze Indices. Market and Industry Indices n.d. History of the Dow Jones Industrial Average: 1900–2007. URL: http://www.analyzeindices.com/dowhistory/djia-100.txt.

Berry, B. J. L. 1991. Long-Wave Rhythms in Economic Development and Political Behavior. Baltimore—London: Johns Hopkins University Press.

Berry, B. J. L., Kim, H., and Baker, E. S. 2001. Low Frequency Waves of Inflation and Economic Growth: Digital Spectra Analysis. Technological Forecasting and Social Change 68: 63–74.

Boretos, G. P. 2009. The Future of the Global Economy. Technological Forecasting and Social Change 76: 316–326.

Bouchaud, J. P. 2008. Economics Needs a Scientific Revolution. Nature 455: 1181–1183.

2009. The (Unfortunate) Complexity of the Economy. arXiv: 0904.0805v1 [q-fin.GN].

Bruner, R., and Carr, S. 2007. The Panic of 1907—Lessons Learned from the Market's Perfect Storm. Hoboken, NJ: John Wiley & Sons.

Coccia, M. 2010. The Asymmetric Path of Long Waves and Business Cycles Behavior. Technological Forecasting and Social Change 77: 730–738.

Devezas, T. C. 1997. The Impact of Major Innovations: Guesswork or Forecasting. Journal of Futures Studies 1(2): 33–50. 2009. The Evolutionary Trajectory of the World System toward an Age of Transition. In Thompson, W. R. (ed.), Systemic Transitions (pp. 223–240). New York: Palgrave MacMillan.

Devezas, T., and Corredine, J. 2001. The Biological Determinants of Long Wave Behavior in Socioeconomic Growth and Development. Technological Forecasting and Social Change 68: 1–57.

Devezas, T., and Corredine, J. 2002. The Nonlinear Dynamics of Technoeconomic Systems: An Informational Interpretation. Technological Forecasting and Social Change 69: 317–357.

Devezas, T. C., LePoire, D., Matias, J. C. O., and Silva, A. M. P. 2008. Energy Scenarios: Toward a New Energy Paradigm. Futures 40: 1–16.

Devezas, T. C., and Modelski, G. 2003. Power Law Behavior and World System Evolution: A Millennial Learning Process. Technological Forecasting and Social Change 70: 819–859.

Hewitt, M., and Petrov, K. 2009. Money Supply and Purchasing Power. URL: http://www.financialsense.com/fsu/ editorials/dollardaze/2009/0223.html.

IMF 2009. International Monetary Fund. World Economic Outlook Update, July 8th. URL: http://www.imf.org/external/pubs/ft/weo/2009/update/02/pdf/0709.pdf.

IIASA 2011. International Institute of Applied Systems Analyses, Logistic Substitution Model II. URL: http://www.iiasa.ac.at/Research/TNT/WEB/Software/LSM2/lsm2-index.vhtml.

Jastram, R. W. 1977. The Golden Constant—The English and American Experience, 1560–1976. New York: Wiley.

Johansen, A., and Dornette, D. 2001. Finite-Time Singularity in the Dynamics of the World Population, Economic and Financial Indices. Physica A 294: 465–502.

Jastram, R. W., and Leyland, J. 2009. The Golden Constant—The English and American Experience, 1560–2007. Northampton: Edgard Elgar.

Kaufman, G. G., Krueger, T. H., and Hunter, W. C. 1999. The Asian Financial Crisis: Origins, Implications, and Solutions. Berlin: Springer.

Kitco n.d. Daily Gold Charts. URL: http://www.kitco.com/charts/historicalgold.html/.

Kondratieff, N. D. 2004. The World Economy and its Conjunctures during and after the War. Moscow: International Kondratieff Foundation [first English translation of the original from 1922].

Korotayev, A. V., and Tsirel, S. V. 2010. A Spectral Analysis of World GDP Dynamics: Kondratieff Waves, Kuznets Swings, Juglar and Kitchin Cycles in Global Economic Development, and the 2008–2009. Economic Crisis, Structure and Dynamics 4(1): 3–57. URL: http://www.escholarship. org/uc/item/9jv108xp.

Maddison, A. n.d. Historical Statistics, Statistics on World Population, GDP, and Per Capita GDP, 1–2006 AD, Horizontal and Vertical Files. URL: http://www.ggdc.net/maddison/.

Maddison, A. 2007a. Fluctuations in the Momentum of Growth within the Capitalist Epoch. Cliometrica 1: 145–175.

Maddison, A. 2007b. Contours of the World Economy 1–2030 AD—Essays in Macro-Economic History. New York: Oxford University Pess.

Marone, H. 2009. Economic Growth in the Transition from the 20th to the 21st Century. UNDP/ODS Working Paper. New York: United Nations Development Programme.

NBER n.d. National Bureau of Economic Research. US Business Cycles Expansions and Contractions. URL: http://www.nber.org/cycles/cyclesmain.html.

Nelson, S. R. 2008. Panic of 1873 Revisited: The Real Great Depression. The Chronicle Review October 17. URL: http://chronicle.com/weekly/v55/i08/08b09801.htm.

Reinhart, C. M., and Rogoff, K. S. 2009. This Time is Different—Eight Centuries of Financial Folly. Princeton—Oxford: Princeton University Press.

Stiglitz, J. E. 2003. The Roaring Nineties—A New History of the World's Most Prosperous Decade. New York—London: W.W. Norton & Co.

Stiglitz, J. E. 2009. GDP Fetishism. The Economics Voice 6(8). DOI: 10.2202/1553-3832.1651.

US Census Bureau n.d. International Data Base. World Population Growth Rates: 1950–2050. URL: http://www.census.gov/population/international/data/idb/worldgrgraph.php.

Wiegand, R. 2009. Gold Prepares for the Big One. URL: http://www.kitco.com/ind/Wieg_cor/roger_jun252009.html.

Chapter 14

LOCAL SOLUTIONS IN A GLOBAL ENVIRONMENT: FACILITATING NATIONAL STRATEGIES IN NEW ZEALAND

Jim Sheffield

How should New Zealand respond to the multiple, intertwined and fast-changing impacts of globalization? What strategies are available to this small South Pacific country and how may these be facilitated? This empirical research frames the facilitation of selected local solutions in a global environment within the theoretical perspective of pluralism and communicative action. The facilitation of aspects of national policies in the domains of science funding, economic development and regional growth is reviewed. Electronic meeting technology was employed. The focus question is: 'Does electronic discourse increase the success of local solutions in a global environment?'

Keywords: New Zealand, local solutions, global environment, pluralism, communicative action, electronic discourse.

1. A NEW ZEALAND RESPONSE TO GLOBALIZATION

Productivity isn't everything, but in the long run, it is almost everything. A country's ability to improve its standard of living over time depends almost entirely on its ability to raise its output per worker.

Krugman 1997

Raising productivity is the core economic challenge for New Zealand over the medium term. Small, high-productivity economies rely heavily on international connections—the flows of people, capital, trade and ideas between countries around the world (New Zealand Government, 2009). In the current era of globalization, New Zealand's combined lack of any major home market effect, small population and lack of major agglomeration effects, and the extreme geographical isolation, breaks the usual link between entrepreneurship, innovation and growth (McCann, 2009). Domestic policy settings in science funding, economic development, and regional planning are critical to making the most of international opportunities. A well-funded science sector encourages entrepreneurial and innovative activity to be located in New Zealand and facilitates international knowledge transfer. Economic development improves competitiveness in global markets, including those in the Asia-Pacific region. Regional planning in Auckland, New Zealand's major growth area, attracts skilled migrants and reduces the loss of New Zealand-born citizens to Australia and other countries (Cheshire, 2012) (Figure 1).

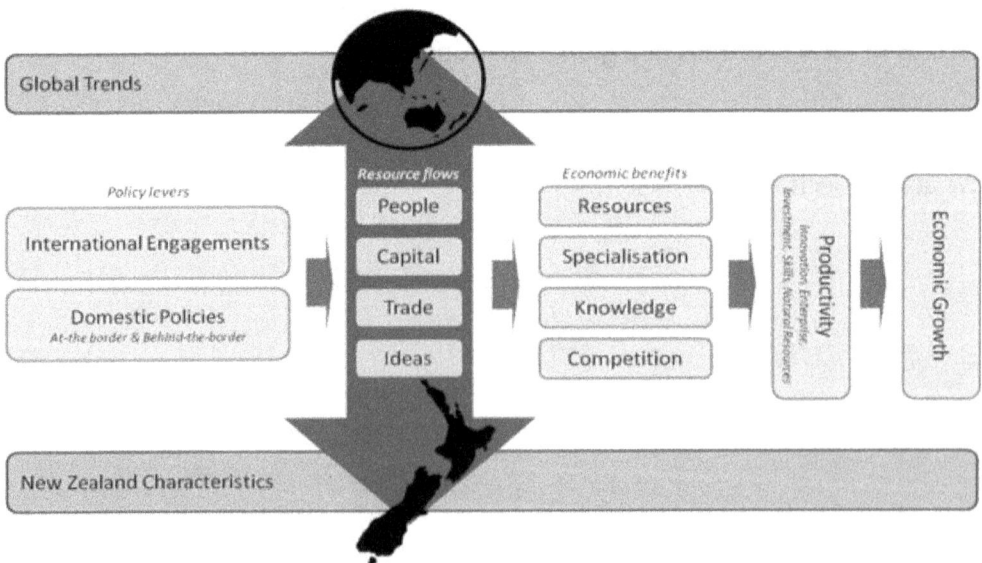

Figure 1 *Some aspects of a New Zealand response to globalization*

Science funding		
Sponsor: New Zealand Ministry of Research, Science and Technology		
Task: Allocation of the US(2012)$2 Billion Public Good Science Fund across all 40 areas of NZ science		
Role/process/group: Design of a 5-day group decision process for a 5-year planning and budgeting period. Implementation of the process with the national Science and Technology Expert Panel		
Goal: Legitimacy in science governance. A national consensus on priorities and transparency in funding		
Economic development		
Sponsor: New Zealand Trade Development Board		
Task: To upgrade New Zealand's competitive position in global markets		
Role/process/group: Design of 70 industry-wide strategic planning interventions conducted with the assistance of Harvard's Michael Porter. Implementation with 1,000+ industry leaders		
Goal: Improved relationships among industry stakeholders and formation of joint action groups		
Regional planning		
Sponsor: Auckland Regional Council		
Task: Strategic evaluation of long-term plans for the Auckland region, NZ's main growth area		
Role/process/group: Design of a group decision process to close out a 7-year planning cycle. Implementation with representatives of the 7 territorial authorities and the Auckland Regional Council		
Goal: Improved trust and understanding among decision makers. Support for a consensus spatial plan		

Table 1 *Facilitating national strategies in New Zealand*

This article reviews the facilitation of aspects of national policies in the domains of science funding, economic development and regional planning (Table 1). Electronic meeting technology was employed. The focus question is: 'Does electronic discourse increase the success of local solutions in a global environment?' The remainder of the article is structured as follows. Section 2 develops a theoretical framework. Section 3 describes the methodology for gathering empirical evidence. Sections 4–6 review the facilitation cases. Section 7 discusses the findings in the light of the theoretical framework. Section 8, which considers the lessons learned, concludes the article.

2. THEORY DEVELOPMENT

Facilitating national policies required extensive consultation among a large number of stakeholders in different organizations. The context was pluralistic—the objectives of social actors were divergent and power was diffused (Jarzabkowski & Fenton, 2006; Denis *et al.*, 2007). A modern information and communication technology—electronic meeting systems—has been found useful in supporting organizational groups engaged in strategic planning activities within an established power structure (Fjermestad & Hiltz, 2001; Shaw *et al.*, 2003).Yet research on electronic support in the context of pluralism and inter-organizational meetings suggests that the role of electronic meeting systems is unclear. For example, if electronic technology is employed in a meeting sponsored by one organization but attended by members of other organizations, whose interpretation of the ends served by the electronically-supported meeting should determine success? Who is the client? (Ackermann *et al.*, 2005). What roles and responsibilities will be recognized? (Franco, 2008). Is it sensible to expect powerful stakeholders to use collaborative technologies when these introduce unwanted accountability and make the exercise of power more difficult? (Schultze & Leidner, 2002; Lewis *et al.*, 2007). What type of model should drive the facilitation process? (Morton, Ackermann, & Belton, 2003). By what concept(s) of rationality or validity should the facilitator be held accountable for a positive outcome? (Kolfschoten *et al.*, 2007). Inter-organizational meetings require the surfacing and testing of assumptions from opposing perspectives (Mitroff & Linstone, 1993). In dialectical terms a pair of opposing perspectives is seen as a Hegelian thesis and antithesis (Millet & Gogan, 2005). Ignorance is reduced via active engagement with the conflict and confusion that accompany surfacing and reconciling opposing (multiple or pluralistic) perspectives, and giving birth to a new, more current synthesis.

Habermas (1984) provides a theory about how claims to pluralistic knowledge should best emerge from the communicative process. In Habermas's theory of communicative action, an ideal speech situation is defined as one in which all participants are free to question any utterance on the basis of its claims to objective truth, rightness for the context, and sincerity of the speaker. The speaker must be open to hearing and rationally responding to the questions that are asked. Power relations, that in other circumstances might allow some participants to ignore the perspectives of others, are set aside in favour of genuine dialogue.

In the theory of communicative action knowledge is evaluated from three perspectives (Habermas, 1984: 100):

- Personal perspective ('why I feel, and would be'). The personal or subjective world that is the totality of the experiences to which the speaker or actor has privileged access (because it is the speaker or actor that experienced them). Claims to subjective truth are evaluated in terms of the sincerity of the speaker or actor.
- Interpersonal perspective ('what we say, and should be'). The totality of interpersonal relations legitimately regulated by contextual expectations or norms. Claims to interpersonal norms are evaluated in terms of the rightness of the speakers or actors.

- Technical Perspective ('how it is, and could be'). The technical world of material fact that is the totality of all entities about which objectively true statements are possible, or could be bought about by purposeful intervention. Claims to facts and technical expertise are evaluated in terms of objective truth.

The ideal speech situation provides a standard of excellence for the reflective communicative action undertaken by two or more stakeholders in order to stabilize mutual understanding. Similarly group decision is seen as a collaborative process that seeks 'rightness' in the fit (coherence) between personal values, interpersonal objectives and technical decision criteria (Shakun, 2003). This requires participants to develop and integrate perspectives from generic roles that Churchman terms system designer (more technical/task oriented), decision maker (more interpersonal/consensus oriented) and client (more subjective/value oriented) (Churchman 1971: 200). Five facilitation principles based on pluralism and communicative action are presented in Table 2.

Integration of the Habermasian perspectives on knowledge is an exercise in sensemaking (Weick, 1979). Themes are detected both prospectively and retrospectively and emerge from communicative acts in a somewhat unpredictable manner. Nevertheless it is common for discourse on intentions to proceed from the personal to the technical, followed by discourse on outcomes that proceed from the technical to the personal (Shakun, 2003). Each pair of discourses (intention and outcome) in the same knowledge perspective develops mutual understanding via one of the principles in Table 2 and evaluates rationality via the relevant Habermasian knowledge claim (Sheffield 2005). The standard of excellence for communicative action can be stated as follows: personal commitment (validated by sincerity) to an interpersonal consensus (validated by rightness) for technical excellence (validated by objective truth). Each aspect of excellence is associated with Principle 1, 2 or 3 and the collective value of all three principles is evaluated in terms of Principle 4 and Principle 5 (Table 2). In the current research pluralism and electronic discourse are evaluated via qualitative measures of the impact on overall success of the facilitation principles and associated framework (Figure 2).

Pluralism is a notable feature not only of communicative action but of research in areas as diverse as neuroscience (Lehrer, 2009), knowledge management (Sheffield & Guo, 2007a, 2007b; Sheffield, 2008b), organizational sensemaking (Weick, 1979; Snowden & Boone, 2007) and systemic development (Sheffield, 2008a, 2009a; Midgley & Pinzo, 2011). Recent advances in neuroscience ground pluralism in the biology of decision behavior (Lehrer, 2009; Sheffield, 2012). Various scanning devices reveal

Principle 1. Personal commitment
Express claims to **sincerity** by free and open disclosure of participants' subjectivity (identity, experience and values)
Ensure that participants give voice to their personal commitments and multiple identities and that periods of silence are provided as an aid to ethical self-reflection
The procedure for evaluating the evidence should be validated by expressing beliefs and aspirations, voices and images ('story telling') that are unconstrained by technical issues and unrestrained by the inter-personal context

Principle 2. Interpersonal agreement
Enact claims to **rightness** via discussion among all those who are entitled to be represented
Ensure that the discussion addresses the role-based needs of stakeholders
The procedure for evaluating the evidence should be validated by full participation in a debate conducted under the norms of established legitimate inter-personal relationships

Principle 3. Technical excellence
Present claims to **objective truth** via research evidence
Ensure that the findings by technical experts are examined critically and the findings documented
The procedure for evaluating the evidence should be validated by a willingness to adopt a cognitive, objectivating attitude towards the facts. *Listen to the evidence, look at the facts—avoid partisan delusions*

Principle 4. Coherence
Assuming that claims for valid personal, interpersonal and technical knowledge have been surfaced, ensure that they are **coherent**. An apparent contradiction (thesis and antithesis) should serve as a precursor to a **Hegelian** synthesis. *Oh my God, I was wrong! We were all wrong!*
The procedure for evaluating coherence should be validated by a willingness to probe the evidence from all three perspectives, to identify strengths and weaknesses in the evidence, and to identify tradeoffs

Principle 5. Overall Success J
Success is conceptualised in **Churchmanian** terms as a meeting of the minds about intertwined relational and task issues that creates the capability of choosing the right means for one's desired ends
This requires participants to develop and integrate perspectives from generic roles that Churchman terms **system designer** (more technical/task oriented), **decision maker** (more interpersonal/consensus oriented) and **client** (more subjective/value oriented)
More specifically, success is indicated by insight leading to a consensus model that provides decision makers with a rationale for action

Table 2 *Five facilitation principles based on pluralism and communicative action. Adapted from Churchman, 1971; Habermas (1984)*

	Principle 4	Principle 5
1. Express values Goal: Expression of values and issues motivating each stakeholder	**Principle 1** Personal commitment (validated by sincerity) to.. ⟷	**6. Commit to action** Goal: Expression of commitment to action by each stakeholder
2. Explore objectives Goal: Obtain consensus on scope of legitimate values and issues	**Principle 2** ..an interpersonal agreement (validated by rightness).. ⟷	**5. Evaluate strategies** Goal: Obtain consensus on the right way forward
3. Obtain evidence Goal: Obtain research evidence for rival strategies	**Principle 3** ..for technical excellence (validated by objective truth) ⟷	**4. Evaluate evidence** Goal: Evaluate research support for rival strategies
Intentions	**Validity Claims**	**Outcomes**

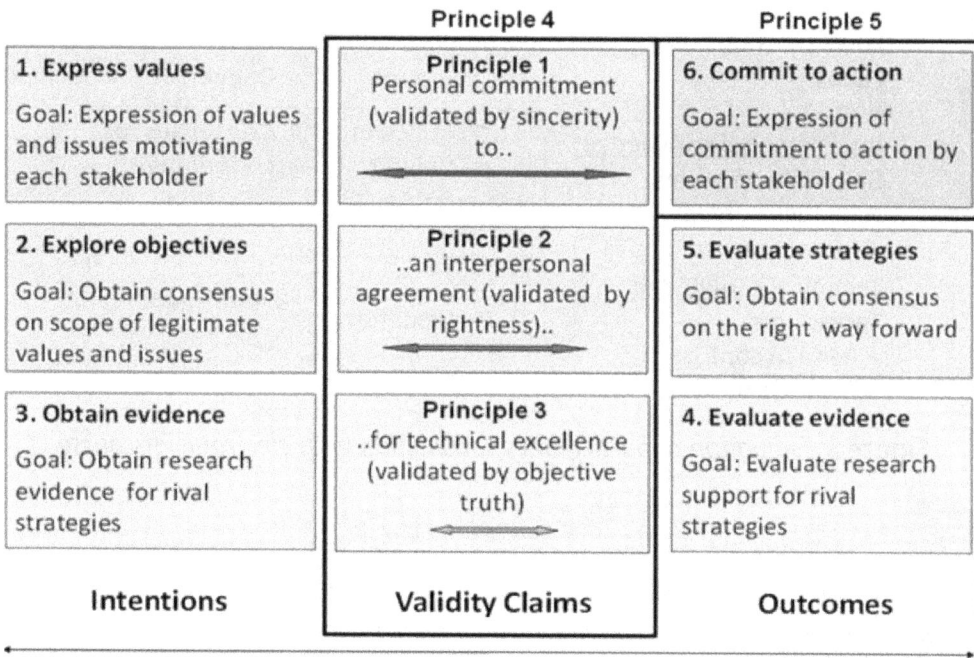

Figure 2 *A framework for facilitating national strategies in New Zealand based on pluralism and communicative action (Habermas 1984). Adapted from Sheffield (2004, 2009b)*

that the brain is an argument between neural regions dealing with emotion, morality and reason. Seen through the perspective of neuroscience the standard of excellence in group decision making becomes the pursuit of success through emotional commitment to a moral agreement for reasoned excellence.

Pluralism can be viewed as a consequence of intertwined relationship and task issues, and intertwined divergent and convergent thinking. The electronic discourse and supporting technology employed in the current research supported pluralism via two key attributes. Firstly, the technology provided a degree of anonymity that reduced the anxiety about surfacing opposing perspectives. This reduced participants' conflict about personal (emotional) commitments and interpersonal (moral) issues. Secondly, the technology reduced confusion by providing automatic recording of all electronic discourse ('group memory'). This enhanced participants' technical (reasoning) capabilities. Together these attributes allowed procedures for idea generation (divergent thinking) to be separated in time from procedures for information analysis (convergent thinking). This in turn enabled a separate focus on interlocked issues about relationships (trust) and cognition (understanding). In the current research all of these concepts are included in the evaluation of satisfaction with electronic discourse (Figure 3).

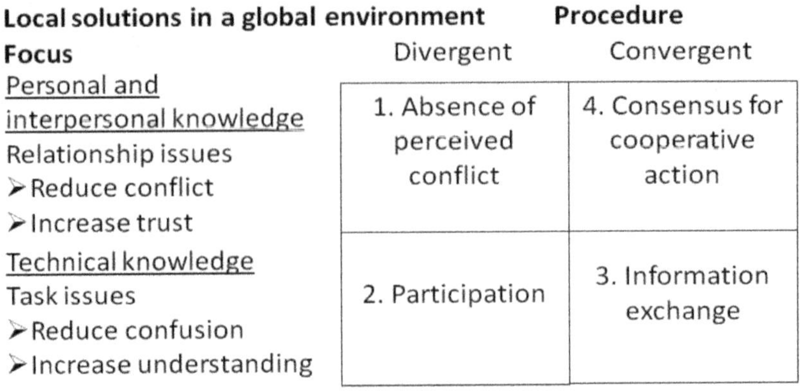

Local solutions in a global environment	Procedure	
Focus	Divergent	Convergent
Personal and <u>interpersonal knowledge</u> Relationship issues ➤Reduce conflict ➤Increase trust	1. Absence of perceived conflict	4. Consensus for cooperative action
<u>Technical knowledge</u> Task issues ➤Reduce confusion ➤Increase understanding	2. Participation	3. Information exchange

Figure 3 *Evaluation of participant's satisfaction with electronic discourse.*

3. METHODOLOGY

A multiple case study approach was adopted. The unit of analysis was a meeting (or series of meetings) facilitated by a leader in the domain of either science funding, economic development or regional planning. The facilitator was not part of the research team. The research team consisted of two academics and two assistants. The role of the research team was primarily one of data gathering and analysis. The data gathering techniques that were used included direct observation, interviews with the facilitator and his staff, interviews with meeting participants, analysis of meeting reports and computer files, and a questionnaire that was administered to participants at the end of their meeting.

All meetings were conducted in an electronic meeting facility at the University of Auckland. This facility, called the Decision Support Centre (DSC), consists of a large room containing 20 computers set out on an elongated table. In addition, the DSC contains a set of four large, moveable whiteboards for more traditional methods of recording the group's activities. The purpose of the computer facilities is to run Ventana Corporation's GroupSystems, a text-based electronic meeting support system (Sheffield & Gallupe, 1994; Fjermestad & Hiltz, 2001; Ackermann *et al.*, 2005). GroupSystems supports processes that include the anonymous and simultaneous individual generation of ideas and the prioritization and brief discussion of key findings (Van de Ven & Delbecq, 1971). GroupSystems also supports the anonymous and simultaneous individual allocation of budget amounts and the amalgamation and analysis of a group budget (Figure 4). In the following three sections the facilitation cases are reviewed.

Figure 4 *Electronic meeting technology.*

4. FACILITATING SCIENCE FUNDING

The clashing point of two subjects, two disciplines, two cultures of two galaxies, so far as that goes ought to produce creative chances.

(Snow 1959: 16)

There was such a huge diversity of people on the panel, from "pure research" oriented scientists to hard-headed business people, that significant political differences were inevitable. "(Electronic discourse) put the politics in a black box, to be dealt with later."

(Participant in a science funding meeting)

Bednarek (2011) analyses the strategizing process in New Zealand's science sector. She found that the context was pluralistic—the objectives of social actors were divergent and power was diffused. In this context institutions found legitimacy to be a powerful determinant of success. Legitimacy was found to comprise aspects which included the cognitive, normative/moral/regulative and sociopolitical. Organizations in New Zealand's science sector were characterized by multiple embedded tensions and complex diffused power structures. The author's analysis demonstrated both the creative potential and challenges in strategizing for legitimacy amidst pluralism.

The facilitation of aspects of science funding starts with the theoretical perspective that objective facts, societal norms, and personal values are intertwined. Objectivism, social constructionism and subjectivism are viewed as emergent perspectives in a broader and more critical discourse. The chief scientist of New Zealand, Sir Peter Gluckman, emphasizes that science is no longer linear, authoritative and definitive, provided only by a domain-specific expert. Rather science is increasingly characterized by complexity, where multiple perspectives on knowledge are required to address the asymmetric payoffs associated with various policy options (Gluckman, 2011).

The chief executive of New Zealand's Ministry of Research, Science and Technology (MORST) and staff spent four days in the Decision Support Centre at the University of Auckland (Figure 4) with the panel appointed to allocate the Public Good Science Fund. The panel distributed US(2012) $2Billion across all 40 areas of New Zealand science. This is by far the largest contestable fund in New Zealand and funding decisions directly or indirectly impact most of the New Zealand economy. The technical (cognitive) issues were complex—each of the twenty panel members had received approximately 1,000 pages of briefing papers. A group memory device would clearly be required to support deliberation. The personal and interpersonal (sociopolitical) issues were perhaps more difficult to ignore—many of the panel were scientists, and nobody wanted reductions in areas dear to them. The decision process was designed to reduce politics about divergent objectives to a manageable level, so that attention could be directed to the more technical, task-oriented aspects of the decision process.

One member of the panel was the chief executive of the New Zealand Trade Development Board, Rick Christie. He reported that electronic discourse "tends to be fairer—more objective—it draws on a different range of skills. But there's no question of not being heard—which can be a problem in meetings where there's just verbal interaction...If you are seeking ideas on something not identified with the contributor, then it's a great leveller..." Another member of the panel was John Butcher, director of the Forest Research Institute's Wood Technology Division. He reported that there was such a huge diversity of people on the panel, from "pure research" oriented scientists to hard-headed business people, that significant political differences were inevitable, and that "(electronic discourse) put the politics in a black box, to be dealt with later" (Sheffield 1993).

Quantitative evidence on the efficiency and effectiveness of facilitating science funding was obtained via a survey instrument (Appendix). The instrument was administered to all participants at the end of the final day of the electronically-supported meetings. Participants' satisfaction with electronic discourse averaged 5.9 on a 7 point scale (1 = low satisfaction, 7 = high satisfaction). Participants were satisfied with the focus on personal and interpersonal knowledge and the management of relationship

Science funding Focus	Procedure	
	Divergent	Convergent
Personal and interpersonal knowledge Relationship issues ➢Reduce conflict ➢Increase trust	1. Absence of perceived conflict 6.1	4. Consensus for cooperative action 6.0
Technical knowledge Task issues ➢Reduce confusion ➢Increase understanding	2. Participation 5.9	3. Information exchange 5.8

Figure 5 *Science funding. Participants' satisfaction with electronic discourse averaged 5.9 (1 = Low satisfaction; 7 = High satisfaction).*

issues—absence of perceived conflict (6.1) and consensus for cooperative action (6.0) received the highest ratings. Participants were also satisfied with the focus on technical knowledge—ratings for participation (5.9) and information exchange (5.8) were also high (Figure 5).

5. FACILITATING ECONOMIC DEVELOPMENT

Sheffield and Gallupe (1994, 1995) describe an application of electronic meeting technology to a series of economic policy-making meetings sponsored by the New Zealand Trade Development Board. The meetings were part of a national study aiming to upgrade New Zealand's competitive position in global markets. They were held in Auckland, the main economic region of New Zealand, and were branded 'Advantage Auckland'. The aim of the research was to determine if electronic meeting technology could support an economic development process where participants came from a variety of backgrounds (e.g., business competitors, different ethnic groups) and where meeting urgency and efficiency were of prime importance.

The national study was implemented with the assistance of Harvard's Michael Porter and was framed by his book The Competitive Advantage of Nations (Porter 1990). It started with the application of Porter's Diamond Model of industry-based competitiveness to analyze the New Zealand economy and to develop recommendations for improvement. Case studies were completed on 20 economic sectors which in total comprised 85 percent of New Zealand's exports. The results were published in an influential book entitled Upgrading New Zealand's Competitive Advantage (Crocombe, Enright & Porter, 1991). It was intended to serve as a basis for positive action by individuals, companies, unions, industry groups, and government. It sought to explain why New Zealand needed:

- A new, more comprehensive economic framework
- A fundamental re-engineering of attitudes, strategies and institutions
- Systematic upgrading of sources of competitive advantage.

At the time of the study, however, the New Zealand economy was in recession. Most businesses were dependent on the shrinking local market and as a consequence faced severe competition on price and high levels of business failure. Growth in export earnings became the primary goal of government economic policy. Cooperative efforts to upgrade competitive advantage were urgently required—yet were expected to be difficult to arrange.

The Advantage Auckland meetings had four key objectives:

1. To involve a large number of business leaders with a variety of backgrounds in sector and enterprise planning;
2. To assist those who were business competitors to move beyond price completion in local markets and seek opportunities for joint action to upgrade industry competitiveness in world markets;
3. To develop business opportunities for ethnic groups such as Maori who were suffering from high rates of unemployment;
4. To develop a collaborative action plan containing five initiatives that the meeting participants were committed to implement.

The final design of the meetings reflected the assumptions of the research team and facilitator:

- That some participants would require 'unfreezing' from their initial viewpoints (Lewin, 1947; Schein, 1993);
- That anonymous brainstorming on carefully selected topics would build opportunities for collaborative action although brief oral discussions would be required for agreement on key ideas;
- That building commitment to implement the action plans was primarily a social process that could best be supported in a rich communication medium (Daft and Lengel, 1986; Sheffield, 1995a).

There were five stages in each meeting. The purpose was to obtain working agreement on: meeting objectives, industry competitive advantages and disadvantages, actions to enhance competitive advantage, detailed action plans, and commitment to implementation. Earlier stages featured anonymous brainstorming within a strong organizing structure. In the last two stages, structure was not imposed—it emerged largely from the direct face-to-face interaction of the participants. In these stages the facilitator served primarily as coach and the electronic support served primarily as a memory aid. The design and evaluation of meeting discourse reflected elements of the task (Porter's Diamond Model) and four recommendations for 'unfreezing' (Lewin, 1947):

1. Participants feel psychologically safe;
2. Participants step outside existing cultural norms;
3. Participants (especially the leaders) learn something new;
4. A formal change process is implemented.

A series of 12 meetings were attended by 250 business leaders with a variety of backgrounds (Sheffield & Gallupe, 1994). The primary result for each participant from their meeting was a 50- to 80-page bound transcript. Quantitative evidence about meeting effectiveness and participant satisfaction was obtained via a survey instrument administered at the end of each meeting. The results of the questionnaire (Appendix) indicated that participants felt that the meetings were both very effective and efficient. Answers to questionnaire item 1 indicated that participants felt that if the meetings were held using conventional meeting support each would have taken three times as long. Average effectiveness (measured via the average of items 3b-24) was 6.1 (1=Low satisfaction; 7=High satisfaction). Participants felt that the way the session was run by the facilitator was excellent (6.3) and the technology was very easy and fun to use (6.3).

Participants' satisfaction with electronic discourse was measured via four measures that are numbered so as to match the four recommendations for unfreezing:

1. Absence of perceived conflict;
2. Participation;
3. Information exchange;
4. Consensus for cooperative action.

As demonstrated in Figure 6, these measures of the meeting process are conceptually related to procedure (either divergent or convergent) and focus (either relationship or task). For the 12 Advantage Auckland meetings, the average of these four measures was 6.1 (Figure 6) (1=Low satisfaction; 7=High satisfaction).

The follow-up study two years after the meetings revealed that the success of the action plans varied considerably. Some were discontinued within months. Others such as the Marine Exporters Group (Marex) remain in existence and have become central to their industries. The most successful action plans were those in industries where previous meetings had been marked by dysfunctional conflict. Individuals in these meetings collectively possessed resources which, when shared and focused in the absence of perceived conflict, were sufficient to support successful initiatives. Subsequently a further 58 meetings were held in Auckland that were attended by approximately 1,000 business leaders. The Advantage Auckland meetings led directly to the establishment of a group support facility at Victoria University in the capital city of Wellington. The Wellington facility has supported many campaigns, most of which are sponsored by national government, some with the goal of upgrading New Zealand's competitive position in global markets.

6. FACILITATING REGIONAL PLANNING

6.1 Introduction

At the time of this research study, the governance of the Auckland region was characterized by divergent objectives (politics) and diffuse power structures (decentralised governance) (Healey 1997). Planners from seven territorial authorities met on occasion with the planning team from the regional council to develop comprehensive urban growth plans. They negotiated shared meaning about facts (at-

Economic development	Procedure	
Focus	Divergent	Convergent
Personal and interpersonal knowledge Relationship issues ➤ Reduce conflict ➤ Increase trust	1. Absence of perceived conflict 6.4	4. Consensus for cooperative action 6.2
Technical knowledge Task issues ➤ Reduce confusion ➤ Increase understanding	2. Participation 5.9	3. Information exchange 5.7

Figure 6 *Economic development. Participants' satisfaction with electronic discourse averaged 6.1 (1=Low satisfaction; 7=High satisfaction).*

tributes of Auckland), norms (mutual expectations), and personal commitments (to one's own visions—and how they should be funded). Comprehensive scenarios for rival strategies were iteratively developed and evaluated throughout lengthy planning cycles. The process was complex and politics, confusion, and conflict were accepted as the norm.

Political differences in the Auckland region had been exacerbated by a combination of limited resources and population growth from internal and external migration. The politics around transportation were particularly difficult. Trip times were increasing and transportation costs, which included lost productivity, were increasing. While transportation modelling had been extensively used, issues of governance, funding, and collaborative planning remained. In the absence of a robust and responsive governance structure, deliberations about managing population growth were marked by political differences (Royal Commission on Auckland Governance, 2007; New Zealand Council for Infrastructure Development, 2008).

Confusion arose from the limited role of a single decision maker and the complexity of the substantive factual issues. For example, multiple organizations were involved in transportation governance—their roles were specialised and included control, participation, planning, funding, and operation/management. While each organization managed part of the transport system, none was responsible for the system as a whole. Region-wide or comprehensive urban planning necessitated a critical evaluation of conflicting claims about intertwined criteria related to transportation, housing, workplaces, amenities, etc, by individuals primarily situated within organizations with divergent objectives. To a greater or lesser extent, all social actors suffered from confusion.

Conflict arose from the complexity of the power relationships among decision makers. Local Government legislation conferred powers on the regional council to plan for the region 'in consultation with' territorial authorities. Each authority maintained a planning office responsible to its own council. Each was empowered to serve its own constituency and expected the comprehensive urban plan to serve its own interest. To a greater or lesser extent, all social actors were embroiled in power conflicts.

In Table 2 overall success required participants to develop and integrate perspectives from generic roles that Churchman terms system designer (more technical/task oriented), decision maker (more interpersonal/consensus oriented) and client (more subjective/value oriented). In the regional planning meeting, each participant was primarily a designer of an urban area for which the elected council was the decision maker, and those who lived in the area were clients (Churchman, 1971: 200).

The current research explores the practical value of electronic discourse in regional governance and comprehensive urban planning (Tables 1 and 2, Figure 2). Because of the complexity of the issues, and the importance of power relations, and the emergent nature of their interactions, and the historical context a non-positivist method of inquiry was adopted. The aim was to describe the general nature of the phenomena observed and to interpret actions, events, and consequences. The evolution of quality measures (validity claims) during the pre-meeting, meeting, and post-meeting phases of decision making was observed. Data was gathered before, during, and after an electronically-supported meeting.

The purpose of the facilitated electronically-supported meeting was the strategic evaluation of a comprehensive 30-year plan for the Auckland region. This plan, known as the Auckland Strategic Planning Model, had been constructed over a seven-year period. The plan described two strategies for an increase in population from 1m to 1.5m. Consolidation drove strategy one. More controls, particularly environmental controls, would be imposed to limit the spread of population into rural areas. The result would be higher population density and increased use of passenger transportation (buses, light rail). Expansion drove strategy two. Planning controls would be relaxed, allowing the spread of population into rural areas. The result would be lower population density and increased use of private transport (cars, freeways) (Sheffield, 2009b).

In summary regional planning in Auckland, New Zealand was subject to political differences, confusion, and conflict. Regional planning was informed not by a search for a purely technical solution but by communication within a diffuse power structure about divergent objectives. Inter-organizational planning meetings were the exercise of technical skills on behalf of constituencies with a history of conflict, confusion, and the exercise of power. An open dialogue across planning organizations was required to resolve contradictions among competing perspectives. Facilitating such a dialogue presents conceptual and practical difficulties that motivated the research reviewed below.

6.2 Before the Meeting

The evidence gathered in the pre-meeting phase revealed that the 16 participants in the electronically supported regional planning meeting were there to represent seven territorial authorities (four cities and three districts) and the Auckland Regional Council (ARC). Each was a professional planner responsible for advising his/her own (elected) council. Each territorial authority constituted one part of the whole of the Auckland region. The issues associated with embedding 'one part' of an urban region in 'the whole' were complex. The chief planner for the ARC advised that most participants had been involved in prior consultations marked to some degree by politics, confusion, and conflict. Participants recognized the difficulties in achieving the goals of their respective councils and engaging in consultations about comprehensive region-wide plans with planners from other councils. Perceptions of costs and benefits varied with the allegiance of the participant and the history of his or her interactions. As the day of the focal electronically-supported meeting approached it became apparent that considerable difficulties were being experienced by ARC planners, and that these were directly related to unresolved technical, interpersonal and personal issues.

Technical perspective. Technical difficulties were encountered in discovering an analytically sound method of combining knowledge from the acknowledged experts. Urban planning is a pluralistic area that Banville & Landry (1989) would describe as 'lacking conceptual integration'. For example, traffic engineers focused on access and transportation and developed estimates of trip times under each strategy. Biologists studied coastal water quality and developed estimates of pollutants in parts per million. Financial analysts focusing on economic values developed quantitative estimates of costs. Other planning consultants developed qualitative assessments of amenity, landscape values and housing choice. Scientific methods were applied by the experts who developed sub-models in sub-disciplines embedded within urban planning. Yet measures such as trip times, pollutants and implementation costs were, by themselves, conceptually unrelated and could not rigorously be compared. Claims to objective truth were diminished by the lack of an analytically sound method of combining knowledge from different sub-specialties.

Interpersonal perspective. The traditional urban planning triple-bottom-line categories of economic, social and environmental concerns appeared to be interlinked in a way that made the separate evaluation of any one category or subcategory impossible. It became clear that there were complex, dynamic and recursive ('chicken and egg') or self-referential (Müller et al., 2005) interdependencies among stakeholder's beliefs, potentially right strategies and available objective facts. These emergent properties of regional planning could only be resolved by discourse.

Personal perspective. The third set of problems was associated with personal commitments. Planners from one major territorial authority (a city of 300,000) were reluctant to attend because they were committed to a city plan based on presuppositions that differed from those of the regional council.

Summary. Analysis from the perspective of pluralism and communicative action (Table 2, Figure 2) provides qualitative evidence suggesting that the observed levels of guarantors (objective truth, rightness and sincerity) immediately before the focal electronically-supported meeting were low.

6.3 During the Meeting

To evaluate rival strategies for the Auckland region the facilitator of the focal electronically-supported meeting chose to apply the five facilitation principles (Table 2) and framework (Figure 2). The first part of the meeting focused on the expression of concerns and issues motivating each stakeholder. The last part of the meeting focused on expressions of degrees of commitment to action, for and against, rival strategies. More than half of the agenda items were devoted to electronically-supported discourse about a decision matrix. Two strategies (columns) were evaluated against five classes of criteria (rows)—cost, amenity and landscape, housing choice, access and transportation, and water quality. Each row of the decision matrix was the subject of a 50-minute session that included the anonymous individual generation of ideas and the prioritization and brief discussion of key findings (Sheffield, 2004). This 50-minute session included the private ordering by each participant of his or her preference for each strategy (Dias & Climaco, 2005). In the following subsections evidence is presented about participant satisfaction with electronic discourse and claims to emergent personal, interpersonal and technical knowledge.

Participant satisfaction with electronic discourse. Participants' satisfaction with electronic discourse averaged 6.0 on a 7 point scale (1 = low satisfaction, 7 = high satisfaction) (Figure 7). Participants (some of who were initially unwilling to attend the meeting) were particularly satisfied with participation (6.2) and the management of relationship issues - absence of perceived conflict (6.1) and consensus for cooperative action (6.1) also received high ratings. The relatively lower rating for information exchange (5.5) reflects most participants' familiarity with the issues. Unstructured comments were collected anonymously from participants by means of the GroupSystems software. The responses were overwhelmingly positive. Participants remarked that the meeting generated intense participation, goodwill and momentum. Many people expressed surprise that the technology existed and stated that the meeting outcomes would not have been possible without electronic support.

Evaluation of claims to objective truth. Through the use of the electronic meetings technology participants produced ten pages of text on each of the five criteria (Sheffield, 2004). This text or 'frozen discourse' includes key issues that were prioritized

Regional planning	Procedure	
Focus	Divergent	Convergent
Personal and interpersonal knowledge Relationship issues ➢Reduce conflict ➢Increase trust	1. Absence of perceived conflict 6.1	4. Consensus for cooperative action 6.1
Technical knowledge Task issues ➢Reduce confusion ➢Increase understanding	2. Participation 6.2	3. Information exchange 5.5

Figure 7 *Regional planning. Participants' satisfaction with electronic discourse averaged 6.0 (1 = Low satisfaction; 7 = High satisfaction).*

via a weighted voting procedure (Van de Ven & Delbecq, 1971). Participants cast a total of 240 votes for each criterion. The key issues were expressed in a manner that was exploratory rather than evaluative. For example, the issue of the extent to which population density must increase to make public transportation sufficiently viable is central to the choice between strategy one (consolidation) and strategy two (expansion). Yet at the end of a seven-year planning exercise that included extensive traffic modeling, the issue was raised as a question rather than as the evaluation of a factual proposition supported by expert analysis. This supports the conclusion that under the norms of a cognitive, objectivating attitude towards the facts, the 'truth' was that neither strategy was superior.

Evaluation of claims to rightness. At the end of the discourse on a criterion, each participant privately recorded how well each strategy performed against the five criteria in Table 1. This enabled participants to interpret technical findings from the perspective of their own organization's norms and values. Each of the 16 participants anonymously rated the two strategies on each of the 5 criterion. The aggregated ratings for each strategy and criterion were made accessible to each participant. On one criterion (housing choice), strategy one and strategy two were rated equally. On the remaining four criteria (cost, amenity and landscape, access and transportation, and water quality) strategy one performed distinctly better than strategy two. The strategies and criteria had been developed through a consultative process over a seven-year period. This supports the conclusion that under the norms of established legitimate inter-personal relationships, strategy one is more 'right' than strategy two.

Evaluation of claims to sincerity. Electronic meeting technology supported sharing personal visions prompted by the question 'What is it like to live in Auckland under strategies 1 and 2?' From the perspective of Churchman's inquiring system, participants were being asked to drop their usual role of designer and adopt the role of client (Churchman, 1971: 200). The goal was disclosure of speaker's subjectivity, unconstrained by the (technical) structure of the model and unrestrained by the interpersonal context. The strategy was to get each individual to: (i) write a personalized account of what it would be like to live in Auckland 30 years hence under each of strategies 1 and 2; (ii) read the accounts of others to identify the most valuable visions. The procedure was a 60-minute silent envisioning exercise in which each account was identified only by a code. Anonymity was almost complete. The most valued visions of what it would be like to live in Auckland 30 years hence showed intense personal support for strategy one, and a willingness to work against strategy two. This supports the conclusion that under the norms of disclosure of speakers' subjectivity, 14 of the 16 participants would, in all sincerity, only have supported strategy one.

Summary. The positive results obtained from the meeting are in strong contrast to the confusion and conflict that existed at the end of the pre-meeting phase. While some participants had been reluctant to attend the focal meeting, and expressed negative views at the beginning of the meeting, all participants provided positive evaluations at the end of the meeting. The functionality of the electronic meeting technology was supportive of an overall positive result. Participation by all participants was intense. By the end of the meeting, electronic discourse produced 80 pages of text. Intense participation in electronic discourse resulted in extensive documentation of claims to objective truth, rightness, and sincerity. The data gathered during the focal meeting support the claim that electronic discourse had successfully reduced conflict and confusion. It is not clear, however, that the decision outcomes integrated the technical, inter-personal, and personal perspectives into a consensus model that provided a rationale for action.

6.4 After the Meeting

We have yet to consider the degree of coherence among the three perspectives. Participants found no difference between the strategies on the basis of technical knowledge. Moderate claims in favour of strategy one were made based on interpersonal knowledge. Strong claims in favour of strategy one were made based on personal knowledge.

The degree of coherence among the decision outcomes at different levels was poor. There was a major discrepancy in preferences at various stages of the decision process. The 80-page report generated by electronic meeting technology (from which the findings were extracted) was circulated to all participants immediately after the meeting. The introductory section of the report highlighted the fact that the participants were strongly supportive of a strategy that lacked factual support. The report became subject to intense scrutiny. Regional planners met repeatedly among themselves about the report and consulted other meeting participants. Support grew for the interpretation that the strategic options were not extreme enough. In Hegelian terms, the dialectical logic (synthesis) of this interpretation was initially lost on the regional planners because they were so firmly wedded to their decision framework (thesis) that they experienced profound difficulty in recognizing that the framework was flawed (antithesis). An abbreviated planning round was subsequently undertaken with more extreme versions of strategies one and two (based on a 100 % increase in population). Support that integrated the technical, interpersonal and personal levels of the facilitation framework was then found for strategy one.

6.5 Summary of Findings

The results showed that the pre-meeting phase was fraught with technical, interpersonal and personal problems. Both the observations during the meeting and the satisfaction reported by participants (Figure 7) demonstrated that the facilitated electronically-supported meeting had increased participant's trust and understanding. During the meeting participants found no difference between the strategies on the basis of technical knowledge, a moderate preference for scenario one on the basis of interpersonal knowledge, and a strong preference for scenario one on the basis of personal knowledge. Reflection after the meeting produced sudden insights that dissolved the perceived lack of coherence. The final analysis integrated technical, inter-personal, and personal perspectives into a consensus model that provided a rationale for action. Empirical evidence was therefore found for the importance of the facilitation framework (Figure 2) and all five principles (Table 2).

7. DISCUSSION

The meeting made it easy to lay your thoughts out without putting your neck on the line.

(Participant in an economic development meeting)

The current research described local solutions implemented as part of New Zealand response to impacts of globalization. Inter-organizational meetings were conducted in the domains of science funding, economic development and regional planning. The importance of pluralism and electronic discourse to the successful facilitation of these meetings was evaluated via quantitative and qualitative measures. Evidence from the quantitative measures indicated that participants found the meetings very efficient and effective and were very satisfied with electronic discourse. Averages across all three cases are reported in Figure 8. Evidence from the qualitative measures indicated that the facilitation principles (Table 2) and framework (Figure 2) were closely associated with overall success. These findings are briefly discussed.

The strategies implemented were developed in inter-organizational meetings attended by a large number of stakeholders with divergent objectives. Because each participant was very busy meeting the demands of their own organization it was imperative that the inter-organizational meetings were efficient and effective. In traditional inter-organizational meetings, even when participants desire to work in a relatively democratic way, the limited airtime creates conflict. In a one hour meeting of 15 people, each must compete to get more than four minutes of airtime. Quite literally it is the sender not the message that is visible. Critical analysis invites interpersonal conflict. But in an electronic meeting all participants can input and read information at the same time (Sheffield, 1995b).

Because everyone can 'talk' at once and still be heard the work was completed two to three times faster. Because it was difficult to identify who has proposed a particular idea, rank and personality differences among participants were less pronounced. Advocacy, coalitions and infighting were less necessary. According to participants, facilitated electronically-supported meetings provided an efficient and effective method of generating informed consensus for action (Figure 8).

The quantitative evidence indicated that participants were particularly satisfied with the focus on personal and interpersonal knowledge and the management of re-

All three cases	Procedure	
Focus	Divergent	Convergent
Personal and interpersonal knowledge Relationship issues ➢Reduce conflict ➢Increase trust	1. Absence of perceived conflict 6.2	4. Consensus for cooperative action 6.1
Technical knowledge Task issues ➢Reduce confusion ➢Increase understanding	2. Participation 6.0	3. Information exchange 5.7

Figure 8 *All three cases. Participants' satisfaction with electronic discourse averaged 6.0 (1 = Low satisfaction; 7 = High satisfaction).*

lationship issues—across all three cases absence of perceived conflict (6.2) and consensus for cooperative action (6.1) received the highest ratings. Participants were also satisfied with the focus on technical knowledge—ratings for participation (6.0) and information exchange (5.7) were also high. This suggests that the anonymity provided by electronic meeting technology was perceived as more important than the raw power associated with the simultaneous use of keyboards. This was particularly apparent in the economic development meetings.

In the 12 Advantage Auckland economic development meetings the absence of perceived conflict (6.4) and consensus for cooperative action (6.2) received the highest ratings. The electronically supported meetings were held when the economy was in recession. Because the level of pain was high and some participants were business competitors the potential for conflict was high. In many industry sectors diminished disposable income and deregulation had led to oversupply, competition on price, heavy discounting, and persistent infighting. Participants indicated that the meeting created a dialogue, and the exchange of valuable information fostered openness and trust. Interviews conducted one to two years afterwards as part of a follow-up study (Sheffield & Gallupe, 1995) confirmed that the meetings had been a catalyst for industry wide change. Participants commented that the anonymous and simultaneous use of the keyboards aided creativity and allowed everybody's comments to be treated fairly.

Our ideas were stimulated, shared and focused.

Domination by individuals whose solutions were not of great quality had often destroyed meetings in the past. Anonymity was essential to get rid of personality clashes. The (electronically-supported) meeting was memorable for the variety of participants, its quietness and structure—nobody dominated. It delivered an action plan that was solid enough to cope with the infighting.

Before the meeting a lot of us didn't believe in talking to the opposition. There's a lot more talking together, pulling together now.

The meeting was definitely the catalyst. Absolutely! Why? Because the computer medium allowed people to feel that their contributions were being treated fairly.

The meeting made it easy to lay your thoughts out without putting your neck on the line.

Empirical support was found for the facilitation framework and all five principles (Table 2, Figure 2). This suggests that, in facilitating local solutions in a global environment, the benefits of electronic discourse are three-fold:

Technical perspective: Electronic discourse provided support for the development and documentation of validity claims about objective truth, rightness and sincerity, and the degree of coherence among them.

Interpersonal perspective: Electronic discourse provided support for discourse that interweaves evidence (experience and reflection, decision and action, theory and practice, individual feeling and objective fact) from multiple, intertwined, conflicting yet mutually supportive evaluative frames.

Personal perspective: Electronic discourse provided support for the 'psychological safety' and 'trust' needed for direct and unreserved expressions of multiple, conflicting individual perspectives.

In totality, the empirical evidence enables the focus question 'Does electronic discourse increase the success of local solutions in a global environment?' to be answered in the affirmative.

8. CONCLUSION

Several lessons have been learned. Firstly, facilitating local solutions in a global environment was a pluralistic endeavour—the objectives of social actors were divergent and power was diffused. Often the goal was a legitimate consensus among diverse stakeholders so that scarce resources could be combined/leveraged for national advantage. Secondly, the theoretical perspective of communicative action was useful in separating out intertwined but quite different types of knowledge. The standard of excellence in communicative action can be stated as follows: personal commitment (validated by sincerity) to an interpersonal consensus (validated by rightness) for technical excellence (validated by objective truth). Thirdly, individual and institutional knowledge was inherently mediated and situated, provisional and pragmatic, aspirational and contested. In an environment of diffuse power relationships, inter-organizational meetings were essential in gaining legitimacy. Fourthly, electronic meeting technology has a raw power that leads to efficient and effective inter-organizational meetings. Excellent performance was observed in the application of electronic meeting technology in science funding, economic development, and regional planning meetings. Fifthly, the findings reported in the current research suggested that the facilitation principles and framework developed in this article may be routinely applied in various other domains. Seen from a Hegelian perspective, the power of pluralism and communicative action lies not in achievement of enlightenment, but in appreciation of the nature of three types of ignorance and the practical consequences of belief.

REFERENCES

Ackermann, F., Franco, L. A., Gallupe, B., and Parent, M. 2005. GSS for multi-organizational collaboration: Reflections on process and content, Group Decision and Negotiation 14(4): 307–331.

Banville, C. & Landry, M. 1989. Can the field of MIS be disciplined? Communications of the ACM 32(1): 48–60.

Bednarek, R. S. 2011. Strategizing for legitimacy in pluralistic contexts: New Zealand's Science Sector, unpublished PhD, Victoria University Wellington.

Cheshire, P. 2012. Cities: the need to understand better before we mess with them. Presentation held at New Zealand Government—Auckland Policy Office on 14 December 2012.

Churchman, C.W. 1971. The Design of Inquiring Systems. New York: Basic Books.

Crocombe, G. T., Enright, M. J., and Porter, M. E. 1991. Upgrading New Zealand's Competitive Advantage. Auckland: Oxford University Press.

Daft, R. L., and Lengel, R. H. 1986. Organizational information requirements, media richness and structural design. Management Science 32(5): 554–571.

Denis, J., Langley, A., and Rouleau, L. 2007. Strategizing in pluralistic contexts: rethinking theoretical frames. Human Relations 60(1): 179–215.

Dias, L., & Climaco, J. 2005. Dealing with imprecise information in group multicriteria decisions: A methodology and a GDSS architecture. European Journal of Operational Research 160(2): 291–307.

Fjermestad, J., and Hiltz, S. R. 2001. Group Support Systems: A descriptive evaluation of case and field studies. Journal of Management Information Systems 17(3): 115–159.

Franco, L. A. 2008. Facilitating collaboration with problem structuring methods: A case study of an inter-organizational construction partnership. Group Decision and Negotiation 17(4): 267–286.

Gluckman, P. 2011. Towards better use of evidence in policy formation: a discussion paper, Office of the Prime minister's Science Advisory Committee. URL: http://www.pmcsa.org.nz/ publications/

Habermas, J. 1984. The Theory of Communicative Action Volume 1: Reason and the Rationalization of Society. Boston: Beacon Press.

Healey, P. 1997. Collaborative Planning: Shaping Places in Fragmented Societies. New York: Palgrave.

Jarzabkowski, P., and Fenton, E. 2006. Strategizing and organizing in pluralistic contexts. Long Range Planning 39: 631–648.

Kolfschoten, G. L., Den Hengst-Bruggeling, M., and De Vreede, G. J. 2007. Issues in the design of facilitated collaboration processes. Group Decision and Negotiation 16(4): 347–361.

Krugman, P. 1997. The Age Of Diminished Expectations. Third Edition, Boston: The MIT Press.

Lehrer, J. 2009. The Decisive Moment: How the Brain Makes Up Its Mind. Melbourne: The Text Publishing Company.

Lewin, K. 1947. Frontiers of group dynamics: concept, method and reality in social science. Human Relations 1(11): 5–41.

Lewis, F. L., Bajwa, D. S., Pervan, G., King, V. L. K., and Munkvold, B. E. 2007. A cross-regional exploration of barriers to the adoption and use of electronic meeting systems. Group Decision and Negotiation 16(4): 381–398.

McCann, P. 2009. Economic Geography, Globalization, and New Zealand's Productivity Paradox. New Zealand Economic Papers 43(3): 279–314.

Midgley, G., and Pinzo´, L. A. 2011. Boundary critique and its implications for conflict prevention. Journal of the Operational Research Society 62: 1543–1554.

Millet, I., and Gogan, J. 2006. A dialectical framework for problem structuring and information technology. Journal of the Operational Research Society 57: 434–442.

Mitroff, I. I., and Linstone, H. A. 1993. The Unbounded Mind: Breaking the Chains of Traditional Business Thinking, New York: Oxford University Press.

Morton, A., Ackermann, F. and Belton, V. 2003. Technology driven and model driven approaches to group decision support: focus, research philosophy and key concepts, European Journal of Information Systems, 12(2), 110-126.

Müller, D. B., Tjallingii, S. P., and Cantor, K. J. 2005. A transdisciplinary learning approach to foster convergence of design, science and deliberation in urban and regional planning. Systems Research and Behavioral Science 22(3): 193–209.

New Zealand Council for Infrastructure Development 2008. Strengthening Auckland Governance, a submission to the Royal Commission on Auckland Governance.

New Zealand Government. 2009. International Connections and Productivity: Making Globalization Work for New Zealand. New Zealand Treasury Productivity Paper 09/01.

Porter, M. E. 1990. The Competitive Advantage of Nations. New York: The Free Press.

Royal Commission on Auckland Governance 2007. Downloaded 10 December 2012 from http://www.royalcommission.govt.nz

Schein, E. H. 1993. How can organizations learn faster? the challenge of entering the green room. Sloan Management Review, Winter 1993: 85–93.

Schultze, U., and Leidner, D. E. 2002. Studying knowledge management in information systems research: Discourses and theoretical assumptions. MIS Quarterly 26(3): 213–242.

Shakun, M. F. 2003. Right problem solving: doing the right thing right. Group Decision and Negotiation 12(6): 463–476.

Shaw, D., Ackermann, F., and Eden. C. 2003. Approaches to sharing knowledge in group problem structuring. Journal of the Operational Research Society 54(9): 936–948.

Sheffield, J. 1993. Silent decision making. The University of Auckland Alumni News 3(1): 16–17.

Sheffield, J. 1995a. The effect of communication medium on negotiation performance. Group Decision and Negotiation 4(2): 159–179.

Sheffield, J. 1995b. Using electronic meeting technologies. People and Performance 3(3): 18–21.

Sheffield, J. 2004. The design of GSS-enabled interventions: a Habermasian Perspective. Group Decision and Negotiation 13(5): 415–436.

Sheffield, J. 2005. Systemic knowledge and the V-model. International Journal of Business Information Systems 1(1/2): 83–101.

Sheffield, J. 2008a. Does health care for systemic development? Systems Research and Behavioral Science 25(2): 283–290.

Sheffield, J. 2008b. Inquiry in health knowledge management. Journal of Knowledge Management 12(4): 160–172.

Sheffield, J. 2009a. Systemic Development: Local Solutions in a Global Environment, Goodyear, Arizona: ISCE Publishing. 2009b. Towards a design theory for collaborative technologies: Habermasian analysis of comprehensive urban planning, Proceedings of the 42nd Annual Hawaii International Conference on System Sciences, 5th—8th January. IEEE Computer Society Press.

Sheffield, J. 2012. My Decisive Moment. Auckland: Pagination Publishers.

Sheffield, J., and Gallupe, B. 1994. Using electronic meeting technology to support economic policy development in New Zealand: short term results. Journal of Management Information Systems. Winter 1993–94. Vol. 10. No 3: 97–116.

Sheffield, J., and Gallupe, B. 1995. Using group support systems to improve the New Zealand economy, part II: follow-up results. Journal of Management Information Systems 11(3): 135–153.

Sheffield, J., and Guo, Z. 2007a. Critical heuristics: a contribution to addressing the vexed question of so-called knowledge management. Systems Research and Behavioral Science 24(6): 613–626.

Sheffield, J., and Guo, Z. 2007b. Ethical inquiry in knowledge management. International Journal of Applied Systemic Studies 1(1): 68–81.

Snow, C. P. 1959. The Two Cultures. London: Cambridge University Press.

Snowden, D. J., and Boone, M. 2007. 'A leader's framework for decision making'. Harvard Business Review, November 2007, pp. 69–76.

Van de Ven, A. H., and Delbecq, A. L. 1971. The effectiveness of Nominal, Delphi, and interacting group decision making processes. Academy of Management Journal 17: 605–621.

Weick, Karl E. 1979. The social psychology of organizing. New York: Random House.

APPENDIX

Decision Support Centre session for _____ (group) on _____ (date)

*Efficiency (Q1-2), effectiveness (Q3a-5), facilitator (Q6-7), technology (Q8-11), reduced barriers to communication (Q12-14), participation (Q15-17), information exchange (Q18-21), meeting outcomes (Q22-24)

Directions: Your opinions are important to us! Please take the time to answer the questions on the front of this sheet. We will use your responses to this questionnaire to upgrade future workshops in the Decision Support Centre. Thank you! Jim Sheffield, Research Director, Decision Support Centre.

1.   You spent ____ hours in the Decision Support Centre to achieve this result. How many hours would you expect to spend to achieve the same result by conventional means? ____ hours

2.   Using conventional means the process would most likely have spread over _____ days

3a.   In the next three months I expect to use/study the report of this session for a total of _____ hours

For questions 3b through 24 indicate your level of agreement with the statement using the following scheme:

(1)	(2)	(3)	(4)	(5)	(6)	(7)
Strongly Disagree	Mostly Disagree	Somewhat Disagree	Neutral	Somewhat Agree	Mostly Agree	Strongly Agree

All questions are answered by circling a number. There are no right or wrong answers.

3b. Overall, I thought the workshop was excellent: 1 2 3 4 5 6 7

4. I enjoyed being a member of this group: 1 2 3 4 5 6 7

5. The report containing all contributions to this session will be highly valuable: 1 2 3 4 5 6 7

6. The way the session was run by the facilitator was excellent: 1 2 3 4 5 6 7

7. The facilitator's use of the whiteboards was highly effective:

                                                                              1 2 3 4 5 6 7

8. The computer facilities were easy to use: 1 2 3 4 5 6 7

9. The computer facilities were highly effective: 1 2 3 4 5 6 7

10. Typing enabled me to focus and refine my ideas before going public:

                                                                              1 2 3 4 5 6 7

11. The Decision Support Centre technology is fun to use:

                                                                              1 2 3 4 5 6 7

12. Internal politics were largely absent from today's meeting:

                                                                              1 2 3 4 5 6 7

13. The rank of participants did not inhibit the free flow of ideas:

                                                                              1 2 3 4 5 6 7

14. The personality of participants did not inhibit the free flow of ideas:

                                                                              1 2 3 4 5 6 7

15. I felt actively involved throughout the session: 1 2 3 4 5 6 7

16. All group members participated equally: 1 2 3 4 5 6 7

17. Participants, both as individuals and as a group, were creative:

                                                                              1 2 3 4 5 6 7

18. I was willing to give valuable information to others in the group:

                                                                              1 2 3 4 5 6 7

19. I was able to give valuable information to others in the group:

                                                                              1 2 3 4 5 6 7

20. I received valuable ideas from others on issues of significance to me:

                                                                              1 2 3 4 5 6 7

21. I received support from others on issues of significance to me:

                                                                              1 2 3 4 5 6 7

22. The issues surfaced during the brainstorming are important:

                                                                              1 2 3 4 5 6 7

22b. I strongly recommend that this and similar groups use the Decision Support Centre for future planning tasks   1 2 3 4 5 6 7

23. The summary of key issues developed on the whiteboards are important:

                                                                              1 2 3 4 5 6 7

24. Participants, both as individuals and as a group, were productive:

                                                                              1 2 3 4 5 6 7

Quotable comment. Please quote me on the following comment:
Please use the back of the sheet for further comments.

Chapter 15

GLOBAL BIFURCATION: THE DECISION WINDOW

Ervin Laszlo

It has been said that our generation is the first in history that can decide whether it is the last in history. We need to add that our generation is also the first in history that can decide whether it will be the first generation of a new phase in history. We have reached a watershed in our social and cultural evolution. The sciences of systems tell us that when complex open systems, such as living organisms, and also ecologies and societies of organisms, approach a condition of critical instability, they face a moment of truth: they either transform, or break down.

Keywords: complex systems, scenario, bifurcation, global trends.

THE BAU (BUSINESS AS USUAL) SCENARIO

Humankind is approaching a critical instability—a global bifurcation. The following scenarios illustrate the nature of the choice at this critical point.

- There is no real change in the world in the way business is conducted, resources are exploited and energy is produced. This leads, on the one hand, to a worsening global economic crisis, and, on the other—to major climate change due to the accelerated warming of the Earth's atmosphere.

- In some regions global warming produces drought, in others devastating storms, and in many areas it leads to harvest failures. In coastal areas vast tracts of productive land are flooded, together with cities, towns and villages. Hundreds of millions are homeless and face starvation.

- Massive waves of destitute migrants flow from coastal regions and areas afflicted with lack of food and water, above all in Africa, Asia, and Latin America, toward inland regions where the basic resources of life are more assured. The migrants overload the human and natural resources of the receiving countries and create conflict with the local populations. International relief efforts provide emergency supplies for thousands, but are helpless when confronted with millions.

- In futile attempts to stem the tidal wave of destitute people India builds up its wall along the border with Bangladesh, the U.S. along the Mexican border, and both Italy and Spain build walls to protect their northern regions from their overrun southern regions.

- The world's population fragments into states and populations intent on protecting themselves, and masses of desperate people facing imminent famine and disease. The conflicts create unsustainable stresses and strains in the structure of international relations. Social and economic integration groups and political alliances break apart. Relations break down between the U.S. and its southern neighbors, the European Union and the Mediterranean countries, and India and China and the hard-hit Southeast Asian states.

- Global military spending rises exponentially as governments attempt to protect their territories and reestablish a level of order. Strong-arm régimes come to power in the traditional hot-spots and local food- and water-wars erupt between states and populations pressed to the edge of physical survival.

- Terrorist groups, nuclear proliferators, narco-traffickers, and organized crime syndicates form alliances with unscrupulous entrepreneurs to sell arms, drugs, and essential goods at exorbitant prices. Governments target the terrorists and attack the countries suspected of harboring them, but more terrorists take the place of those that are rounded up and killed or imprisoned.

- Hawks and armaments lobbies press for the use of powerful weapons to defend the territories and interests of the better-off states. Regional wars fought initially with conventional arms escalate into wars conducted with weapons of mass destruction.

- The world's interdependent and critically destabilized economic, financial and political system collapses. The environment, its productive processes and vital heat balance impaired, is no longer capable of providing food and water for more than a fraction of the surviving populations. Chaos and violence engulfs peoples and countries both rich and poor.

Here, however, is another scenario.

The TT (Timely Transformation) Scenario

- The experience of terrorism and war, together with rising poverty and the threats posed by a changing climate, trigger a widespread recognition that the time to change has come. In country after country, an initially small but soon rapidly growing nucleus of people pull together to confront the dangers of the global crisis and seize the opportunity it offers for change.

- The rise of popular movements for sustainability and peace leads to the election of political leaders who support economic cooperation and social solidarity projects. Forward-looking states monitor the dangerous trends and provide financing for the urgently needed economic, ecological, and humanitarian initiatives.

- Non-governmental organizations link up to undertake projects to revitalize regions ravaged by environmental degradation. Emergency funds are provided for countries and regions afflicted by drought, violent storms, coastal flooding, and failures of the harvest.

- Military budgets are reduced and in some states eliminated, and the resulting 'peace-dividends' are assigned to increase the production of staple foods, safe water, basic supplies of energy, and essential sanitation and health services for the needy disadvantaged populations.

- Country after country shifts from fossil-fuel based energy-production to alternative fuels, reducing the release of greenhouse gases into the atmosphere and slowing the process of global warming. A globally networked renewable energy system comes on line, contributing to food production, providing energy for desalinizing and filtering sea-water, and helping to lift marginalized populations from the vicious cycles of poverty.

- Leading business companies join the classical pursuit of profit and growth with the quest for social and ecological responsibility. On the initiative of enlightened managers a voluntarily self-regulating social market economy is put in place, and the newly elected forward-looking political leaders give it full support.

- As the new energy system and the self-regulating social market economy begins to function, access to economic activity and technical and financial resources becomes available to all countries and economies. Frustration, resentment, animosity and distrust give way to a spirit of cooperation, liberating the spirit and enhancing the creativity of a new generation of locally active and globally thinking people. Humanity is on the way to a peaceful and sustainable, diverse yet cooperative planet-wide civilization.

The choice between these scenarios is not yet made. As of today, we are moving along the path of the BAU scenario, but more and more people are waking up and searching for ways to move to a scenario of timely transformation. The question is, how much time is there for this shift? The window of time is finite: when conditions in a complex open system reach a critical point the system becomes chaotic, and it either transforms, or breaks down. The longer the transformation is delayed, the more difficult it becomes to carry it out.

To define the feasible decision-window we must take into account both the time by which individual trends reach a critical phase, and cross-impacts and feedbacks among the trends.

1. The unfolding of individual trends. Time estimates of when individual life-threatening trends would reach points of criticality have been reduced from the end of the century to mid-century, and for some trends to the next ten to twenty years.]

For example, the sea level has been rising one and a half times faster than predicted in the IPCC's Third Assessment Report published in 2001. Forecasts published at the end of 2008 project global sea-level rise that is more than double the 0.59 meter rise forecast even by the Fourth Assessment Report.

Carbon dioxide emissions and global warming have likewise outpaced expectations. The rate of increase of CO_2 emissions rose from 1.1 percent between 1990 and

1999 to over 3 percent between 2000 and 2004. Since 2000 the growth-rate of emission has been greater than in any of the scenarios used by the IPCC in both the Third and Fourth Assessment Reports.

The warming of the atmosphere progressed faster than expected as well. In the 1990s forecasts spoke of an overall warming of maximum 3 degrees Celsius by the end of the century. Then the time-horizon for this level of increase was reduced to the middle of the century, and presently some experts predict that it could occur within a decade. At the same time, the prediction for the maximum level of global warming rose from 3 to 6 degrees. The difference is not negligible. A three degree warming would cause serious disruption in human life and economic activity, while a six degree warming would make most of the planet unsuitable for food production and large-scale human habitation.

2.   Feedbacks and cross-impacts. Most predictions of points of criticality take into consideration only one trend—the global warming and attendant climate change; water quality and availability; food production and self-reliance; urban viability, poverty, and population pressure; air quality and minimal health standards, or others. They fail to consider the possibility that a criticality in one trend could accelerate the unfolding of other trends toward a point of criticality.

There are multiple feedbacks and cross-impacts among the relevant trends, both in regard to the biosphere and conditions in the human world.

In the biosphere, all the trends that affect human life and well-being also impact on the cycles that maintain the planet's ecology within a humanly viable range. This is the case in regard to the global water and the global carbon cycle: the alteration of these cycles by any one trend affects the way the other trends unfold. For example, an increase of carbon dioxide in the atmosphere leads to global warming and that affects rainfall and the growth of forests. That, in turn, reduces the biosphere's carbon absorption capacity. Feedbacks are also conveyed by air and ocean currents. Warmer water in the oceans triggers hurricanes and other violent storms alters the course of major ocean currents, such as the Gulf and the Humbold. And that triggers further changes in the climate.

Feedbacks also obtain between ecological and societal trends. For example: The warming of the atmosphere produces prolonged drought in some areas and coastal flooding in others. Starving and homeless masses are impelled to migrate from the highly impacted areas to less hard-hit regions creating critical conditions in those regions as well. A drop in the quality of the air in urban and industrial megacomplexes below the minimum required for health creates a breakdown in public health, with epidemics spreading to vast areas. A breakdown of the financial system would impact not only on banks and stock markets, but would interfere with industrial output and trade the world over, creating critical conditions first of all for the poorest countries and economies.

Cross-impacts among accelerating global trends reduces the feasible decision-window. The precise time for effecting meaningful change is not predictable with certainty, but due to feedback and cross-impacts among the trends, it is likely to be less than the forecasts of critical points for individual trends. The decision-window may close within ten years and possibly sooner.

Chapter 16

TOMORROW'S TOURIST: FLUID AND SIMPLE IDENTITIES

Ian Yeoman

The globalization of tourism and increases in real wealth have meant tourists can take a holiday anywhere in the world, whether it is the North Pole or the South Pole and everywhere in between including a day trip into outer space with Virgin Galactic. Increases in disposal income allow a real change in social order, living standards and the desire for quality of life with tourism at the heart of this change. Against this background the concept of a fluid identity emerges. This trend is about the concept of self which is fluid and malleable in which self can not be defined by boundaries, within which the choice and the desire for self and new experiences drive tourist consumption. However, as wealth decreases this identity becomes simpler a new thriftiness and desire for simplicity emerge. This paper examines the values, behaviors, trends and thinking of the future tourist, whether it is a fluid or simple identity.

Keywords: tourism, tourist, attitudes, behaviors, futures.

INTRODUCTION: WHICH IDENTITY

Rising incomes and wealth accumulation distributed in new ways alter the balance of power in tourism. The tourist is the power base which has shifted from the institution of the travel agent through the opaqueness of online booking for holidays and travel to an individual. At the same time, the age is rich for new forms of connection and association, allowing a liberated pursuit of personal identity which is fluid and much less restricted by influence of background or geography. The society of networks in turn has facilitated and innovated a mass of options provided by communication channels leading to the paradox of choice. In the future market place, a tourist can holiday anywhere in the world whether it is Afghanistan or Las Vegas, to the extent the tourist can take a holiday at the North Pole or the South Pole and everywhere in between including a day trip into outer space with Virgin Galactic (Yeoman, 2008). If 25 million tourists took an international holiday in 1950, 903 million took a holiday in 2008 (Yeoman, 2008). Why? The growth in world tourism is founded on increase in real household income per head, which doubles every 25 years in OECD countries. This increase in disposal income, allows a real change in social order, living standards and the desire for quality of life with tourism at the heart of that change. Effectively, consumers want improvement year by year, as if it was a wholly natural process like aging. That change in disposable income has meant a greater and enhanced choice for tourists.

This tourist has demanded better experiences, faster service, multiple choice, social responsibility and greater satisfaction. Against this background, as the world has moved to an experienced economy in which an endless choice through competition and accessibility because of the low cost carrier, and what has emerged is the concept of fluid identity. This trend is about the concept of self which is fluid and malleable in which self can not be defined by boundaries, within which the choice and the desire for self and new experiences drive tourist consumption. The symbol of this identity is the fact that a consumer on average changes their hairstyle every 18 months according to the research by the Future Foundation (2007), from a tourist perspective it is about collecting countries, trying new things and the desire for constant change. It means the tourist is both comfortable with a hedonistic short break in Las Vegas or a six month eco-tourism adventure across Africa. This fluid identity makes it difficult for destinations to segment tourists by behavior or attitude as it is constant and fluid. However, as wealth decreases that identity becomes more simpler a new thriftiness and desire for simplicity emerge (Flatters & Wilmott, 2009). This desire for simplicity is driven by inflationary pressures and falling levels of disposable incomes, squeezing the middle class consumer. As the economies of wealth slow down, whatever the

Figure 1 *The Author's Fluid Identity and the Desire for New Experiences*

reason, new patterns of tourism consumption emerge, whether it is the desire for domestic rather than international travel or what some call a stay-vacation. A fluid identity means tourists can afford enriching new experiences and indulge themselves at premium 5 star resorts. They can afford to pay extra for socially conscious consumption, whereas a simple identity means these trends have slowed, halted or reversed (maybe reVERSed?). As resources become scarcer, a mind set of a whole generation of tourists change their behavior. Between now and 2050 the world will go through a cycle of economic prosperity and decline which is the nature of the economic order. When wealth is great, a fluid identity is the naked scenario, however, when a recession emerges, belts are tightened, tourists like other consumers search for a simple identity. This chapter examines the values, behaviors and thinking of the future tourist, whether it is a Fluid or Simple Identity.

FLUID IDENTITY

This tourist is both interested in a two week eco-tourism vacation where s/he will undertake an authentic and sustainable experience but at the same time s/he will take a short break in Las Vegas, whether it is a retail therapy, gambling or something more erotic (see Figure 1). Why? Tourists cannot be labelled according to their attitudes and beliefs—what they say and what they do, are two totally different things. They constantly evolve and seek something new, just like David Beckham and his hairstyles (Yeoman, 2008). That is why segmenting tomorrow's tourists is becoming much more difficult. If the future, is rising incomes and wealth accumulation in which individuality is central, the pursuit of personal identity becomes liberated and fluid as boundaries are broken which are not restricted by geography, culture or the past.

Fluid identity produces consumer volatility of proliferated choice and magnesium where a high entropy society exists (Future Foundation, 2007). Tourists have the means for endless choice and creative disorder. They have the power to express opinion and they do so, whether through www.tripadvisor.com or www.youtube.com. In fact, they form their opinion not on trusted sources from the authority but on a peer review, hence the importance of the consumer generated content and the advocacy of local authentic information as provided, for example, by the citizens of Philadelphia at www.uwishunu.com.

They are excellent at using networking tools to get a better deal or complain about poor service. A fluid identity allows tourists to be frivolous, promiscuous and just plain awkward. A fluid identity means tourists want to sample a range of new experiences, hence the rise of the long tail (Anderson, 2008) and emergence of bespoke tourism products i.e. special cruise markets at www.insightcruises.com.

A fluid identity emerges because the society is socially less rigid, that individuals have become less class-defined and human relationships are not restricted by accident of birth, but the combination of breaking class distinction through education, income and mind expanding influence of modern travel and entertainment which broadens preordained identities and choices. The emerging tourists from Brazil, Russia, India and China are the new tourists who are now not restricted to one town, one church, one marriage and one football team, especially generations Y and Z (Mc-Crindle & Wolfinger, 2010) Fluid identity results in massively propelled ad hoc communities of new friends and connections some via social media and others through shared interest activity groups. Ethan Watters (2004) calls this Urban Tribes, groups of like-minded people and friends doing activities together whether it is a girlie weekend of pampering or a boyish rugby game. It is the idea that an infinite number of options are available; this propels the idea of fluid identity.

Globalization shapes people's lives and the mixture of cultures produces exposure to new ideas and different identities. The tourist is the centre of the globalization of experiences, where holidays in exotic locations that are deep inside countries are becoming the norm. No longer is an international holiday confined to a resort, the tourist has become a traveler, staying longer and going deeper into the culture of destinations. Globalization is brought nearer to us all through social media and the world of personalized communications, the society that is fast, instant and networked. No longer is the internet bound to a wire or a desk but is mobile and wireless. Everyone seems to be online 24 hours a day, anywhere, as technology has become more accessible and costs of transactions are falling. The power of personal mobile technology means more features, interactivity and multi-functionality which deliver a different way in which tourist providers have to engage with future tourists. One of the challenges for tourist destinations is how they protect their brand equity when it can be quickly destroyed or poked fun at www.youtube.com or www.facebook.com. It means brands have to work harder to remain an authority with trust as a disruptive discourse which is shaped by the word of month or someone being followed at www.twitter.com.

A tourist's sense of timing and patience is changing; society is now just a click away from a screen and is not the one that likes the notion of delayed satisfaction. Patience is now measured in nanoseconds driving an immediacy culture. The tourist has become programmed to be narcotic, wanting more all the time in an instance. In Tokyo, 30% of hotel reservations are on the day of arrival as smartphone augmented technology allows tourists to look at a hotel through the smartphone camera and gauge availability, then book accommodation through a related website like www.expedia.com (Hatton, 2009), all driven by applications such as the Wikitude AR Travel Guide (www.wikitude.org).

Longevity is a key trend associated with fluid identity, as consumers live longer with wealth they expect richer experiences and more. They visit places and do things that their parents could not afford or would not have heard of. They will search for experiences that hold back the wrinkle of old age, whether it is a spa treatment in Hungary or a medical procedure in South Africa. Health and medical tourism become more important in this scenario along with any service that rejuvenates the soul or a tired body. Longevity also changes life courses, so change becomes the norm and is unpredictable. Although tourists may have their favorite place, they like refreshment and renewal. This means, they ask themselves who they are and a multiplicity of answers suffice. Michael Wilmott (Wilmott & Nelson, 2005) calls this complicated lives, in which the choice explosion of holidays and travel means tourists have brought upon themselves complexity and complications resulting in some anxiety. At one level, this means many tourists are opting out and talking career breaks and travelling the world, on another scale authenticity becomes important as tourists look for simplicity. At another level, destination brands have to find a means to ensure they can help a tourist unclutter this world through a brand search optimization, a high brand value and choice management. Although choice is regarded as a positive value within a consumption culture, choice making support is important, such as a book recommendation service at www.amazon.com.

Tourism destinations need to understand their tourists, not engaging in a relationship which is about mass selling but focusing on what tourists want at the right time and at the right place. To a certain extent, fluid identity is about wealth and a have-it-all society, these tourists can afford holidays several times a year and a multitude of short breaks. This is a tourist that can afford to be concerned about the environment so s/he does not mind paying a little bit extra. In a have-it-all society, the desire is for sociality, economic gain, family involvement, leisure and self improvement which are less delineated by stages of life or gender, all of these desires are reflected in holiday activity, whether it is an extended family holiday at Walt Disney Park or a cultural short break in Paris. The expectation amongst the tourist with a fluid identity is they want a richer and fulfilling life, but at the same time there is pressures of expectation, hence the previous mentioned link to a disruptive discourse in this identity.

Although rising wealth means more opportunities, it also means a fear of loss, in which society is portrayed in decline. Here a consumer turns to therapies and anti depressants, is anxious about the future and thinks society has lost its way. Writers, such as Frank Furedi (2006) label this 'the culture of fear'. From a tourism and media perspective there seems to be a focus on a health scare or terrorism incident which impacts upon destinations. The incident is portrayed as overtly bad news which results in countries issuing travel advisories advising us not to travel to such and such a place.

A heightened sense of personal freedom has undoubtedly increased the growth of world tourism, where identity is built on liberal attitudes reinforced through education and knowledge. The exposure of tourists to a multi cultured society allows a greater expression of individuality, whether this is sexual behavior or unconventional lifestyles, however this degree of liberalism differs around the world. Fundamentally, as economies grow they become more liberal in outlook and seek to push out their identity. As such, they will try new things and visit new places, destinations in the far away places that seemed inaccessible to previous generations.

The manifestations of a fluid identity are wide-ranging, from overt and status-driven to the anonymous and elusive. Yet the common characteristic for the tourists is that they simply do not want to consume but experience the consumption in several ways, increasing aspirations and higher order expectations (Yeoman 2008). One noticeable trend shaping a fluid identity is the movement from conspicuous consumption to inconspicuous consumption, especially amongst tourists from advanced economies of the world who are well versed with travel. It has become the norm not to parade wealth and success in a deliberate ostentation, but to be more conservative, wiser and discreet. From a tourism perspective, inconspicuous consumption has developed as the experience economy has matured from theatre to the desire for authenticity, where tourists search for deeper and more meaningful experiences. This trend has changed the meaning of luxury in society away from materialism to more about enrichment and personal development, for a tourist it is about the point of self-actualization in Maslow hierarchy of needs (Maslow, 1998). Luxury has therefore become more accessible to the growing middle classes of the world, in which they can hire a Ferrari for the weekend (www.gothamdreamcars.com) or even hire the latest designer handbags (www.bagborroworsteal.com).

Related to the changing nature of luxury is the importance of cultural capital, that is how tourists talk about destinations and experiences. The importance of cultural capital defines identity and status, it becomes the critical currency of conversation i.e., 'have you been to South Africa', 'I swam with the dolphins in New Zealand' or 'I built a bridge for a community in India'. It is the knowledge and experiences of the arts, culture and hobbies that help define who people are rather than their socioeconomic grouping. Sociologists such as Rifkin (1984) and Bourdieu (Bourdieu & Nice, 1987) argue that consumers are moving from the era of industrial to cultural capitalism, where cultural production is increasingly becoming the dominant form of the economic activity and securing access to many cultural resources and experiences. This means that the definition of culture changes, the tourist is both happy with a high-brow opera and low-brow comedy, hence the rise of the creative class and no-brow culture associated for example, with the success of Edinburgh's festivals which embody the diversity of cultural capital and the breath of experiences.

The emergence of a fluid identity means tourists are genetically programmed to be suspicious and rather cynical of all marketing and advertising. As if the tourist that is instinctively mobilized to mount resistance and rebuke. It also becomes increasingly difficult to label and segment tourists by demographics, attitudes and economic well-being as fluidity becomes the norm in this scenario. A fluid identity represents a challenge for tourism destinations because of the constant change and resistance.

...therefore, a fluid identity is represented in the following scenario

Michael Hay is a 28-year-old business executive from London. Michael is a seasoned traveller, who likes to take two long haul holidays a year and several short breaks. This year Michael is visiting Tokyo and wants to climb Mt Fuji and see the

Figure 2

snow monkeys near Nagano. He chose Japan because friends had previously visited the country; they often talked about the food, people, how everyone was so helpful, how safe the country is and what a wonderful experience it was. Michael had considered China, but he had watched so many viral videos, that he was put off from visiting China at the present. Prior to visiting Japan he had read a couple of guide books which formed the basis of a vague itinerary. He looked at the destinations website for information and could vaguely recall Tourism Japan had sponsored some sort of sporting event.

His seven day vacation to Japan begins in Tokyo, he has not booked an accommodation and is relying on his Nikon 300UXP contact lenses[1], such is the speed of the technology that at the flick of the eye, details of the JP five star hotel is sought using the latest augmented reality technology and its availability is confirmed and a reservation made. In addition, a five day tour to central Japan is organized by his online travel agent based upon his requisites, attraction booking, hotel accommodation and transport connections. Japan is a place known for its organizational efficiency and excellent transport infrastructure making it easy to get around. Michael is even going to road test a classic 2020 Ferrari, something that requires manual control and skill not like today's automatic personal vehicles. Each day Michael tries something new, whether it is a Japanese spa treatment, staying in a traditional Ryokan or hiking up Mt Fuji. All in all, a wonderful action packed holiday, everything from adventure to tranquillity.

---

1. A new generation of contact lenses built with very small circuits and LEDs promises bionic eyesight. The University of Washington, in Seattle has engineered a lens akin to Terminator movies. Arnold Schwarzenegger's character sees the world with data superimposed on his visual field—virtual captions that enhance the cyborg's scan of a scene. In stories by the science fiction author Vernor Vinge, characters rely on electronic contact lenses, rather than smartphones or brain implants, for seamless access to information that appears right before their eyes. These lenses do not give us the vision but have the potential to deliver the vision of an eagle and the benefit of running subtitles http://www.spectrum.ieee.org/biomedical/bionics/augmented-reality-in-a-contact-lens/0.

This scenario is shaped by many of the trends associated with a fluid identity including, wealth, a networked society, resistance to marketing, strong brand image, culture of fear, choice management, personal recommendation, variety of experiences and its cultural capital. The importance of the scenario highlights how individuals shape their life using technologies as short-cuts and choice managers, however the biggest influence choice is personal recommendation and the ability to lead a fluid identity depends on wealth.

SIMPLE IDENTIFY

The Global Financial Crisis (GFC) plummeted the value of the High Net Worth population by US $32.8 trillion or 19.5 % according to the World Wealth Report (2009) published by Capgemini and Merrill Lynch, so the rich are less rich. Flatters and Wilmott (2009) argue that in most developed economies pre-GFC the precession consumer behavior was the product of 15 years of uninterrupted prosperity, driven by growth in real levels of disposal incomes, low inflation, stable employment and booming property prices. Therefore, new consumer appetites emerged in which a consumer could afford to be curious about gadgets and technology, in which tourists shelled out for enriching and fun experiences on exotic locations. The GFC changed that, propelling tourist trends into slowdown, halting or even reversing the trajectory of growth in world tourism. So, is this a sample of the future, the era of the pension crisis, scarcity of oil, inflation and falling levels of disposal income in which tourism expenditure falls year by year? If so, what will the future tourist look like? Rather than having a fluid identity it will be more akin to simplicity.

During an economic slowdown, tourists tend to travel less, stay nearer home (increase in domestic tourism) and seek simplicity such as www.exploreworldwide.com' value based holidays focusing on basic facilities, meeting locals, lots of free time and cheap. This trend is accelerated in the scenario of falling incomes as a simple and functional product that will suffice. A simple identify means offering advice becomes extremely important, whether its website's www.farecast.com which advises travellers when the optimal time to purchase an airline ticket is or price comparison technologies which are found on many online booking services.

When simplicity is combined with thrift, tourists trade down. The Pod Hotel in Manhattan (www.thepodhotel.com), where accommodation usually costs US $300 a night, offers single beds from US $89 a night including bunk beds. The use of technology and social media assists tourists in the search for bargains, whether it is the use of augmented technologies in smartphones or contact lens which view availability and prices as we view them in the street or recommendations from a network of friends on social media sites. Thrift and simplicity also combine to drive the trend of Visiting Friends and Relatives (VFR), as incomes fall getting back to basics and developing human relationships are very important, and the most important aspects of tourists' lives are friends and relatives.

Research by the Trajectory Group (Flatters & Wilmott, 2009) highlights that affluent consumers have revealed mounting dissatisfaction with excessive consumption. Many desire a wholesome and less wasteful life. Hence there is a desire to get back to nature, something that is tranquil, basic, rooted, human and simple (Yeoman, 2008). As a consequence, the desire for more authentic and simple luxury experiences accelerates. An example of simple luxury is a tree house hotel which offers a unique experience in a natural setting. A new experience is seen not as conspicuous consumption, but as overtly inconspicuous. For instance, the Costa Rica Jungle Hotel is situated in the rainforest around Arenal Volcano, surrounded by wildlife and birds (treehouse-shotelcostarica.com/). Another example is hay-vacations, where holiday-makers pay to stay and work on farms. Holiday-makers are turning to haycations to experience a world far removed from their daily life. At Stoney Creek Farm (www.stonycreekfarm.org), tourists are charged up to US $300 a night to work on the farm. This is an experience where tourists pick there own food, then cook it that evening and in a location with no cellphone reach. During the times of recession tourists are searching for back to basic experiences that are simple, with a sense of community and authenticity. About 50 % of the tourists to Stoney Creek Farm are locals from the same county. This is a typical example of inconspicuous consumption and a desire for a simple identity.

In a simple identity, ethical consumption declines as paying a premium for a Starbucks coffee falls by the wayside, even if they use organic coffee which supports children in a third world country. From a tourism perspective, many of the ethical tourism projects in third world countries, such as Africa and India, which depend on independent travellers, will suffer.

Tourists also have become canny at searching for bargains, which economists call mercurial consumption, whether it is using price comparison software, or grapping last minute offers from websites such as www.grabaseat.co.nz which offer last minute air travel deals to New Zealand consumers, or www.5pm.co.uk which offers diners a chance of discounted meals after 5pm that evening. The dominance of technology and social media has changed consumers purchasing behavior to something more mercurial in which they actively search for bargains through price comparison websites.

Attitudes to travel also change, as tourism has to compete with other forms of leisure expenditure, whether it is the latest technology gadgets or virtual holidays. There is a generation of Japan youth who prefer their X-Box to climbing Mt Fuji. The desire for new experiences is more about 'insperience' (Trend Hunter, 2008), where technology provides a better experience than the one, in which consumers desire to bring top level experiences into their domestic domain.

There have been many predictions about the end of the high street travel agent in the last decade, but in fact during times of economic slowdown, when tourists are trying to unravel complexity and give up excess, they go back to travel agencies to reduce choice through an efficient filtering process and maximise time management. In addition, the desire for new experiences slows down as a number of simple repeat trips to usual places also increases (Buhalis, 2009).

In an economic slowdown, the role of authority changes as governments intervene to stabilize markets, bring assurance and confidence to markets, create jobs and increase public expenditure. Therefore many countries have increased marketing expenses, particularly in domestic markets to entice tourists to stay at home this year (hence the term stay-vacation) as international markets fall. The tourism industry in particular, will turn to government to offer support and strategic leadership when the private sector is failing. Therefore, trust in authority increases and destination brands that offer value, honesty and can deliver on brand promise become more important.

New Zealand is the adventure capital of the world, whether it is a bungee jumping, jet boating, bugging or skydiving. During an economic slowdown extreme-experience seeking stalls, as they are seen as expensive, frivolous, risky and environmentally destructive. Extreme adventure is partially about how a tourist differentiates themselves. But conspicuous consumption is out of favour and the trend of simplicity and discretionary spend is in. So for destinations like New Zealand, extreme sports like bungee jumping and jet boating will be curtailed.

The GFC has focused the consumer mind on the boardroom, in particular the executive bonuses of companies such as AIG, Royal Bank of Scotland or General Motors. Excess has become a dirty word, as such travel and the meetings industry have taken a hit as too many think that this sector is about excessive and unnecessary expenditure.

If the future is a simple identity, the key words are simplicity, thrift and mercurial consumption which leads to a scenario in the following manner:

Sheena Michaels is 68, lives in London and is a part-time social worker. She is well travelled, has just completed an Open University degree in Technology and is a volunteer with a number of local community projects. Circumstances force Sheena to work part time because of her pension shortfall and she thinks that this will continue until her health dictates otherwise. Since London is recognized as a cultural centre for tourism. Lack of monies means Sheena has to watch how she spends her money. Websites such as www.culturalprice.com tell Sheena in advance when it is the right time to book a theatre ticket and Sheena's social media network of friends advise on special deals etc. Sheena would like to travel but nowadays tends to stay in the local region doing day trips in the surrounding hinterland. When she does go on holiday, it is staying with friends and families. This year she managed to take a short break in Barcelona, staying with friends and capturing much of the city culture, especially the galleries. Today, Sheena is travelling to the Soho Theatre Quarter in central London as it is the opening day of the Quarter's Festival and many of the acts are performing free street shows. She manages to take in several short acts including, an eight minute performance of all Will Shakespeare's plays by the Royal Company and a lunchtime comedy performance by Leo Blair on 'the exploits of a Prime Minister's Son'. Eventually, Sheena and friends find a café for a cup of tea and just watch the world go by.

This scenario is shaped by many of the trends associated with a simple identity including, a networked society, simplicity and thrift, pricing technologies, highly educated, community, use of leisure time and personal recommendation. The importance of the scenario highlights how individuals trade down and are thrifty with spending.

CONCLUSION

Tourism is an unpredictable industry, shaped by events, world economy and the sociopolitical environment. Tourists are fickle and when times are good will spend large amounts of disposal income on tourism. To a certain extent, tourists retrench and focus on lower order basic needs when times are hard, so tourism declines. Given the Global Financial Crisis and the forthcoming demographic and pensions time bomb we could see year by year decline in tourism expenditure with 2050 being the flip point. When tourists do have money, they possess a fluid identity of constant change in a fast moving world, in which they are easily bored, seek novelty, desire thrill, something new, aspiration and enrichment. Tourism has always been about fun, relaxation, entertainment, enrichment and enjoyment, but will it be simple or fluid, only time will tell.

REFERENCES

Anderson, C. 2008. Long Tail: The Revised and Updated Edition: Why the Future of Business is Selling Less of More. London: Hyperion.

Bourdieu, P., and Nice, R. 1987. Distinction: A Social Critique of the Judgment of Taste. Boston: Harvard Business Press.

Buhalis, D. 2009. Personal communication on owner-trinet-l@hawaii.edu discussion board.

Flatters, P., and Wilmott, M. 2009. Understanding the Post Recession Consumer. Harvard Business Review. July–August: 106–112.

Furedi, F. 2006. Culture of Fear Revisited: Risk-taking and the Morality of Low Expectation. London: Continuum.

Future Foundation, nVision Central Scenario UK. London: Future Foundation, 2007.

Hatton, C 2009. The Future of Technology. Tourism Futures Proceeding. Goldcoast 18th August.

Maslow, A. 1998. Toward a Psychology of Being. New York: Wiley.

McCrindle, M. and Wolfinger, E. 2010. The ABC of XYZ: Understanding the Global Generations. Sydney: UNSW Press,

Rifkin, J. 1984. The Age of Access: The New Culture of Hypercapitalism, Where all of Life is a Paid-For Experience. London: Tarcher.

Trend Hunter (2008) Insperience Economy. Accessed on the 16th September 2009 at www.trendhunter.com

Watters, E. 2004. Urban Tribes: Are Friends the New Family? New York: Bloomsbury.

Wilmott, M., and Nelson, W. 2005. Complicated Lives: The Malaise of Modernity. Chichester: Wiley.

World Wealth Report 2009. Capgemini and Merrill Lynch, 2009. Available at http://www.capgemini.com/resources/thought_leadership/2009_world_wealth_report/

Yeoman, I. 2008. Tomorrows Tourists: Scenarios & Trends. Oxford: Elsevier.

Yeoman, I., Munro, C., and McMahon-Beattie, U. 2006. Tomorrows World, Consumer & Tourist. Journal of Vacation Marketing 12(2): 174–190.

CHAPTER 17

WORLD ENERGY AND CLIMATE IN THE TWENTY-FIRST CENTURY IN THE CONTEXT OF HISTORICAL TRENDS: CLEAR CONSTRAINTS TO FUTURE GROWTH

Vladimir V. Klimenko and Alexey G. Tereshin

The paper deals with global energy perspectives and forthcoming changes in the atmosphere and climate under the influence of anthropogenic and natural factors. In the framework of the historical approach to energy development the forecast of the future global energy consumption for the present century is elaborated, and its resource base and the global impact of the power sector on the atmosphere and climate against the background of natural factors influence are studied. It is shown that, following the historical path of global energy evolution, the global energy consumption will remain within 28–29 billion tons of coal equivalent (tce) by the end of the century, with CO_2 emissions peaking in the middle of this century. In this scenario, the CO_2 concentrations will not exceed 500 ppm, and the global temperature should rise by 1.5 °C by 2100, with the growth rate not exceeding the adaptation limits of the biosphere.

Keywords: energy, climate change, carbon dioxide emission.

INTRODUCTION

Energy is a fundamental base of the evolution of civilization, and the twenty-first century poses for the world energy sector a challenging task of ensuring sustainable development of human society. The progressing growth of population will undeniably lead to the necessity of accelerated development of many regions of the world, and, as a result, to enhanced demand for energy in the nearest decades. Thus, to provide fuel and energy resources to the world economy is one of the principal problems posed to humanity. On the other hand, today energy sector is considered as one of the principal factors entailing global environmental change, which overrides all other anthropogenic factors and compares with powerful natural forces in its impact on the climate of the planet (see Solomon *et al.*, 2007). The concern about the scale of observed climate changes (in particular, an increase of the mean global temperature by 0.8 °C over the past 120 years) and anxious projections of further warming (up to 5 °C over the current century) make the ecological policy, along with the state of the resource base, one of the principal regulators of the world energy development.

There have been a great number of publications concerning the above-mentioned problems over the past decades, and these problems have been in the centre of attention of leading national and international institutions. However, great controversy still exists in the opinions as to the global energy perspectives and the scale of their associated environmental and climate changes.

At the same time, it is quite understandable that without a more or less clear view of the future energy use, one cannot build realistic scenarios of its impact on the environment and climate and develop an efficient adaptation policy. A question thus arises of whether long-term forecasts of energy demand are feasible at all? Many experts, bearing in mind an extensive negative experience in this field (see, e.g., a review of scenarios of global energy consumption in Klimenko *et al.*, 2001) tend to give a negative answer to this question.

In our opinion, the situation may prove to be not so hopeless if one resorts to a historical extrapolation approach that is widely known in contemporary sociology and economy as the theory of institutional changes (see North, 1990), which is based on the concept that the history of complex systems development predetermines their future behavior for many years ahead. In the present work we set ourselves the task to outline the direction of the world energy development, based on the principal trends in its historical evolution, and to assess, from the same standpoint, resource availability and the most probable impact on the global climatic system.

The suggested assessments are based on the so-called genetic forecast of global energy consumption developed at the Moscow Energy Institute (MEI) over 20 years ago (see Klimenko & Klimenko, 1990; Snytin *et al.*, 1994), and which has shown a remarkable correspondence to the observed data over the past decades. The deviation of projected values from the global energy statistics data was within 2 %, which, in our opinion, makes it possible to build up a super long-term energy forecast with an accuracy sufficient for climatic assessments. A consistent application of the genetic approach to energy use forecasting (identification and extrapolation of historical tendencies to the future) allowed drawing two basic conclusions as to the development of world energy demand in the nearest decades:

1. Stabilization of the national per capita energy consumption at the level primarily determined by climatic and geographic factors (see Klimenko, 1994). This process has already been completed in the most developed countries (see Klimenko *et al.*, 2001; Energy Statistics Yearbook, 2009; BP Statistical Review..., 2009; International Energy Annual, 2009; Demographic Yearbook, 2009).

2. Steady and nearly linear decrease of the carbon intensity of global energy as a result of structural changes in the world fuel mix that lingered for more than 100 years (see Klimenko, V. V., Klimenko, A. V., Andreichenko *et al.* 1997).

The realization of the former tendency should lead to the fact that the global per capita energy consumption will reach 2.9–3.0 tce/year, which, by the way, is quite close to the present level (2.6 tce per capita in 2009); as a result, since the Earth's population is expected to reach 9.5 billion by 2100 (see World Population..., 2009), the energy consumption will make 28–29 billion tce/year, which is 1.6 times above the present level. Thus, the historical approach prohibits the energy consumption increase to 60, 100, and, all the more, 200 billion tce/year over the present century, which frequently conceded by the authors of the most radical energy scenarios (see Nakicenovic & Swart, 2000).

The preservation of the latter tendency means that the growth of the anthropogenic impact on the climatic system steadily slows down and, therefore, one can expect that the anthropogenic emission of CO_2 will fairly soon, within the next quarter of century, reach its maximum. There is not any mysticism in the steady and, apparently, irreversible decrease of the carbon intensity. Moreover, this fact can be philosophically substantiated in terms of the principle of progressive simplification, a phenomenon of widespread occurrence in nature and social life and observed not only in engineering, but also in science, art, philosophy, and theology (see Toynbee, 1988). As applied to the energy sector, this principle is manifested in a gradual transfer from more complex, 'conserved' energy sources to more elementary, natural. Such is the trend of global energy development from coal to oil, then to gas, and, finally, to renewable (solar, wind, tidal etc.) sources.

GLOBAL RESOURCES OF FOSSIL FUELS AND RENEWABLE ENERGY SOURCES AND PROSPECTIVE ENERGY BALANCE IN THE TWENTY-FIRST CENTURY

A necessary test for consistency of any energy scenario includes assessment of its fuel and energy reserves' availability. Surprisingly as it may seem, the amount of fossil fuels expected to be consumed in many previous 'high' energy consumption scenarios did exceed not only its proven recoverable reserves, but often also hypothetical additional resources.

In the present work to estimate the consumption of hydrocarbon fuels (oil and natural gas) we made use of the so-called 'depleting resource consumption concept' (Energy and Nuclear Power…, 1985) which suggests declining production rates of a resource as its stocks are gradually depleted. In this case, the cumulative consumption trend of the resource is described by a logistic function with an exponential initial portion and an asymptote defined by the ultimate amount of recoverable reserves. For the latter we accept the sum of discovered recoverable reserves and prospective additional resources, which poses the theoretical limit of the availability of this kind of fuel from the geologic and economic viewpoints (in terms of the World Energy Council (WEC) [see WEC Survey…, 2001]). In this respect the present research differs from our previous work (see Snytin, Klimenko, & Fedorov, 1994) where we took no account of additional oil and gas resources, which resulted in a slightly distorted projection of the structure of global energy balance, envisioning continuous increase of the coal share and, vice versa, decrease of oil and gas share already from the beginning of the current century. In reality, the last decade showed that oil has preserved its leading position in the world fuel mix, while coal consumption, increasing with the annual rate of 4 %, left natural gas behind.

Figure 1 shows a record of changes of discovered recoverable reserves of hydrocarbon fuels over the past 60 years (according to BP Statistical Review…, 2009; International Energy Annual…, 2009; Energy and Nuclear Power…, 1985; WEC Survey, 2001). Evidently, the estimates for oil, as well as for gas resources have changed considerably: as compared with 1950, they have increased by an order of magnitude, regardless of current high production rates (about 5 and 4 billion tce/year for oil and gas,

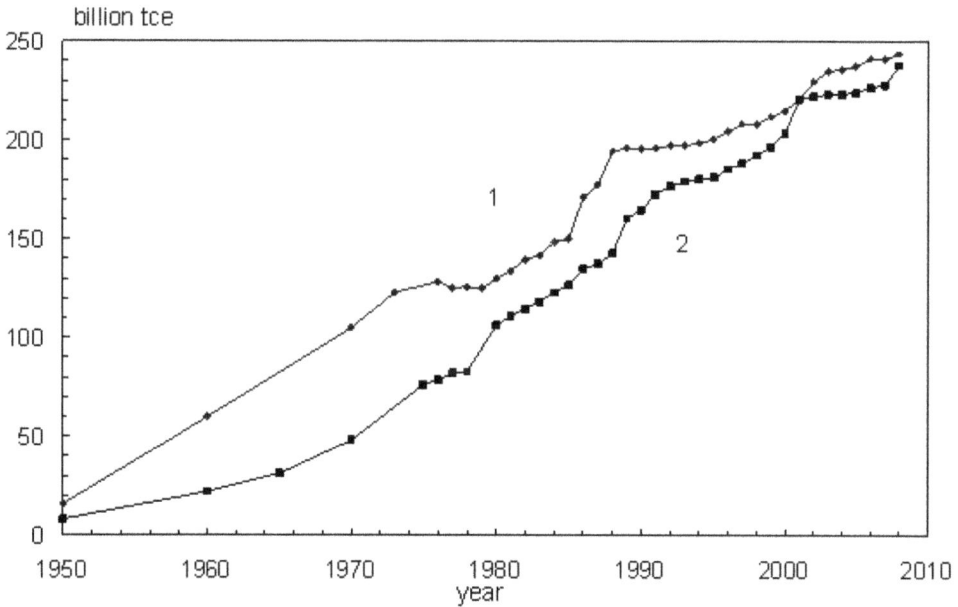

Figure 1 *Record of the estimated proven recoverable reserves of hydrocarbons: (1) oil and (2) natural gas.*

respectively). However, it is quite clear that this situation cannot last indefinitely long, and the curve shapes in Figure 1 show that annual build-up of oil resources today nearly match oil production rate, and the situation with gas will obviously become the same over the nearest decade or two. Thus, the global proven recoverable oil and gas reserves are currently about 240 billion tce each, and the ultimate recovery (including additional recoverable resources (see WEC Survey..., 2001), comprises 620 and 490 billion tce, respectively.

Fitting the historical series of the cumulative oil and gas production (see Energy Statistics Yearbook, 2009; BP Statistical Review..., 2009; International Energy..., 2009) by a logistic function with ultimate oil and gas reserves as asymptotes determines the trend in the annual production of hydrocarbon fuels for the nearest decades (Figure 2 and 3). The genetic forecast assumes that these kinds of fuel will cover some 40 % of global energy demand by 2050, but less than 10 % by the end of the century. For comparison in Figure 2 and 3 the principal scenario of the WEC and International Institute of Applied System Analysis (WEC/IIASA) (see Nakicenovic, Grubler, and Mc-Donald 1998) is shown, according to which the total consumption of oil will fully deplete its resources by 2100, and the total consumption of natural gas will even exceed its ultimate resources. The same features are characteristic for the scenario B2 of the Intergovernmental Panel on Climate Changes (IPCC) (see Nakicenovic & Swart, 2000). Although this scenario is not specified as a basic one, it assumes moderate demo-

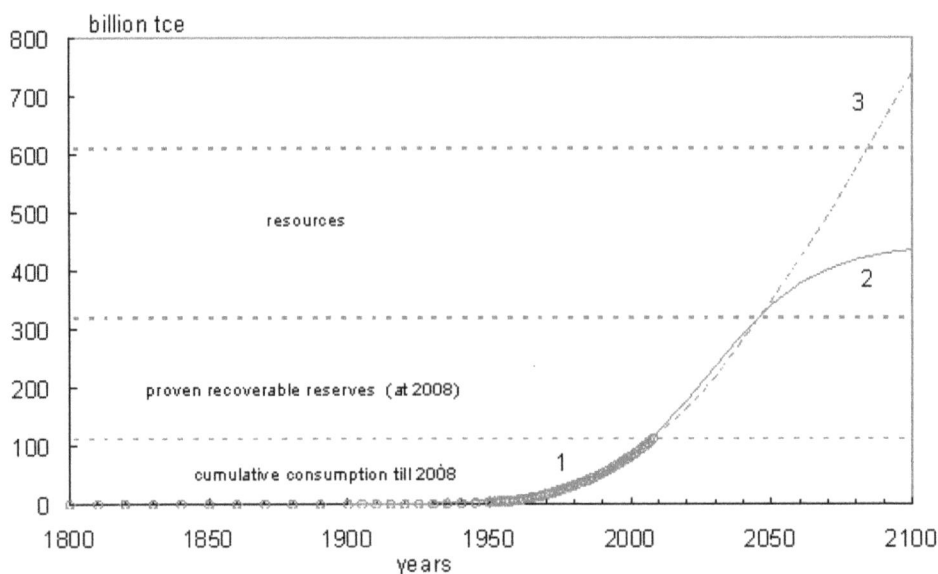

Figure 2 *Cumulative global gas consumption: (1) historical data (Energy Statistics Yearbook 2006; BP Statistical Review... 2009; Energy Statistics Yearbook 2009); (2) forecast of the present work; and (3) WEC/IIASA Reference Case scenario (Nakicenovic, Grubler, and McDonald 1998).*

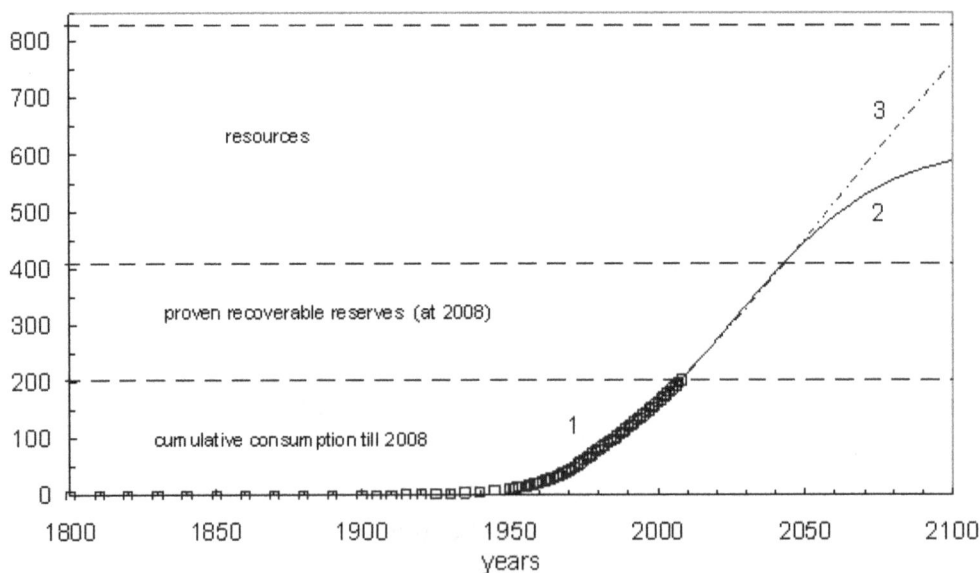

Figure 3. *Cumulative global oil consumption: (1) historical data (see International Energy Annual 2009; BP Statistical Review... 2009; Energy Statistics Yearbook 2009); (2) forecast of the present work; and (3) WEC/IIASA Reference Case scenario (see Nakicenovic, Grubler, and McDonald 1998).*

graphic and economic growth parameters and is placed in the centre of the spectrum of forty alternative energy development scenarios presented in (Ibid.).

To keep the tendency for specific CO2 emission decrease with the growing energy production, the suggested genetic forecast requires that the share of coal in the global energy balance be maintained at a level of 15–20 %. Thus, this 'clean energy' scenario assumes that non-fossil energy sources will cover about 30 % of the energy demand by 2050 and up to 65 % by 2100 against present 20 % (Table 1).

Scenario	Energy source	2000	2020	2050	2100
'Clean energy' (present work)	Coal	3,2 (23 %)	3,1 (15 %)	3,7 (14 %)	5,5 (19 %)
	Oil & gas	8,0 (56 %)	11,1 (54 %)	10,2 (39 %)	2,0 (7 %)
	Non-CO$_2$ commercial	1,7 (12 %)	4,2 (20 %)	9,8 (38 %)	18,4 (65 %)
	Non-commercial	1,3 (9 %)	2,1 (10 %)	2,3 (9 %)	2,4 (9 %)
	Total	14,2 (100 %)	20,5 (100 %)	26,1 (100 %)	28,3 (100 %)
'Coal energy' (present work)	Coal	3,2 (23 %)	3,3 (16 %)	6,7 (26 %)	8,1 (29 %)
	Oil & gas	8,0 (56 %)	11,1 (54 %)	10,2 (39 %)	2,0 (7 %)
	Non-CO$_2$ commercial	1,7 (13 %)	4,0 (20 %))	6,9 (26 %)	15,8 (56 %)
	Non-commercial	1,3 (9 %)	2,1 (10 %)	2,3 (9 %)	2,4 (9 %)
	Total	14,2 (100 %)	20,5 (100 %)	26,1 (100 %)	28,3 (100 %)
Reference case WEC/ IIASA (Nakicenovic et al. 1998)	Coal	3,2 (22 %)	5,1 (26 %)	6,5 (23 %)	12,6 (26 %)
	Oil & gas	8,0 (56 %)	10,5 (54 %)	13,4 (47 %)	14,4 (29 %)
	Non-CO$_2$ commercial	1,7 (13 %)	2,6 (13 %)	7,2 (25 %)	21,6 (44 %)
	Non-commercial	1,3 (9 %)	1,2 (6 %)	1,2 (4 %)	0,6 (1 %)
	Total	14,2 (100 %)	19,4 (100 %)	28,3 (100 %)	49,2 (100 %)
Scenario	Energy source	2000	2020	2050	2100
B2 scenario IPCC (Nakicenovic and Swart 2000)	Coal	3,2 (22 %)	3,3 (17 %)	3,0 (10 %)	10,2 (22 %)
	Oil & gas	8,0 (56 %)	12,6 (65 %)	17,8 (60 %)	13,4 (29 %)
	Non-CO$_2$ commercial	1,7 (13 %)	2,2 (11 %)	7,6 (26 %)	21,4 (46 %)
	Non-commercial	1,3 (9 %)	1,3 (7 %)	1,3 (4 %)	1,3 (3 %)
	Total	14,2 (100 %)	19,3 (100 %)	29,7 (100 %)	46,3 (100 %)

Table 1 *Global fuel mix for different energy scenarios, billion tce*

However, the energy consumption parameters over the last decade point to prevailing growth rates of global coal consumption, primarily due to China and India. This tendency provides some evidence in favor of the so-called 'coal bridge' theory formulated three decades ago, according to which this kind of fuel should fill the gap between depleting hydrocarbon reserves and slowly developing renewable energy sources. To account for this tendency, an alternative scenario was developed, that put emphasis on coal whose annual consumption was estimated by a procedure analogous to that used with oil and gas (Figure 4). The shape of fuel mix for this scenario ('coal energy'), which expects that by 2100 the share of coal will increase to 30 % and non-CO2 emission energy sources will make about a half, is also presented in Table 1, along with the WEC/IIASA and IPCC data (see Nakicenovic, Grubler, & McDonald, 1998; Nakicenovic & Swart, 2000).

Considering the dynamics of the world fuel mix, one can note that the long-term expectations associated with the development of nuclear energy technologies (specifically, the WEC/IIASA scenarios [see Nakicenovic, Grubler, & McDonald, 1998]) projecting a growth of the annual nuclear power production over the current century up to 25–40 trillion kWh, which is equivalent to annual combustion of 8–13 billion tce at thermal power plants[1]) have not come true: most experts (see International Energy Outlook 2009; World Energy..., 2008) do not see the possibility that the present nuclear electricity production (about 2.8 trillion kWh/year) will increase considerably. Thus, the basic scenarios both of the US Department of Energy (see International Energy Outlook..., 2009) and International Energy Agency (see World Energy Outlook, 2008) suggest that the annual nuclear electricity production will span the range 3.5–3.8 trillion kWh over the period to 2030. Thus, the share of nuclear energy in global energy consumption will comprise no more than several percent. Hydroenergy, regardless of expected increase of its production rate (at present one third of the economic global hydro potential is already harnessed), too, will be able to cover no more than 10 % of the total energy demand. As a result, by 2100, to implement the genetic scenario will require the energy production from non-traditional renewable sources to increase to 16–18 billion tce/year or about 50 trillion kWh/year, which is quite possible, since these production rates are well below the technical potential (and just about three times as high as the economic potential calculated for the conditions at the beginning of this century) of both solar and other kinds of renewable energy (Table 2) whose utilization rates have grown consistently by 8 % a year over the past three decades (see Energy Statistics 2009; International Energy Annual..., 2009).

---

1. Recalculation of the so-called primary electricity, *i.e.* non-fossil electric power, is performed by the equation 1 kWh = 0.319 kgce, with the global average efficiency of thermal power plants taken to be equal to 0.385.

Renewables	Theoretical Potential	Technical Potential	Economical Potential
solar	8700	720	5,3
hydro	40	15	8,0
Renewables	Theoretical Potential	Technical Potential	Economical Potential
wind	500	53	2,4
wave and tides	22	6	0,6
geothermal	5,000,000	6	1,0
TOTAL	5,009,262	800	17,0

Table 2 *Potential of renewable energy sources, trillion kWh/year (see Energy and Nuclear Power 1985; WEC Survey 2001; International Energy Outlook 2009; World Energy Outlook 2008)*

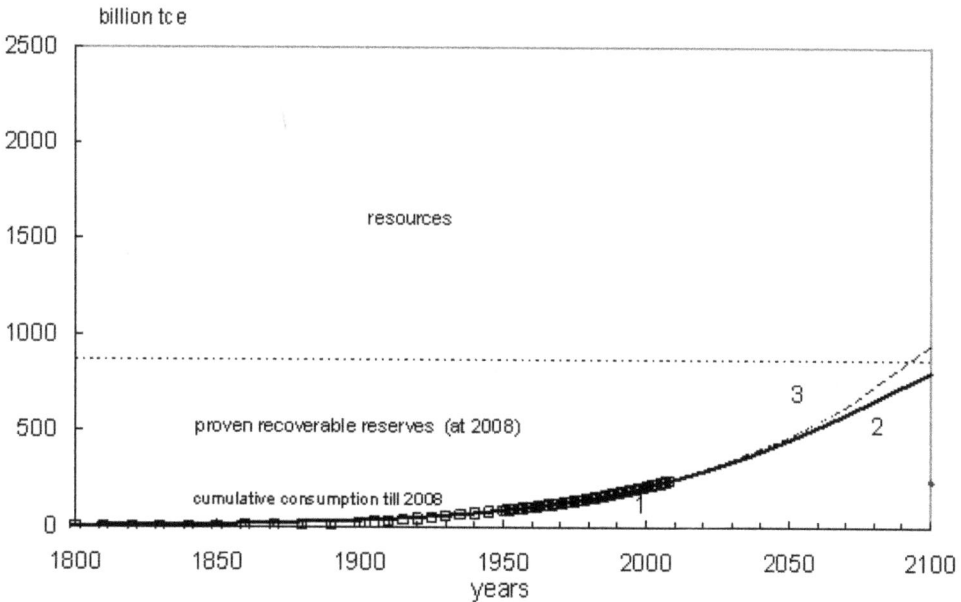

Figure 4 *Cumulative global coal consumption: (1) historical data (see International Energy Annual 2009; BP Statistical Review... 2009; Energy Statistics Yearbook 2009); (2) forecast of the present work; and (3) WEC/IIASA Reference Case scenario (see Nakicenovic, Grubler, and McDonald 1998).*

The structure of the world energy balance suggested in the present work for the forecast of global energy consumption up to 2050 is quite similar to that in (Nakicenovic, Grubler, & McDonald, 1998; Nakicenovic & Swart, 2000) (Table 1). Appreciable differences in the estimates for the shares of hydrocarbon fuels and non-CO2 emission energy sources arise closer to 2100, resulting from the fact that the status of energy technologies by this time is still difficult to predict. Nevertheless, the share of non-fossil energy sources that we expect by 2100 (55–65 %) is considered in a number of IPCC scenarios (i.e., A1T, A1B, and B1 scenarios) (see Nakicenovic & Swart, 2000). Thus, even though we made use of quite a different approach to assess global energy perspectives, our suggested structure of global fuel balance does not generally contradict expert assessments for the development of energy technologies and, with regard to fossil fuels, it is fully provided by available natural resources.

Environmental characteristics of the suggested scenarios are determined by the carbon coefficient of global energy consumption (Figure 5). One can see that, with the recent growing share of coal in the global commercial energy consumption, the long-term tendency for a decrease of carbon intensity reverses, approaching the current 1.9 ton CO2/tce from the minimum of 1.8 ton CO2/tce in 2000 but the subsequent decrease to the 'clean' scenario level by the end of the century is anticipated. Surely, such changes should appreciably enhance emissions of CO2 (and other greenhouse

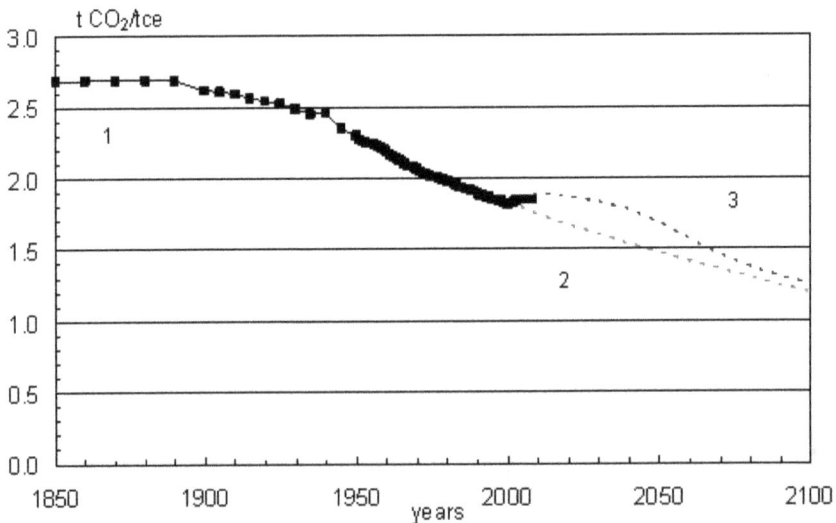

Figure 5 *Dynamics of the carbon coefficient of the global commercial energy consumption: (1) historical data (see International Energy Annual 2009; BP Statistical Review... 2009; Energy Statistics Yearbook 2009); (2) 'clean energy' and (3) 'coal energy' scenarios of the present work.*

gases). These consequences and their associated global climate changes are considered below.

ATMOSPHERE AND CLIMATE CHANGES IN THE TWENTY-FIRST CENTURY

Dramatic scenarios of future global warming (see Solomon *et al.* 2007) are based on models of general atmosphere and ocean circulation simulations.2 As was repeatedly shown (e.g., see Klimenko, V. V., Klimenko, A. V., and Tereshin, 2001; Klimenko, V. V., Klimenko, A. V., Andreichenko *et al.*, 1997), these models, while getting more and more complicated, are still incapable of adequately representing the observed climate changes and give widely scattered estimates for such an important parameter of a climatic system as the sensitivity to the content of greenhouse gases in the atmosphere, which, according to various estimates (their review is given in Solomon *et al.* 2007), varies in the range 1.5–5.5 deg. with doubling CO2 concentration. To overcome these difficulties, we have developed a more simple regression analytical climatic model (RACM) (see Klimenko, 1997, 2007; Klimenko, V. V., Klimenko, A. V., Andreichenko *et al.*, 1997; Klimenko *et al.*, 1994; Klimenko, Mikushina, & Tereshin, 1999; Klimenko & Mikushina, 2005; Khrustalev *et al.*, 2008) which combines physical methods for representing thermodynamic processes in the ocean – atmosphere system and statistical methods for correlating their impact (temperature responses) with external perturbing factors. With a correct account for the effect of a few major natural climate forcing (solar, volcanic etc.) we estimate the sensitivity of the global climatic system at about 1.9 °C for doubled CO2 concentration, which falls into the lower range of estimates for this parameter (see Solomon *et al.*, 2007).

To estimate the changes in atmospheric CO2 content, associated with anthropogenic emission, we made use of the box diffusion model of global carbon cycle, developed at the Moscow Energy Institute (see Klimenko, V. V., Klimenko, A. V., Andreichenko *et al.*, 1997). The CO2 concentrations calculated by greenhouse emission scenarios for the 'clean' and 'coal' global energy development models are shown in Figure 6.

According to the RACM global climate change projections, assuming the basic forecast of principal climate forcing factors (Ibid.) and 'coal energy' scenario, the global average temperature will increase by about 1.3 °C within this century. Although this value even exceeds the maximum Holocene mark, it is, in terms of another important

---

2. According to the most extreme of them, the global average temperature will increase by 5 °C over the current century, which has never occurred not only during the history of civilization, but also over the past 40 million years on the whole.

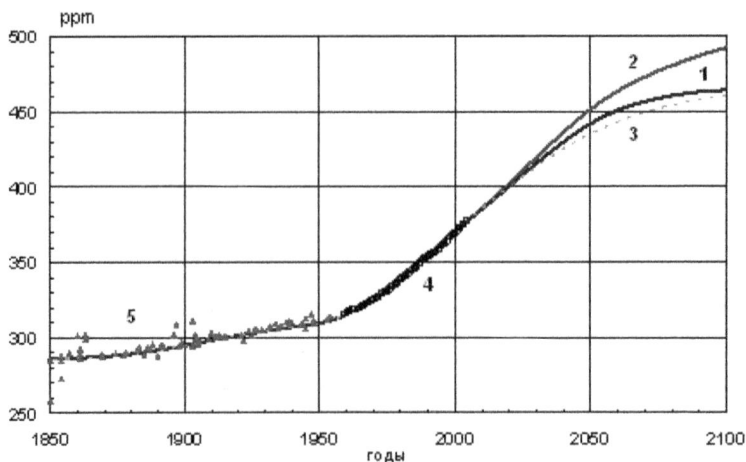

Figure 6 *Carbon dioxide concentration change: model simulations by the emission scenarios of the (1) 'clean' and (2) 'coal' global energy development of the present work, model simulation (3) (from Klimenko, V. V., Klimenko, A. V., and Tereshin, 2001) and data of instrumental measurements (4) and ice-cores (5) (Solomon et al., 2007).*

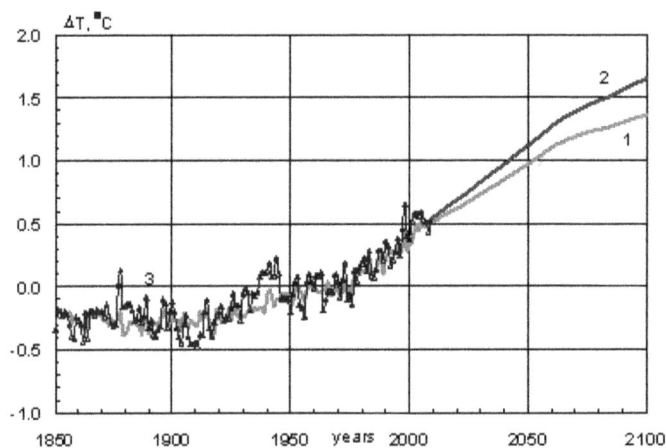

Figure 7 *Global average temperature change (compared with the 1951–1980 mean): model simulation and forecasts for the 'clean' (1) and 'coal' (2) energy scenarios of the present work; and (3) instrumental data (see Solomon et al. 2007).*

criterion (temperature change rate), probably within the adaptation limits of the biosphere. Simulation of the implementation of the 'clean' scenario similar in goals to those of the Kyoto Protocol (Figure 7) shows that the measures suggested by this international agreement for climate stabilization, even though they cannot much affect the dynamics of the global average temperature, will still help to reduce the global warming by 0.3 °C. A comparison of the 'clean' and 'coal energy' scenarios (Figure 7, curves 1 and 2, respectively) shows that the meeting of the Kyoto Protocol targets can favor a more environmentally safe energy development.

Local climate changes are expected to be quite diverse. Our survey for various parts of Russia (see Klimenko, 2007; Khrustalev et al., 2008; Klimenko & Mikushina, 2005) showed that in the nearest decades the average annual, winter, and spring temperatures will appreciably increase, which, in its turn, will affect a number of applied climate characteristics crucial for different economy sectors. Thus, a shorter and warmer cold period will require less fuel for heating (down 15 % from the present level by 2050) (see Klimenko, 2007). Positive changes in transport and agriculture are also expected, which, too, will decrease the required energy consumption. Probably, permafrost areas are the most vulnerable to climate changes and will require huge additional investments in the existing infrastructure (see Khrustalev et al., 2008).

A comparison of temperature and precipitation fields for present warming and other historically warm periods which are useful analogs for the expected warming, such as the Atlantic Holocene Optimum (about 7–6 thousand calendar years ago) and the Medieval Warm Epoch (the late ninth-twelfth centuries) (see Klimenko, 2001, 2004), shows that considerable temperature changes may occur only in several countries of the Northern hemisphere. Thus, considerably increased average annual temperatures are observed, along with Russia, only in Canada, the Northern part of the USA and Middle and Central Asia, whereas the temperature changes in Europe, South-East of the USA, and most part of China and India are either inconsiderable or even negative. These changes will almost everywhere be accompanied by enhanced precipitation, except for the North East of the USA, the Mediterranean, eastern Provinces of China and South-East States of India, where a certain desiccation takes place.

However, the scenarios of future climate changes, presented in other works, are quite different up to catastrophic ones (Figure 8). Thus, the most recent IPCC review (see Solomon et al., 2007) does not exclude that the global average temperature may increase by 5 °C by the end of this century. Such a large-scale global warming will entail irreversible environmental changes in most regions of the world, including Russia, and, as a result, will have an extremely negative impact on all spheres of human activities. Undoubtedly, the ecological pressure on economy, in particular, in terms of the Kyoto Protocol, is much dependent on whether these projections will be proved or disproved. Provided the catastrophic forecasts are disproved, further tightening of Kyoto constraints will be less likely. At the same time, if things follow an unfavorable scenario, further consolidation of the environmental protection community should be expected. However, we have to repeat that the results of our long-term research provide firm evidence in favor of moderate scenarios of global climate changes.

Thus, our early forecasts (see Klimenko et al., 1994; Klimenko, 1997) of global climate changes are still valid, which, by the way, evidenced by the fact that they fully represent the actual data for the past two decades (see Klimenko, V. V., Klimenko, A. V., & Tereshin, 2001; Klimenko & Tereshin, 2010). According to this forecast, we expect that the global average temperature will increase by another 1–1.5 °C over the course of the current century (Figures 7 and 8), which falls below the range of IPCC estimates for possible atmosphere and climate changes (see Solomon et al., 2007), even including scenarios assuming the world population decrease (B1), and is five times below the possible temperature rise due to the extreme scenarios group A1FI, which expects the most intensive growth of fossil fuel consumption. Nevertheless, the expected warming is far beyond the range of the natural variability of global climate, recorded in palaeoclimatic data over the past 2.5 thousand years (see Klimenko, 2001, 2004); however, the warming rate (about 0.1 °C per decade) appears to be within the adaptation limits of the biosphere (Klimenko & Tereshin, 2010). It can be concluded that both the warming expected to occur by the late twenty-first century and the increase in the atmospheric CO2 concentration will only slightly exceed the scale of global changes that have already occurred over the past century.

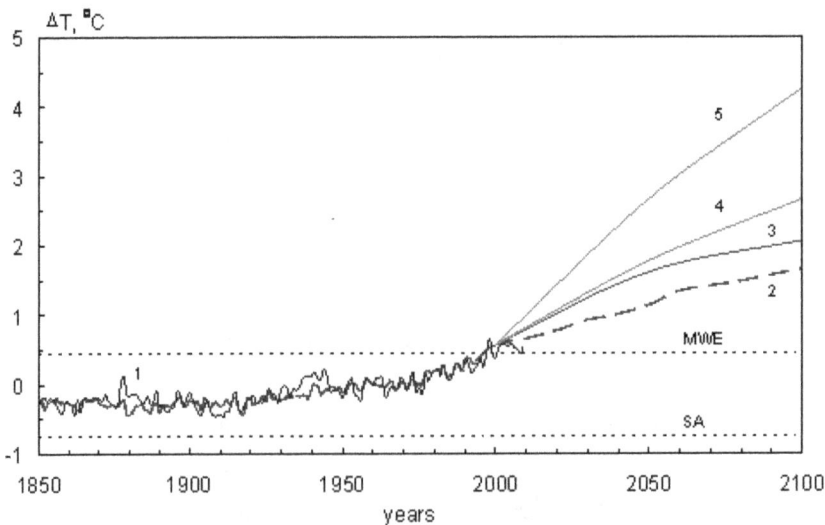

Figure 8 *Global average temperature change (compared with the 1951–1980 mean): (1) instrumental observations data (see Solomon et al. 2007); (2) model simulation and forecasts by the (2) 'coal' energy scenario of the present work and IPCC scenarios (Ibid.): (3) B1, (4) B2, and (5) A1FI. Temperature levels of the Medieval Warm Epoch (MWE) (see Klimenko 2001) and Cold Subatlantic Epoch (SA) (see Klimenko, 2004) are also indicated.*

CONCLUSIONS

The time passed after the publication of the first results of the application of the genetic approach to forecasting future energy use showed that this approach gives encouraging results. Our early forecasts of world energy consumption represent the actual data for the last 20 years remarkably well.

The development of the method of historical extrapolation to assess the future global fuel mix allowed us to develop the perspective energy balance for the current century, in which the key role of fossil fuels will hold up to at least 2060–2065.

This historical scenario of global energy consumption is completely provided by the available resources of fossil fuel and does not contradict the assessments of prospective development of non-fossil energy sources.

An implementation of the historical scenario of energy development is expected to cause moderate atmosphere and climate changes which are quite comparable with the scale of global changes that have already occurred over the past century.

Local manifestations of expected climate changes will be very diverse. Thus, in the nearest decades in moderate and high latitudes we can expect a shorter and warmer cold period, as well as appreciable destructive phenomena in the permafrost zone of the Russian territory.

ACKNOWLEDGEMENT

The study was supported by the Russian Ministry of Education and Science in the framework of the Federal Program 'Scientific and Educational Staff of Innovative Russia'.

REFERENCES

BP Statistical Review of World Energy 2009. London: BP, 2009. URL: http://www.bp.com/ statistical review.

Demographic Yearbook 2006. New York: UN, 2009.

Energy and Nuclear Power Planning in Developing Countries. Tech. Rep. Ser. 245. Vienna: Int. Atomic Energy Agency, 1985.

Energy Statistics Yearbook 2006. New York: UN, 2009.

International Energy Annual 2006. Energy Information Administration. Washington, D. C.: Department of Energy, 2009.

International Energy Outlook 2009. Washington, D.C.: DOE/EIA, 2009.

Khrustalev, L. N., Klimenko, V. V., Emelyanova, L. V. *et al.* 2008. Dynamics of Permafrost Temperature in Southern Regions of Cryolithozone under Different Scenarios of Climate Change. Kriosfera Zemli 12(1): 3–11. In Russian.

Klimenko, V. V. 1994. An Influence of Climatic and Geographical Conditions on the Level of Energy Consumption. Physics. Doklady 39(11): 797–800.

Klimenko, V. V. 1997. Why Carbon Dioxide Emission should not be Controlled. Thermal Engineering 44(2): 85–89.

Klimenko, V. V. 2001. Climate of the Medieval Warm Epoch in the Northern Hemisphere. Moscow: MEI Publ. In Russian.

Klimenko, V. V. 2004. Cold Climate of the Early Sub-Atlantic Epoch in the Northern Hemisphere. Moscow: MEI Publ. In Russian.

Klimenko, V. V. 2007. Climate Change Impact on the Heat Demand in Russia. Energiya 2: 2–8. In Russian.

Klimenko, V. V., Fedorov, M. V., Andreichenko, T. N., and Mikushina, O. V. 1994. Climate on the Border of Millennia. Vestnik MEI 3: 103–108. In Russian.

Klimenko, V. V., and Klimenko, A. V. 1990. Will Energy Development Result in Climatic Collapse? Teploenergetika 10: 6–11. In Russian.

Klimenko, V. V., Klimenko, A. V., Andreichenko, T. N. *et al.* 1997. Energy, Nature, and Climate. Moscow: MEI Publ. In Russian.

Klimenko, V. V., Klimenko, A. V., and Tereshin, A. G. 2001. Power Engineering and the Climate on the Eve of the New Century: Forecasts and Reality. Thermal Engineering 48(10): 854–861.

Klimenko, V. V., and Mikushina, O. V. 2005. History and Projection of Climate Change in the Barents and Kara Seas Basin. Geoecologiya 1: 43–49. In Russian.

Klimenko, V. V., Mikushina, O. V., and Tereshin, A. G. 1999. Do We Really Need a Carbon Tax? Applied Energy 64: 311–316.

Klimenko, V. V., and Tereshin, A. G. 2010. World Energy and Global Climate Beyond 2100. Teploenergetika 12: 38–44. In Russian.

Nakicenovic, N., Grubler, A., and McDonald, A. 1998. (Eds.). Global Energy Perspectives. Cambridge: Cambridge University Press.

Nakicenovic, N., and Swart, R. 2000. (Eds.). Special Report on Emissions Scenarios. Cambridge: Cambridge University Press.

North, D. C. 1990. Institutions, Institutional Change and Economic Performance. Cambridge: Cambridge University Press.

Snytin, S. Yu., Klimenko, V. V., and Fedorov, M. V. 1994. A Forecast of Energy Consumption and Carbon Dioxide Emission into the Atmosphere for the Period until 2100. Physics – Reports 336(4): 457–460.

Solomon, S., Qin, D., Manning, M. *et al.* 2007. Climate Change 2007: The Physical Science Basis. Contribution of Working Group I to the Fourth Assessment Report of the Intergovernmental Panel on Climate Change. Cambridge: Cambridge University Press.

Toynbee, A. J. 1988. A Study of History. London: Oxford University Press.

World Energy Outlook 2008. Paris: OECD/IEA, 2008.

World Population Prospects: The 2008 Revision. New York: UN, 2009.

WEC Survey of Energy Resources. London: World Energy Council, 2001.

Chapter 18

WILL THE GLOBAL CRISIS LEAD TO GLOBAL TRANSFORMATIONS?

Leonid Grinin & Andrey Korotayev

This article analyzes some important aspects of the world socioeconomic and political development in the near future. The future always stems from the present. The first part of the article analyzes the global causes of the contemporary crisis and the possibilities to eliminate the most acute problems that have generated this crisis. The authors believe that in some respects the global financial system, notwithstanding all its negative points, still performs certain important positive functions including the 'insurance' of social guaranties on a global scale. New financial technologies decrease the risks in a rather effective way, they expand possibilities to attract and accumulate enormous capitals, actors, and markets. The modern financial sector also contributes to the insurance for social funds on a global scale. The participation of pension and insurance funds in financial operations leads to the globalization of social sphere. The countries poor in capital, but with large cohorts of young population, are involved more and more in a very important (though not quite apparent) process of supporting the elderly portion of population in the West through the vigorous unification of the world's financial flows, their standardization, and by increasing global mobility and anonymity.

The second part of the article considers some global scenarios of the World System's future and describes several characteristics and forecasts of the forthcoming 'Epoch of New Coalitions'. Among the problems analyzed in this paper are the following: What are the implications of the economic weakening of the USA as the World System center? Will the future World System have a leader? Will it experience a global governance deficit? Will the world fragmentation increase?

Keywords: global crisis, global financial system, financial revolution, financial technologies, pension funds, social funds, medium-length economic cycles, Juglar cycles,

---

international order, the World System, World System leader, global hegemony, center, periphery, global governance, national sovereignty.

GLOBAL CHARACTERISTICS OF THE CRISIS AND THE NECESSITY OF CHANGES

The global crisis that has somehow sobered down (at least for some time) those who believed that in this century global development would proceed without crises, appears to be fading away (though the threat of one more wave of crisis does not seem to have disappeared entirely). The world economy in general has moved from the phase of recession to the phase of depression, and there seem to appear some indications suggesting certain movement toward the recovery phase in a number of countries. However, in Europe one can hardly notice any signs of recovery. To a certain degree the growth rates slow down also in China, India and some other developing countries for which the global crisis was not as difficult as it was for the countries of the World-System core. This implies that we can look at the causes (as well as proximate and ultimate consequences) of the deepest (within the last 75 years) economic crisis in a way somehow different from that of four years ago.

The history of economic crises suggests that each of them was connected with the type of relationships within the World System.[1] However, the strongest crises also changed in a rather significant way the World System structure, the connecting lines of this system. The current global financial-economic crisis is also likely to contribute to the beginning of the process of major changes in the World System structure and functioning, as well as in the principles of the international relations in the forthcoming decades. In the first part of article we analyze the global causes of the contemporary crisis and both the negative role of the world financial flows and their important positive functions including the 'insurance' of social guaranties at the global scale. The second part of the paper is devoted to the analysis of probable future transformations in connection with the crisis and to estimating the probabilities of various scenarios in the development of the World System during the forthcoming decades.

---

1. The world-system approach originated in the 1960s and 1970s due to the work by Fernand Braudel, Andre Gunder Frank, Immanuel Wallerstein, Samir Amin, and Giovanni Arrighi (Braudel, 1973; Frank, 1990, 1993; Frank & Gills, 1993; Wallerstein, 1987; Chase-Dunn & Hall, 1994, 1997; Arrighi & Silver, 1999; Amin et al., 2006). The term World System/world-system is used rather widely and not by world-system analysts only. For more detail on the history and contents of 'the World System' notion see Grinin & Korotayev, 2009a: 9–19; 2009b.

1. Global Causes of the Global Crisis

The growth and deepening of financial-economic globalization has led to the unprecedented development of a number of countries and regions in the last decade (see e.g., Maddison, 2007, 2010; World Bank, 2012); yet, it has also caused some crises. That is why the current crisis may be considered as the reverse side of globalization (see Grinin, 2008c; 2012).

It is quite natural that the causes and character of the current crisis will be a subject of attentive research for quite a long time. However, it is quite evident that the main factors causing the crisis have not disappeared. Also many problems have been just temporarily dampened by an unprecedented pumping of funds that can be only justified as an extraordinary measure that can worsen the situation in the future. That is why there are some grounds to expect a new outburst of the crisis in the near future (within 3–5 years). In the meantime, there is a considerable probability that the strongest manifestations of the crisis will be felt most distinctly in the fast growing Asian economies that have suffered rather moderately from the current crisis. Similar situations, with a similar asynchrony during the strongest crises with respect to Europe and North America, were observed in the late 19th century and the early 20th century (see e.g., Lescure, 1907; Tugan-Baranovsky, 1954, 2008 [1913]).

The major changes in the global division of labor between countries are associated with the most important causes of the crisis.[2] One of the most salient points here is that the countries of the World System center (especially the UK and USA) have developed their financial sector in the most active way.[3] In the meantime, the semiperipheral countries have been more actively developing the 'real economy'. As a result, in the Western economies the GDP share produced by the financial sector reaches between a quarter and a third of the total GDP, exceeding the share produced by industry. In general, within the world economy (due to the West's ability to accumulate the world capitals, as well as because of the formation and diffusion of new financial technologies) the financial sector has been growing faster than the other sectors during the last three decades. As a result, the financial sector has transformed from a sector serving the economy, into a sector producing the main vector of its development; thus it has become a sector where an immense share of added value is produced. Such a division of labor has a number of important consequences. Western countries

---

2. For more detail on the causes of the crisis, as well as on the development of new financial technologies see Grinin, 2008c, 2009a, 2009c; Grinin & Korotayev, 2010; Grinin, Korotayev, & Malkov, 2010.

3. The World System core countries may be identified as the high income OECD countries that include 24 out of 30 members of this organization and that produce 60 % of all the world GDP.

become not only the world capital accumulation center, they also become net importers of capitals.

In these countries one can observe a phenomenon of deindustrialization. On the
contrary, one can observe a fast industrial growth in the semiperipheral countries.[4]
This has appreciably contributed to the development of the situation when the growth
rates in such developing economies—e.g., the BRICS members and some other semiperipheral countries—are significantly higher than in the West. In the semiperipheral
countries one can observe a particularly fast growth of the export sectors, whereas
the USA and some other core countries become more and more the world center of
consumption whose demand determines to a considerable extent the prosperity of
semiperipheral and peripheral economies. Thus, in general we can observe the decline
of the role of the West as an industrial-economic center of the World System. On the
other hand, this is accompanied by the growth of its importance as an importer of
commodities and capitals; correspondingly, the economic role of the semiperiphery
(in general, and certain semiperipheral centers, in particular) grows; yet, their economy becomes more and more dependent on the ability of the West to consume. The
consumption economy has become an imperative not only for the West, but for the
whole World System (see e.g., Wolf, 2005).

An anarchic and extremely rapid development of new financial centers, financial
currents and technologies (that has secured a fast growth of the financial sector) has
also contributed in an extremely significant way to the genesis of the global financial-
economic crisis. Their negative role has been amplified by the lack of transparency
with respect to many financial instruments and institutes, which led to the actual ob-
scuring of risks and to the general underestimation of global risks (Kudrin 2009: 9–10;
see also Suetin, 2009; Grigoriev & Salikhov, 2008).

It should be noted that the aspiration for risk (which is usually characterized as a
positive quality feature of an entrepreneur's psychology) should be reconsidered in
the context of globalization. If the financiers (and finally other businessmen) consider
the whole world to be a sphere for possible investment, and thus, given this condition, risks are counted in trillions of dollars, then to risk or not to risk stops being just
a question of personal choice for individual entrepreneurs and firms. An adventurous
inclination for risk (whose consequences could produce a fatal influence on the whole
global economy) becomes a very dangerous feature. Consequently, it becomes nec-

---

4. Some analysts maintain that relative wealth is flowing now from the World System center to its semiperiphery (National Intelligence Council, 2008), however, the opposite view is more wide-spread.

essary to control activities of such global entrepreneurs (for more detail on the crisis psychology see Grinin, 2009b).

2. Why Have Classical Features of Previous Economic Crises Manifested in the Current Crisis?

The global causes of the contemporary crisis have led to an unexpected effect—we observe within it some classical features of the cyclical crises of the 19th and early 20th centuries that appeared to have been eliminated. Crises in their classical form (as unexpected and even unexplainable economic collapses occurring against the backdrop of unprecedented florescence, growth of profits and prices) were typical for that period of time. Later, in the second half of the 20th century (in a direct connection with an active countercyclical interference of the state) the cyclical crises became much weaker and less pronounced.

Let us recollect that medium-length economic cycles with a characteristic period of 7–11 years (that go through the upswing phase turning into the overheating sub-phase, and ending with a crisis/collapse/recession and depression) are also known as Juglar cycles.[5] Such cycles were typically characterized by fast (sometimes even explosive) booms (that implied a great strain on the economic system) followed by even faster collapses. The period of upswing, followed by boom and overheating, was accompanied (a) by a fast and inadequate growth of prices of raw materials and real estate objects; (b) by an excessive demand for credit funds and investment expansion beyond any reasonable limits; (c) by outbursts of speculations with commodity and stock assets; (d) by enormous increase in risky operations. All these are vivid features of the Juglar cycle that were described many times in studies produced by representatives of various schools of economic thought (see e.g., Juglar, 1862, 1889; Lescure, 1907; Tugan-Baranovsky, 1954, 2008 [1913]; Marx, 1993 [1893, 1894]; Mendelson, 1959–1964; Hilferding, 1981 [1910]; Keynes, 1936; Hicks, 1946 [1939]; Minsky, 2005; Samuelson & Nordhaus, 2005, 2009; Haberler, 1964 [1937]; see also Grinin, Korotayev, &Malkov, 2010; Grinin & Korotayev, 2010). All these features have been observed in the current crisis.

Our analysis has also demonstrated that almost at every upswing phase a new financial technology (or a new type of financial assets[6]) acquires a special significance

---

5. They were denoted as Juglar cycles after Clément Juglar (1819–1905), who was one of the first to demonstrate the periodical, regular, cyclical character of economic crises; though a number of economists (including Karl Marx) studied the economic cycles simultaneously with Juglar.

6. In the 19th century for some time this role was played by railway shares whose use made it possible

(on the appearance of new financial technologies during new economic cycles see Grinin & Korotayev, 2010). Abrupt transitions from booms to collapses were connected with spontaneous economic development that was regulated by market forces and almost by nothing else, as state interference into the economic development was not sufficient. Under these conditions (against the background of the presence of gold standard) acute crises became inevitable.[7] Karl Marx had already considered the anarchic character of development inherent in a capitalist economy (against the background of the economic agents' urge towards the expansion of supply) as the main cause of the economic crises.

As a result of the Great Depression the role of the state in regulating the economy changed. Due to various direct and indirect ways of influencing the macroeconomic framework of national economic development it became possible for the state to minimize dramatic distortions of booms and busts.[8] As a result, the crises became far less pronounced than before.

However, the global causes of the current crisis have made those Keynesian monetary methods of economic regulation (that are effective at the scale of a single country) ineffective at the global level. The world economy is being transformed into a single system, but technologies of the countercyclical management at the World System scale have not yet been worked out. Nation-states wage a tense competition for higher growth rates (and the question of possible limitation of those rates is not even discussed). Respectively, in the absence of the necessary level of control, the features of anarchic and arrhythmic development of non-regulated market economy become more and more salient at the World System level. This implies a certain systemic similarity between the functioning of unregulated national economy and the one of the modern global economy. We believe that this similarity accounts for the recurrence of

on a number of occasions to expand dramatically credit and speculation, overheating the economy.

7. Thus, with the overexpansion of credit and the swelling of financial assets the amount of money substitutes (shares, promissory notes, bonds, etc.) expanded enormously (numerous proponents of the important role of supernormal credit belong to the so-called Austrian school, e.g., Mises, 1981 [1912]; Hayek, 1931, 1933). As a result, with the decrease of trust in those stocks a sudden demand for gold and cash grew so dramatically that it was able to crash the whole banking system (see e.g., Tooke, 1838–1857; Evans, 1969 [1859]; Juglar, 1862, 1889; Lescure, 1907; Tugan-Baranovsky, 1954, 2008 [1913]).

8. It became possible to put speculation under some control. For example, after the Great Depression in the USA, the Glass-Steagall Act was passed, forbidding banks, investment firms and insurance companies to speculate at stock exchanges (see: Lan, 1976; Samuelson & Nordhaus, 2005, 2009; Suetin, 2009: 41). In 1999 in the USA the law on financial services modernization was passed, which annulled the Glass-Steagall Act that was in force for more than 60 years (see: Suetin, 2009: 41). As a basis for introducing the law on financial services modernization, it has been claimed that American credit organizations are inferior to foreign rivals, especially European and Japanese 'universal banks' which were not subject to such limitations (Greenspan, 2007).

some features of cyclical crises of the earlier epoch (see Grinin, 2009a, 2009c; Grinin & Korotayev, 2010; Grinin, Korotayev, & Malkov, 2010 for more detail).

1. In many respects subjects of the international economy (because of a lack of development in the financial regulation of international law) behave in a similar uncontrolled and anarchic manner as was observed earlier with respect to subjects of a national market (i.e., because of lack of development in economic regulation of national law). As they use floating courses of exchange in their accounts, this inevitably leads to sharp distortions in international trade, devaluations, defaults, bankruptcies, etc.

2. The urge of states and major corporations to attain maximum growth rates in the absence of any effective macroeconomic limitations leads, at the level of the world economy and world financial system, to consequences that are analogous to the ones that were produced by uncontrollable growth and competition for market share in the capitalist economies of the 19th and early 20th centuries: overheating, 'bubbles', and collapse.

3. In recent decades, the movements of capitals between countries have become free (e.g., Held *et al.*, 1999; Held & McGrew, 2003); that is, they are rather weakly regulated by national law and are hardly regulated at all by international law. This causes enormous impetuous movements of capitals that lead to very rapid upswings in some places and later, with crises, to sharp declines.

4. The development of the modern economy not only has been accompanied by the formation of new financial technologies but it has started to produce more and more added value precisely in the financial sphere (as financial services). This led to a sharp increase in the financial component of the crisis (in comparison with earlier decades when the main growth was observed in industry).

3. Financial Speculation: Does It Have a Positive Side?

Financial middlemen were cursed in all epochs. And there were always certain grounds to curse them. But they exist and the modern economic system cannot reproduce itself without them, as the modern market economy depends on financial middlemen in a rather significant way, as they transform households' savings into productive investments (Greenspan, 2007).

The activities of modern financial corporations and funds (which lead to the uncontrolled growth of financial assets and anarchy in their movements) are quite justly criticized (this point will be discussed below; see also Grinin, 2008c, 2009a, 2009c; Grinin & Korotayev, 2010). However, it would not be correct to maintain that modern financial technologies are fundamentally deleterious, that they only lead the world economy to various troubles, that they are only beneficial to the financiers and speculators. On the contrary, both the formation and the current development of the financial sector are connected with the performance of very important functions—and just on a global scale.

Thus, the modern financial globalization should not only be cursed; it also has some positive sides. Summing up the achievements of what is called 'the financial revolution' (see Doronin, 2003; Mikhailov, 2000; see also Held *et al.*, 1999) we would provide our own version of the most important directions of the development of financial engineering (in addition to the computerization of this sphere of business). We shall also try to specify the positive influence produced by them. Those directions can be described as follows:

1. Powerful expansion of nomenclature of financial instruments and products, which leads to the expansion of possibilities to choose the most convenient financial instrument.

2. Standardization of financial instruments and products. This creates the possibility to calculate an abstract (that is, an aggregate, unified measure based on a standard package of shares and other stocks) base (and not just concrete prices of concrete securities). This secures a considerable economy of time for those who use financial instruments; it makes it possible to purchase financial securities without a detailed analysis of particular stocks; this leads to an increase in the number of participants by an order of magnitude.[9]

3. Institutionalization of ways to minimize individual risks. In addition to the above mentioned expansion of nomenclature and assortment of financial products, it appears especially important to mention: first, the development of special institutions—specialized clearing chambers—with their internal regulations (which makes it possible to avoid reliance on courts of law); second, the use of special rules and computer software, various technologies; and, third, new forms of risk hedging. All these changes help to minimize both the individual risks of unfulfilled deals and also of bankruptcies in the framework of certain stock markets.[10]

---

9. This is similar to the situation with a wholesale purchase of a large batch of any standard commodity when the buyer has no need to examine every particular piece.

10. However, the expansion of the operations' volume and their acceleration create a threat of global financial collapses.

4. Increase in number of participants and centers for the trade of financial instruments. Modern financial instruments have made it possible to include a great number of people through various special programs, mediators, and structures.[11] These changes result in the diffusion of technologies among the owners of capitals of various size (this is similar to the development of joint-stock companies that made it possible to accumulate enormous capitals). It is also possible to observe a significant increase in the number of financial centers and their specialization, as well as in the interconnection of national and world financial centers. It is also extremely important that we observe the growth of the number of emitters of various financial derivatives.[12]

The significance of the changes outlined above for the financial sector on the global stage can be described as follows:

*   Enormous new capitals, actors, markets are accumulated and engaged, which creates a 'difference of potentials' which is necessary for the activization of an economy to attract capitals and investors.

*   Due to enormous growth in the volumes of operations, we observe the emergence of possibilities to extract profits from such operations, from which it was impossible to extract any profits earlier. Thus, a firm could earn just 3 cents from re-selling one share, but it may re-sell millions of such shares every day—and what is more, it may re-sell the same shares dozens of times within the same day (see Callahan 2002).[13] One may compare this with the industrial concentration of low-grade ores, whose processing was not profitable before the invention of respective technologies.

*   The growth of diversity of financial products, the development of specialization in the production of financial services, and the increase in nomenclature of those services make it possible to smooth the demand fluctuations and to increase the general volume of sales (in fact, the growth of nomenclature of products gives the same results within any branch of economy).[14]

---

11. The Foreign Exchange Market (FOREX) is the most famous among them.
12. This is similar to the growth of the number of commodity producers with the growth of the network of units selling and servicing respective commodities.
13. It is quite natural that this is most relevant for the upswing phases, whereas this is observed to a much smaller (but not zero) degree during recessions.
14. We believe that the trend toward the maximum standardization of the contract conditions has consequences that are similar to the ones produced by the standardization in manufacturing: in both cases the use of standards expands the sphere of the use of respective technologies and products by an order of magnitude.

- Financial currents and financial centers start to structure the world economy in a new way. The market economy is always structured along certain modes of communications. One may recollect how railway construction not only altered the transportation of commodities but also changed the whole organization of economic life. Within modern information economy the financial currents start playing a role of such system-creating communications. In those zones where we observe the most important financial currents, we may also observe the most intensive economic life. Small financial streams (like, before, small streams of commodities along the railways) create a new economic network.

- The new structure makes it possible for the periphery to participate actively in the world economy. It is quite clear that the spontaneous movement of capital can lead to collapses and global crises; yet, the same was observed in the 19th century when the vigorous railroad construction (accompanied by unprecedented speculations) led first to enormous upswings, and, later, to collapses. Thus, the main task is to put the most dangerous and unpredictable actions under control.

4. Financial Currents as the World Pension Fund?

Our research has made it possible to detect such global functions of the world financial sector that do not seem to have been noticed by analysts. Those functions have developed in conditions of currency not guaranteed by gold and they are connected with the necessity to protect savings in conditions of inflation against losses and risks during long periods of time. They emerged as an unintended consequence of the radical transformations in the world financial system that began in the 1970s. At that time the world financial system finally rejected the gold monetary standard as a result of the double devaluation of dollar and the collapse of the Bretton Woods monetary system. The price of gold was no longer tied to the dollar even nominally, it became free, whereas the currency exchange rate became floating.

However, as a result of the rejection of the golden standard the function of savings' protection moved finally from an 'independent' guarantor (i.e., precious metals) to the state.[15] However, there was no state left, on which the capital owners could rely

---

15. Naturally, the value of gold and silver could fluctuate. One can easily recollect the so-called price revolution of the 16th century, as a result of which the prices grew four times (e.g., Goldstone, 1988, 1991). But there has never been a single case when gold or silver lost their value momentously, or when their prices dropped close to zero (this eventually happens with prices of shares), whereas in the 19th and early 20th centuries (when many states applied the gold standard [Held et al., 1999]) the value of money was sometimes surprisingly stable for long periods of time (the same is true for prices of many key commodities), and this allowed many people to live from the interest rates of their sav-

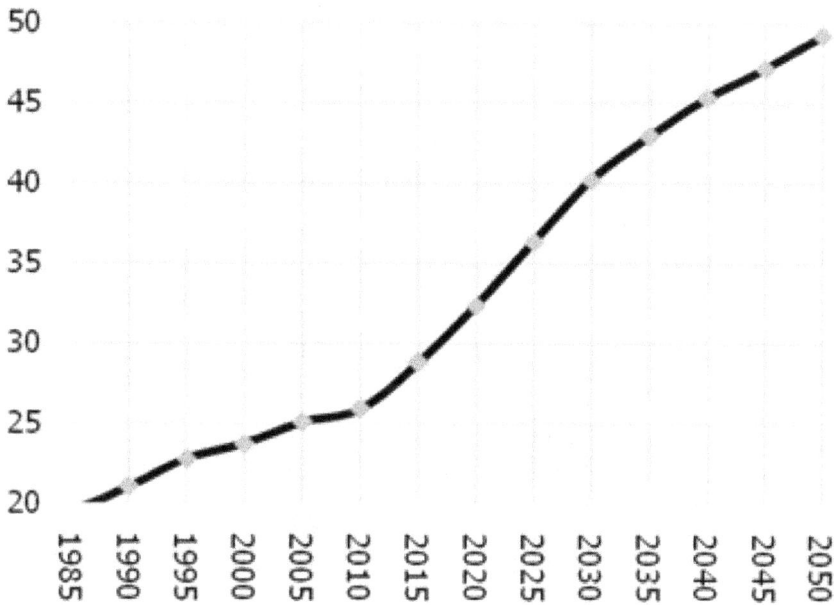

Figure 1 *Number of pensioners per 100 working age adults in developed countries, 1985–2050.*

entirely as on a perfectly secure guarantor. One should add to this, the growth of infla-tion that especially bothered the West in the 1970s and 1980s. One should note that it was during the 1960s and 1970s that the volume of 'social capitals' in the direct sense (i.e., various pension, social, insurance funds, including the medical insurance funds) grew very significantly in direct connection with active social legislation, the growth of the quality of life, and some demographic processes (first of all, the coming to age of the numerous baby-boom generation). There were some other important sources for the growth of capitals in the 1970s and 1980s, in addition to the above-mentioned ones.

The general volume of capitals also grew due to the petrodollars, the increase in the emission of stocks, and borrowing (including the sovereign borrowing).[16] In gen-eral, since that time one may observe the process of rapid growth of the volume of free capitals that should be invested somewhere.

---

ings. It made it possible to rely on savings in the form of gold/silver coins to guarantee one's survival in old age or for any emergencies. Incidentally, this was a very important basis for the development of thriftiness. Today prices of precious metals are as unstable as the ones of any other assets, and the magnitude of their fluctuations is great.

16. Many years later some other sources were added to these; for example, the so called state in-vestment funds (national development funds) that accumulated resources obtained by states through some super-profits (stemming, for example, from the exportation of oil) and invested them in financial markets abroad. At present a few dozen of states have such funds (National Intelligence Council, 2008).

With inflation the question of where to invest capitals and funds not guaranteed by gold or hard currency becomes extremely important. This is especially relevant for capitals accumulated by pension funds, as their designation is to be found dozens of years later preserved and multiplied. Thus, it was necessary to find new ways to guarantee the preservation and multiplication of capitals.

The actual abandonment of the gold standard led to the transformation of not only the world monetary system, but also to the transformation of the financial economy and all financial technology. The sharp increase in the quantity of capitals, the necessity to preserve them from inflation and to find their profitable application objectively pushed the financial market actors to look for new forms of financial activities. As a result, one could observe the rapid growth of volumes of financial operations, the number of financial assets, objects, instruments, and products. Some new instruments were already available at that time, and it became possible to apply them rapidly on a wide scale (see Grinin, 2009a, 2009c; Grinin & Korotayev, 2010 for more detail). A factor greatly contributing to all that, was nothing else but the information-computer revolution that occurred simultaneously with the financial revolution and that became a solid material basis.

Thus, in contrast with precious metals (that retained their value even if they were not invested in anything) the modern capitals do not have such an anchor; no fortune

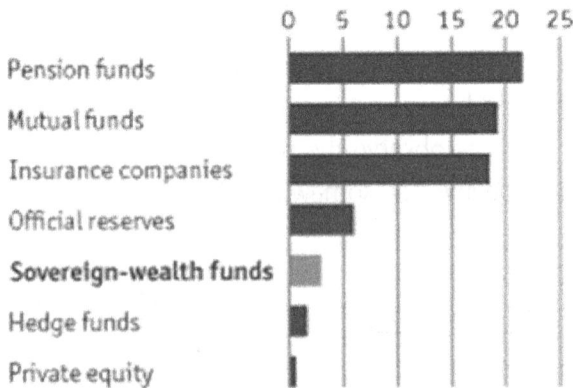

A lot, or a little?
Global assets under management
Latest available, $trm

Pension funds
Mutual funds
Insurance companies
Official reserves
Sovereign-wealth funds
Hedge funds
Private equity

Source: Morgan Stanley

Figure 2 *Assets under management of various types of funds.*

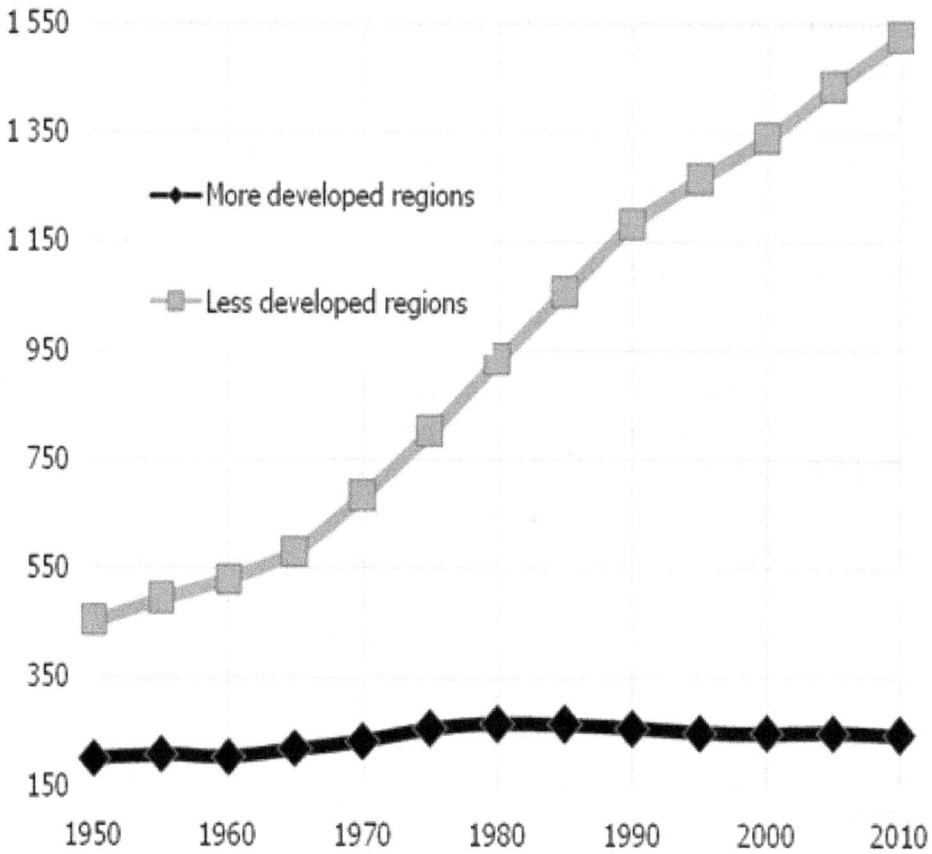

Figure 3 *Young population of more and less developed regions, mlns, 1950–2010.*

can be accumulated in a monetary form without serious risk of a rather fast loss of a substantial part of its value (see e.g., Movchan, 2010: 49). That is why if capitals just lie still (as gold in treasuries), they risk to degrade gradually into 'dust'. What are the possible sources of their preservation and growth—and, hence, what are the possibilities of the fulfillment of respective financial obligations (as well as social obligations connected with them)?

The first point is a system of dynamic movements of capitals, continuous change of their forms, the engagement of new people, mediators, and services that allow them to be preserved and multiplied. The faster the movements and transformations of financial objects, the better the preservation of capitals.

The second point is the distribution of risks at the global scale. We observe the growth of opportunities to distribute risks among a larger number of participants and countries, to transform a relatively small number of initial financial objects into a very large number of financial products. This makes it possible to achieve the maximum diversification by letting people choose convenient forms of financial products and to change them whenever necessary. That is why derivative financial instruments become more and more derivative, they acquire more and more modified forms that become more and more distant from the initial monetary one.[17]

The third point is the growth of specialization (including various forms of deposit insurance) that supports diversification and the possibilities for expansion.

The additional importance of the world finances' functions—with respect to the preservation and multiplication of savings in pension, insurance, and social funds—is amplified every year by a very significant (and, in some sense, global) process of the finalization of the second phase of demographic transition in Western countries and Japan (see e.g., Korotayev, Malkov, & Khaltourina, 2006). It is well known that as a result of this process the natural population growth rates declined in those countries to values that are close or even below zero; depopulation began in a number of developed countries, a rather rapid population aging is observed, whereas the proportion of pensioners in total population tends to increase in a rather dramatic way.[18] The forecasts suggest a further acceleration of this process. In 2010, one can observe 1 pensioner per 4 working-age adults, whereas in 2025, according to forecasts, there will be less than 3 working-age adult per pensioner in the developed countries (National Intelligence Council 2008), and there are even more pessimistic forecasts (see e.g., Meliantsev, 2009: 30).[19] Who will be filling the pension funds in the future, who will fulfill the social obligations with respect to hundreds of millions of elderly voters? Note meanwhile that, in addition to the overall growth of the number of pensioners, one can also observe simultaneously the increase in volume, complexity, and value of respective obligations (in particular, health care services).

Indeed, within such a context, against the background of the slowdown of the economic growth in the West and the threatening growth of state debts in many de-

---

17. One should take into account that pension funds, insurance companies etc. act as institutional investors and owners within many corporations that invest in numerous stocks and projects; this way the finances of the world have been so mixed up that it is difficult to comprehend whom exactly these particular funds belong to, whether they are 'good', or 'toxic'.

18. It is not coincidental that one of the main concerns of Alan Greenspan (about which he writes in his book [2007]) is connected with the point that the numerous generation of baby-boomers will become pensioners soon, whereas the extant financial sources are not sufficient for the complete fulfillment of social obligations with respect to them.

19. The dramatic change of the ratio of pensioners to the working age adults may be illustrated with the following data: in 1950 in the USA the ratio of pensioners to working age adults was 1:16 while in mid-2000s it was 1:3, i.e., it had changed fivefold (Meliantsev, 2009: 30).

veloped countries, the guarantees of pension (and other social insurance) payments do not look perfectly secure.

Here one should take into account the point that most pension funds are concentrated not in the state pension funds, but in thousands of private (non-state) pension funds that are very active as regards the search for the most secure and profitable investments. Pension funds are important shareholders of listed and private companies. They are especially important to the stock market where large institutional investors dominate. The amounts of money concentrated in pension funds are enormous. The Economist (January 17, 2008) reported that Morgan Stanley estimates that (privately managed) pension funds worldwide hold over US $20 trillion in assets, the largest for any category of investor—ahead of mutual funds, insurance companies, currency reserves, sovereign wealth funds, hedge funds, or private equity.[20]

So, in sum, at the world scale pension and other social funds' total volume of money is counted in dozens trillion dollars. Note that we are dealing here not with some financial derivatives or bad debts, but, in general, with honestly earned money accumulated over three to five decades that constitute a working life. Thus, according to some calculations, the volume of the American pension funds can be estimated for the moment of the start of the world financial crisis as about 10 trillion US dollars (Shtefan, 2008), which is comparable with the total size of the US GDP (see Antolin, 2008: 7).

The crisis led to tangible losses and even bankruptcies of many of those funds.[21] How to make the preservation of those resources secure? It is easy to see, that security is a rather relative notion. The best shares can suddenly turn out to be insecure, the same goes with respect to the apparently best shares, real estate, and even state obligations. The OECD estimates the losses of pension funds in OECD countries to be $5.4 trillion or about 20 percent of the value of assets in these countries in 2008 (Hinz *et al.*, 2010: 3; Antolín & Stewart, 2009).

In 2008, the Russian State Pension Fund lost 10 billion roubles because of drop in rate of these obligations (Naumov, 2008). It seems that within a single developed country it becomes more and more difficult to achieve a sufficient level of the security of pension funds.

---

20. By the end of 2005, total assets held by privately managed pension funds in the 23 countries covered in this study amounted to over US$ 15 trillion (Antolin, 2008: 6).

21. At the end of 2008, when economies throughout the world were spinning into recession, many stock markets saw gains of the past decade completely wiped out. The value of pension fund equity holdings in the United States alone fell by $4 trillion over 2009 (Johnson & de Graaf, 2009).

In the meantime in the developing countries we observe enormous numbers of young adults; and it is extremely difficult to provide all of them with jobs and education.

It is impossible to solve this task without an active integration of the peripheral economies into the World System economy, without the diffusion of capitals and technologies from the World System core, whereas such integration cannot be achieved without the development of the world financial system. In the meantime the number of pensioners in the developing countries is still relatively small, the social obligations with respect to them are relatively low, and only after a significant period of time the problem of the pensioners' support will become acute in those countries.

Against this background, the world monetary resources have already begun to take part in solving this social problem (though, naturally, they are unable to solve it completely). It appears that the redistribution of capitals throughout the whole world and the distribution of risks through investments in the diverse assets of developing countries (through numerous mediators and specialized funds) actually creates for the borrowers/recipients from the developing countries (i.e., those countries with a high proportion of young adults in their population) financial obligations that multiply the invested capitals within rather long periods of time. And those multiplied capitals will be potentially used for the payment of pensions and other social obligations in the creditor countries.[22] The developing countries are very interested in attracting capitals that create jobs for the numerous cohorts of young adults. A considerable part of requested capitals come from pension, insurance, and social funds of the developed countries. In other words, to some extent the young adults of India, Indonesia, Brazil, or Egypt will be working indirectly to support the elderly population of the core countries.

Thus, those countries that are rich in demographic resources, but that are poor in capitals are involved more and more in an extremely important (though not quite apparent at the surface) process through which they participate in the support of the elderly population cohorts living in the core countries through the vigorous unification of the world financial currents, its standardization, and increasing global mobility and anonymity.[23]

---

22. One may recall how the financial obligations of the USA (that had been forming for a very long period of time to fund various private projects) became quite unexpectedly an additional factor for the victory of the creditor countries of the Triple Entente in World War I. This was a large debt of the USA with respect to France and Britain that made it possible for them to get vitally important supplies in return for the redemption of that debt (through rather complex financial schemes).

23. This may resemble the situation in Britain in the early 20th century when the revenues derived from

In other words, global finances not only integrate capitals of the various regions of the world, they also contribute to the solution of an extremely important social problem[24]—to support the numerous elderly population of the core countries. Within such a context one can foresee a situation when a failure of one country would be regarded as a common failure. Actually, this outruns (and prepares) political and legal globalization in some very important respects. Such interweaving of interests (as soon as it becomes evident) will make the actors move more actively toward the institutionalization of some financial and social relationships, toward a more rigorous control of financial currents, toward the full security of financial technologies.

In other words, the modern financial assets and currents have become global and international, huge funds are circulating within this system (though, of course, not all its participants extract equal profits). Meanwhile, it is important to understand that a considerable percentage of the circulating sums are social (pension and insurance) money whose loss may lead to disasters with such consequences that are very difficult to forecast.[25]

Thus, a more secure management of the world capital has (in addition to its evident economic and social dimensions) such a dimension as the security of the future of pensioners and those in need of social protection (there are certain respective insurance systems at national levels, but what could they mean in the situation of a global financial collapse?). Hence, the issue of the institutional support of the financial globalization becomes more and more important. However, one could wonder how many new crises are necessary in order that this problem would be solved?

---

the export of capitals helped to sustain a high level of life against the background of falling industrial growth rates. In this period the revenues derived from lands, houses, state loans, foreign and colonial loans constituted just a bit less than a half of all the taxable national income (Tugan-Baranovsky, 2008 [1913]: 321).

24. Note that this problem is apparently internal from the viewpoint of a single country; however, it becomes more and more difficult to solve it in the framework of a single country.

25. The importance of accounting for useful functions (including social ones) of global money should warn us against various extremist statements such as 'working in a bank does not deserve to be excessively well-paid ... the society should not allow for people to become wealthy only because of their redistributing financial means' (from an interview with an eminent French economist Jacques Attali—see Bykov et al., 2009: 103). All these recall hundred year-old declarations that capitalists do not perform useful functions in production. Indeed, as soon as the interest in enrichment through financial operations disappears, who will risk their capitals? And what will happen to them? However, this does not deny the necessity of accurate and consistent limitation of the extremes of speculations and excessive enrichments. And in this respect some ideas of Jacques Attali (in particular, his suggestion about a 'new, this time global, Glass-Steagall Act' [ibid.]) look rather interesting.

Today many specialists see that the main current problems of the world stock markets stem from the defects of their regulatory system (see e.g., Doronin, 2003: 129-130 for an analysis of their views), though many specialists (if not most of them, at least in the United States) still believe that the problems of stock and financial markets stem from defects and imperfections of the national (rather than supranational) regulation systems. One should admit that the United States has derived some conclusions from their crisis experience; in particular, the American actors have started discussing (and taking) measures aimed at tighter regulation. They have also begun cleaning bad and 'toxic' debts. All these are important developments, especially with the account of special and enormous global influence of American financial institutions and instruments. Today the World Monetary Fund has more opportunities to affect global economy as a result of the increase in its credit resources. However, there are strong doubts that the World Monetary Fund will be able to move significantly toward its own transformation into a sort of 'World Central Bank' though such suppositions are sometimes made (see e.g., Zotin, 2010). The world experience demonstrates that the new principles (which should also serve as a basis for the new world financial system) do not develop from or as a result of those institutions which have already realized their functions. These, more established institutions are hardly liable to such a radical transformation.

A few quite reasonable opinions have been expressed recently with respect to the possible directions of the necessary regulation of financial activities. For example, Schäfer maintains the following:

Particularly risky financial products must be prohibited. At present, if one invents a new financial instrument, he can offer it to his clients the next day. For example, an inventor of a new derivative is not obliged to register it in any state agency; he can start selling it immediately. Free market proponents believe that financial markets will regulate everything themselves, that they will sort out and discard bad products by themselves. In reality this does not happen. Banks and funds threw 'toxic waste' amounting to trillions of dollars to the market, and meanwhile they diffused a belief that one can produce really valuable stocks from a large number of dubious assets, whereas nobody felt being responsible for all this. But if the market cannot take responsibility upon itself, it should be assumed by the state. Financial corporations must be obliged to register in advance all the financial products that they invent (similarly to what is observed with the production of medicines in the pharmaceutical industry). A state agency should anticipatorily check and test all the financial instruments before banks get their right to sell them. And if those instruments turn out to be too dangerous, the agency should prohibit them (Schäfer, 2009: 279–280). State agencies controlling financial markets should subdivide rating agencies in such a way that a part of them would calculate ratings, whereas the other part would provide consulting services to banks. In the meantime, rating agencies and their clients should publish all the information that has been used to calculate the rating. In this case, any other rating agency will be able to check the ascribed rating and to publish an alternative calculation if it does not agree (ibid.: 280).

Actually, the business of tax havens consists of the sucking of funds from industrially developed countries. The 'havens' attract them with their extremely low tax rates. They offer absolute secrecy to their depositors and exempt financial corporations from any checks. That is why the industrial countries should coerce 'havens' to abandon bank secrecy and make them inform foreign tax agencies of all the respective capitals and revenues. 'Havens' should raise their tax rates to an internationally acceptable level. They will not do this voluntarily. That is why, if necessary, they should be coerced to do this with economic sanctions (ibid.: 284).[26]

It is easy to notice that this citation (with an important exception in the last section) is addressed to the national government. However, though the role of national regulation still remains very important, today we observe a situation in the financial markets when a single state (in contrast with the previous period) cannot cope with it. Thus, as finances internationalize more and more, respective measures should be taken at the supranational level.

For quite a long time some analysts have been discussing the necessity of the transition from the national level of regulation to the supranational one (see e.g., Van Der Wee, 1990; Soros, 1998). Lester Thurow, as well as some other analysts, finds the cause of the instability of the world stock markets in the contradiction between the in-

---

26. As is well known, in many Western countries high taxes provoked a vigorous growth of the number of those who try to avoid regulations, as well as an increase in the number of off-shore safe havens (see e.g., Cassard, 1994: 22–28; Zorome, 2007: 24–25; Platonova et al., 2009). The G20 London meeting evidenced rather active speeches (especially on the part of Germany and France) against the off-shores. Indeed, a few resolutions aimed against them were taken, some countries (but not all the relevant countries) found themselves in a black list of states putting obstacles in the way to the international control over the tax havens. However, with economic recovery, the anti-off-shore thrust is likely to weaken, especially taking into consideration the point that some G20 countries (e.g., China and Britain) are interested in some off-shores (Bykov et al., 2009: 101). There were also declarations regarding such things as the necessity to tie salaries of the managers of investment banks to mid and long range results. One should also note the pressure to reduce bank secrecy (see e.g., Fokin, 2010), though this cannot be regarded as a purely positive development.

ternational character of operations of the world stock markets and the national nature of the stock markets themselves. However, he has very good grounds to note that, though the epoch of national economic regulation is coming to its end, the epoch of the global economic regulation has not started yet (Thurow, 1996). Will radical changes in this direction take place in the near future?

The transformation of the international order starts to be discussed in an especially urgent way when the world is shaken by global crises. Hence, it is not surprising that the concepts of the 'revision of the world order' (e.g., Tinbergen, 1976) emerged just in 1970s crisis years. The crises of 1969–1971, monetary crises of 1971–1973, but especially the 1973–1975 crisis, were indeed periods of economic chaos without any entirely comparable precedents in the post-war era (*ibid.*). This stimulated the development of new ideologies of global development; particularly with respect to the relationship between developed and developing countries or sustainable development. Many of the questions posed in this period (as well as many recommendations worked out at that time) remain rather valid today. People living in any epoch always believe that their epoch with its reforms and crises is the most unique. Still we do not think that it is an exaggeration to say that the current global crisis has demonstrated, in an especially salient way, the necessity for major changes in the regulation of international economic activities and movements of world financial currents. These changes would include the need for the growth of coordinated actions by governments and unified international legislation regulating financial activities and movements. Actually, the world needs a new system of financial-economic regulation at the global scale.

Comprehension of the causes of the crisis may provide a push to start a new round of global transformations, but the respective path (to effective transformations) appears rather long. However, even the transition to the very initial phase of a new system of supranational-national regulation will imply rather profound changes (whereas many transformations can hardly be predicted today). The point that the political landscape and the balance of world power will change in the forthcoming decades is felt more and more strongly. American analysts believe that 'the international system—as constructed following the Second World War—will be almost unrecognizable by 2025... The transformation is being fueled by a globalizing economy, marked by an historic shift of relative wealth and economic power from West to East, and by the increasing weight of new players—especially China and India' (National Intelligence Council, 2008: 1).

The variations of probable future transformations in connection with the crisis and estimations of the probabilities of various scenarios of the World System development in the forthcoming decades will be presented in the second part of the article. Among the problems which are analyzed in this part are the following: Will there be a leader in the future World System? Will the deficit of global governance and world fragmentation increase? How could national sovereignty be transformed?

ON THE POSSIBLE WAYS OF WORLD SYSTEM DEVELOPMENT

1. Change of Leadership or a Fundamental System Modification?

It is quite obvious that today we observe the weakening of the economic role of the USA as the World System center; in a more general sense we observe the weakening of the World System core countries as a whole.[27] That is why there is no doubt that sooner or later (in any case in the near future) the USA's status as the World System's leader will change and its role will diminish. American analysts are worried by this more than anyone else (see, e.g., Mandelbaum, 2005; National Intelligence Council 2008). The current crisis is an important step toward the present leader's weakening. As we have already spelled out (see Grinin, 2008b, 2009a, 2009d; Grinin & Korotayev, 2010: ch. 4; 2011), the former priorities and foundations of the world economic order which were based on making profits for the USA will sooner or later start to transform into a new order. In the foreseeable future such a transformation will constitute collisions of relations between USA national interests, on the one hand, and the general world interests, on the other (see Grinin, 2008b, 2009e, 2011; Grinin & Korotayev, 2010: Conclusion; 2011 for more detail).

However, such a collision will lead to rather important transformations, which, unfortunately, tend to be ignored. There is a general universal tendency to believe that in the proximate future, the USA's current tenuous leading position will be occupied by the European Union, China, or some other country (starting with India and ending with Russia [see, e.g., Frank, 1997, 1998; Pantin & Lapkin, 2006]). But to model the World System transformations mostly with regard to a change of its leader is a serious mistake. Today we are dealing not only with a crisis in the World System or even with a crisis of the World System core; rather, we are dealing with a crisis of the established

---

27. This is manifested in the slow-down of the economic growth rates in the World System core and their acceleration in the most of the World System periphery (see Grinin & Korotayev, 2010; Khalturina & Korotayev, 2009; Meliantsev, 2009; Inozemtsev, 2008 for more detail), in too rapid and anarchic deindustrialization, in the dependence on cheap imports, budget deficits, general growth of public and private debts, negative demographic developments, etc. This resurrects the ideas of the 'death of the West' (see, e.g., Buchanan, 2002).

model of its structure which is based on having a leader who concentrates many aspects of leadership (political, military, financial, monetary, economic, technological). The USA also acts as an avant-garde of the developed countries as a whole (about some aspects of USA position in the World see, e.g., Renwick, 2000; Nye, 2002; Bacevich, 2002). Thus, we are dealing with a rather complex leadership structure: the USA—leading European countries and Japan—newly industrialized countries of Asia and so on. Besides, a special position is occupied by the USA's neighbors (Mexico, etc.).

When we speak about the USA losing their leadership status, we should not simply surmise that there will be a change in leader of the World System. We should rather presume that there will be a radical transformation of the overall structure of the world economic and political order. The simple change of the World System leader is rather unlikely already due to the fact that no country will be able to occupy the position of leadership in a way that is equivalent to the one held by the USA today, as no country will be able to monopolize so many leadership functions. Only as a result of this factor (although there is a number of other important factors), the loss of the leading role by the USA will mean a radical transformation of the World System as a whole.

First of all one should point out that the USA world leading position is unique in history. We also believe that the concentration of the world's economic, financial, military, political, and innovation-scientific potential in one center (i.e., in the USA) after World War II was a generally positive factor (and the existence of the USSR as the alternative political and military center even amplified its positive value in some respects). Let us recollect that the USA became the World System leader after World War I. But even at that time the United States controlled only economic and financial power, it lacked equivalent political power, and did not even strive for it. Further, it should be emphasized that such a situation, that is the absence of a recognized World System leader, contributed significantly to a very severe economic and political World System crisis during the period between the two World Wars and also to the beginning of the Second World War.

In the 1960s one could observe a decrease in the economic role of the USA in the World System which led to the emergence of a three-center model of economic leadership: the USA—Western Europe—Japan. However, it is important to note that this system was formed under the political and military (recognized and desired) leadership of the USA. This structure turned out to be rather viable for almost four decades. It works even now; yet, if it turns out to be impossible to restore the economic dynamics of the Western economies, its role will weaken (whereas since the early 1970s one can observe a general trend toward the decline of the economic growth rates in all the three centers). Unfortunately, today those centers are not able to give much to each other as they have rather similar problems. The opportunity to strengthen themselves is connected to unifying their forces in order to preserve certain advantages inherent to developed countries (and that are also useful for the World System as a whole, see below). In a way this process would be similar to the one just after World War II through which the West strengthened itself against the backdrop of the expansion of the Communist Block and the simultaneous disintegration of its colonial empires, by uniting militarily, politically, and ideologically (and partly economically).

One cannot exclude the possibility, of course, that the emergence of new revolutionary technologies could give new life to the economic development of the USA (as happened in the late 1980s and 1990s), and the West as a whole; however, first, such technologies do not seem to be likely to be developed in the forthcoming decade, and in this period the problems in the American economy will be aggravated. Second, in order for such new technologies to produce major results a rather long period (at least 15–20 years) would be necessary and over that time many things are bound to change. Third, even such new technologies would be unlikely to help preserve military and political leadership.

Thus, it is evident that a place similar to the one held by the present-day USA cannot be occupied in future, neither by another state nor by an alliance of states. According to Fareed Zakaria (2009), the functions of the World System leader can only be performed by a country that achieves dominance in ideas or ideology, an economic system, and military power. However, in the near future, there will be no country (nor even an alliance of countries) that will be able to concentrate several aspects of the World System leadership.

It is often proposed that China will replace the USA as the new World System leader (for an analysis of such views see, e.g., Wang, 2010). But this function is not likely to be performed by China even if China eventually surpasses the USA with respect to its GDP volume. Those who suggest China as such a leader do not appear to take into consideration that its economy is not adequately innovative, that it does not develop on the basis of technologies of tomorrow (and, to some extent, even of today).[28] This is noted even by those analysts who take rather optimistic view of the Chinese capabilities to sustain extremely high GDP growth rates for long time in fu-

---

28. Meanwhile, China has already moved to the third place in the world (after the USA and Japan) with respect to the absolute size of its R&D expenses (Meliantsev, 2009: 123–124).

ture (e.g., Mikheev, 2008: 311, 319; see also Meliantsev, 2009: 123–124).[29] In addition, the Chinese economy is too much export-oriented. We believe that the economic center of the World System cannot be based on the exporting of non-innovative (and even not sufficiently highly technological) products. Besides, the Chinese model is very resource-intensive which makes it dependent on the opportunities of extending the world raw materials production and their prices, whereas China becomes the leading importer of a number of commodities. And at the same time this makes the world extractive industry extremely dependent on the Chinese economy's growth (Gelbras, 2007: 29–30). And what is more important, the economic growth in China is based to a great extent on an inadequate technological basis (*ibid.*: 30).

In order to perform the World System center role, the Chinese economy should become, on the one hand, innovative and highly technological (which is hardly compatible with heavy industry, or conveyor industries), and ecologically advanced—on the other hand. However, China does not possess necessary conditions for this. One would need no less than 20–25 years to become an innovation pioneer. We believe that it is India that is more likely to become a technological leader (see, e.g., National Intelligence Council, 2008; Meliantsev, 2009: 107, 60), but India does not have many other leadership components that China has. That is why the idea that in 15–20 years many countries will be more attracted by 'China's alternative development model' rather than by the Western models of political and economic development (National Intelligence Council, 2008: iv), provokes serious doubts. The Western models may be criticized, whereas it is quite natural that China's successes could hardly avoid bringing attention. But it is very unlikely that any country (with a possible exception of North Korea) will try to introduce the Chinese model.[30] The point is that this model simply cannot be introduced. In order to do this one would need a totalitarian communist party. Even the USSR was unable to copy the Chinese model. Not to mention that the Chinese polluting model of economy can hardly suit any country.

The issue of the restructuring of the model of the Chinese economy is tightly connected with China's ability to preserve the current high growth rates, whereas the latter is extremely important for the ideological prestige of the Chinese administration (though, we seem to observe the growing influence of the faction that believes that it is necessary to slow down the growth rates in order to decrease the social stratifi-

---

29. However, one should not neglect the explosive growth of patent applications and patent grants in China (see, e.g., Korotayev, Zinkina, & Bogevolnov, 2011), as well as the point that in some areas (e.g., in biotechnologies) China has certain undeniable innovation achievements.

30. Those capable have already introduced it (we mean first of all Vietnam).

cation and tension). The Chinese administration has already announced that it plans to reduce the growth rates to 8–7 % in 2011–2012 and in the period of 12th five-year plan—till 2015 (see, e.g., Beglaryan, 2011), which is connected not only with the increase in inflation but also with the evident difficulty of supporting the previous extremely high growth rates that deform the social system. It is not coincidental that the analysts note overinvestment and the presence of excessive production capacities in the Chinese economy that are caused by fierce competition between provinces that struggle to attract investments and to secure high growth rates in the respective regions. The transition to such an economic model focused on internal consumption and technological innovation is further complicated by the following points: a) the growth of internal consumption implies the acceleration of the growth of the living standards and Chinese labor costs (that are growing anyway); b) the growth of the labor costs is not likely to be compensated by an adequate increase in the labor productivity (as this happened before); c) consequently, the costs of exported commodities may increase, their competitiveness may fall, while the attractiveness of investments in China will then decrease.[31] This may cause a slowdown of growth rates. Thus, for China the transition to a new type of economy with a simultaneous preservation of its leadership in economic growth rates is difficult. Even though home demand will continue to develop, it will either be incapable of a sufficient replacing the export demand or this will mean a profound structural rebuilding of the economy. Investments into infrastructure, housing construction, etc. can be locomotives of development only if sufficient resources obtained through exports are present, but combining the two directions simultaneously does not seem probable in the long run.[32] A decrease in growth rate implies a state revenue decline accompanied by the aggravation of unemployment and an increase in social obligations.

Moreover, it can be assumed that in the coming decade the Chinese economy's growth rates will sooner or later, and inevitably, decline (regardless of whether the attempt to re-orient the economy from an export model to the model of internal consumption will turn out to be successful or not). As a result, the same processes may start that were observed in Japan after 1975 (see, e.g., Karsbol, 2010). However, the slowing down of growth rates, particularly in an authoritarian country (where a remarkable population ageing is to be observed soon), will lead to the aggravation of

---

31. Before the crisis there has already been forecasted some decrease in direct foreign investments into Chinese economy during the next 15 years (Mikheev, 2008: 311).

32. Economy re-orienteering turns out to be very complicated even for such developed countries as Japan. Let us remember that attempts made in Russia during several decades to re-orient the export have been unsuccessful so far.

social conditions and changes in state priorities, which may ultimately weaken China's economic potency. The developing impulse force is still great in China, while inertia is still powerful, but it is quite obvious that both will most probably weaken. At the same time, the idea of higher living standards for the majority of China's population will be spreading at an advanced rate. This has both positive (as this inspires energy and new search motivations in some part of the population) and negative (as it increases ungrounded claims to the state and decreases the competitiveness of the Chinese economy) implications (for more detail about the Chinese economic model and perspective of China world leadership see Grinin, 2011).

2. Hypothetical and Real Alternatives

Thus, the future World System will not be able to possess the same structure as the current one with an equally strong centre. What can be an alternative to the modern 'order' in the world? Here we step upon unsteady and ungrateful soil of forecasts.

Let us first consider the future structure based upon the following probable but still hypothetical suggestion. Objectively, globalization leads to the appearance of some new forms of political and economic establishments of a supranational type. The EU represents just a version of such a type, other types and forms have just been outlined or are currently being outlined. However, they may emerge rather fast under favorable conditions. The largest states (e.g., the USA, China, and India) may rival such supranational establishments for quite a long time, but still the future lies in front of the latter, not the former.

According to this hypothesis, the new World System leader (if it emerges altogether) will hardly be a separate state, but rather a (potentially increasing) block of states. Will this alliance be headed by some of the largest states of the modern world, or will it arise from a coalition of states of medium size and power? Or will such a coalition emerge on some other bases? Obviously, neither China, nor India can unite around themselves a relevant group of countries on their political (for India also civilizational) specific features. In order to become such an integrating center, China must change its political regime.[33] For China, changing its political regime will most probably imply a severe shock (presumably even disintegration, Tibet being the main candi-

---

33. It should be noted that China has got some projections on more active integration with neighbors. In particular, one could mention the idea of free trade area 'China+ASEAN' and a united integrative space in North-East Asia with participation of Japan, China, South Korea, and possibly Russian energetic resources. However, taking into account the tense relations between China and Japan this hardly seems realistic (for further details see Mikheev, 2008: 319).

date for separation[34]), while the preservation of the regime requires relying on its own capacities. The regime in China will remain solid and strong until it is capable to support the process that is quite accurately denoted by Yunxiang Yan (2002) as managed globalization. This implies the ability of the government to control diverse (cultural, in particular) global influences. That is why China is not ready to lead the process of the economic coordination of the region. But, at the same time, it will not agree to perform the role of 'number two' (see, e.g., the opinion of a famous Japanese economist, Richard Koo [Ivanter, 2009: 97]) and clearly increases the economical influence on the ASEAN countries (Kanayev & Kurilko, 2010: 43–44). Still, Richard Koo's statement that China sooner or later will have to burden itself with the leadership, at least in Asia (*ibid.*), does not look entirely convincing.

A more natural integration of the American region under the aegis of the USA (some kind of a pan-America) could theoretically revive the role of the USA as the world center. However, the disposition of political forces in Latin America is too unstable, and the level of development differs greatly among the states. Brazil has already stepped into the first line of the largest countries. Besides, quite a number of regimes are much tempted to play on confronting the USA. A union with Mexico and Canada (NAFTA), though supplying more than 85 % of the whole export for Canada and Mexico, is incapable of fulfilling a function which could solve the above-mentioned task (Kirichenko *et al.*, 2008: 226).

Among all variants of the emergence of such hypothetical leading union the European version has the largest (though on the whole small) probability. Even though the European expansion crosses the natural geographical limitations, the possibility of Turkey with its more than 70 million population entering the EU someday should not be excluded. This would turn the EU into a supra-European union (we should also account for the strengthening ties between the EU and non-European Mediterranean states). If Europe could integrate with Russia, Ukraine, and Belarus, this would give a certain impulse to restructuring the World System relations and even form some strong centre. In terms of practice, this is a highly complicated, but rather possible scenario.[35] In any case the EU must cope with financial problems, periodically encountered by one or another its member.

---

34. The latter, not possessing statehood traditions, most probably will go the way of Kosovo, turning into one more drug state (see, e.g., Bykov *et al.*, 2009: 103).

35. However, the high living standard of the Europeans and the aspiration of the new EU members to immediately attain the same level substantially decrease the impulses to adopt new members (the same, but in lesser scale, occurred during the reunification of Germany).

All the above-mentioned scenarios are rather unlikely. Thus, the most real alternative to the role of the USA is currently... the USA itself. That is why during the nearest one or two decades the USA will remain the most real leader if, of course, the Americans do not undermine their positions themselves (through a sharp change in foreign policy, strong devaluation of dollar, a default, or an economic collapse).[36] In the current absence of an obvious leader counter-weighing the USA, the world will be obliged to support the preservation of the USA as a non-alternative, though getting decrepit, center as any weakening in the USA position can lead to a great extent uncontrolled transformation of the World System. A certain 'imbalance cycle' arises (National Intelligence Council 2008), where imbalances support each other. On the one hand, this plays into the hands of the USA, but on the other hand, the absence of strong competition for leadership greatly weakens the USA's capacities to renewal. There exists an opinion that, though the demand for the USA leadership remains high, the interest and readiness of the USA to play the leading role may decrease, as the American voters will reconsider their attitude towards economic, military and other expenses of American leadership (National Intelligence Council 2008). To put it more exactly, fluctuations in foreign policy are more possible, along with variations of the struggle between isolationism and hegemony, as a result the USA's foreign political activity may decline for some time. However, state transition from the policy of sheer hegemony and external expansion to passive foreign policy took place a number of times in the course of history, in particular, in Japan, Germany, and in most recent times in Russia.

In uncertain conditions the number of probable scenarios can be large. Thus, the document prepared by the National Intelligence Council of the USA Global Trends 2025: A Transformed World (National Intelligence Council, 2008) considers four hypothetical scenarios: 'A World Without the West' when new forces press the West out of its leading positions in the geopolitics;[37] 'October Surprise'—an ecological catastrophe; 'BRICs' Bust-Up'—a conflict between India and China over the access to vitally important resources; 'Politics is not Always Local'—when various non-state structures

---

36. The last two ones will indeed require fast decision-making at the global scale. Spontaneous US dollar collapse can lead to downfall of all national financial and currency systems. Along with a sharp fall of the main global currency rate the whole global economy would devalue (see Platonova *et al.*, 2009: 88).

37. As regards economic ousting of the West, which, of course, results from faster economies' growth in the periphery, there is a fundamental point to note, which will change not so fast, if at all. Currently (and it will stay this way in the foreseeable future) the development vector is still being set by the West, while the fastest-growing economies, with rare exception, are adjuncts to the Western economy. If we try to imagine that only developed countries are left in the world, without the periphery ones, life standard and the level of technology in the West would suffer much less than those in periphery countries in a vice versa situation. Besides, in this fantastic scenario the Western economy would obviously go up rather rapidly, while in the periphery a collapse would occur.

unite in order to develop an international program for the environmental protection and to elect a new Secretary General of the United Nations. All of them, though based on certain trends of the modernity, do not seem sufficiently real, which is admitted by the authors themselves (National Intelligence Council, 2008).

Given the wide range of the variants of future, it is remarkably difficult to consider all variations. That is why it is better to select certain main parameters for the analysis of the hypotheses. Let us take such an important parameter of future development as the degree of suddenness and sharpness of geopolitical and geoeconomic changes. Obviously, if the process proceeds gradually, people get accustomed to it and try to put it under influence, and the system somehow has time to transform. If the changes occur suddenly, for some time there arises a vacuum of the system and order, chaos, hasty building of temporary and thus not always successful constructions. Let us view two such hypothetical scenarios: the one of gradual change and the one of sharp change.[38]

In the first one, the USA power would decrease not sharply, but gradually. In this case the USA, trying to preserve its leading position, would possibly be obliged to maneuver, enter some coalitions, give in sometimes in certain questions, and accept some global ideas in order not to lose leadership and to preserve an acceptable geo-political balance. On the other hand, the USA would aspire to create something at the global scale, try to institutionalize the situation, seeking to strengthen the position of primus inter pares in some commonly accepted international and interstate agree-ments and interaction systems (organizations, consultations, etc.) at the same time not insisting on absolute or even evident hegemony which is present nowadays.[39] Naturally, this would require great skill. This process would be more successful if the USA could, according to Brzezinski's recommendation, unite with Europe and Japan in important directions (Brzezinski, 2004; for the analysis of American foreign policy and the USA position in the world see also Kagan, 2003; Bacevich, 2002; Jervis, 2005). Given the low growth rates in developed countries, the West is objectively interested in creating an order which would institutionally formalize some of its advantages (to

---

38. Though the process will most probably be uneven: slow uncontrolled changes will be succeeded by large but not fatal collapses and crises, inspiring the transformation process and even changing its directions. As the forecasting experience shows, not a single forecasting model could be realized in its pure version; usually something arises, in which different trends can be seen in specific combination.

39. Wilkinson called a somewhat similar scenario 'unipolarity without hegemony', but with probable dominance of the USA (Wilkinson, 1999). In our version some crucially important points have been added, including the necessity of trying to secure de jure the advantages in some relations. Securing the prevailing role of dollar turned out to be exceptionally important for the USA in its time.

some extent this would be useful for the whole World System).[40] It is doubtful whether this necessity could be realized in time, but the success of such an institutionalization greatly depends on whether this realization occurs sooner or later.[41] This would be, so to say, a scenario of 'planned re-building'.

The second scenario will occur in case when the USA changes its position dramatically, i.e., as a result of a sudden dollar collapse and especially as a consequence of an American default (say, at sudden change in the global economy resulting from a crisis sharper that one of fall 2008).

In this case the US public opinion may sway to folding up the global functions of the USA, which will additionally aggravate the vacuum of international governance. In this situation, the possibilities include anarchy (a less likely scenario) or hasty gathering (or 'knocking together') of some system capable of supporting the collapsing world order and solving the momentary tasks, offering certain palliative solutions and agreements which on the whole can turn out to be perspective further on.

However, among all hypothetical variants the two alternative ones seem most probable to us. The first one, which is naturally more preferable, is the expansion of the 'club' of leading global players up to a number which would allow them to influence somehow the course of world development (which will be considered further on). The second variant implies spontaneous uncontrolled development where main players will be mostly concerned with domestic problems, the politicians will mind only the popularity ratings, while the global problems will be solved in passing. The Japanese society is a good example of such self-isolation (though even there some certain trends to integration can be observed [see, e.g., Ivanter, 2009: 99]), the EU also too frequently reveals reticence with their own interests. For Western countries there exists a danger of becoming hostages to a democratic system, in which the position of politicians precludes them from thinking about future, and this could ultimately turn them into demagogues and state-mongers. Besides, incidents of protectionism and other anti-globalization measures are in no way excluded. In this case only unex-

---

40. An example of such securing is the definition of the five leading states having the veto right in the UN Security Council. The disposition of powers in the world is changing, while this order is not easy to alter.

41. One of the possible outcomes for the West implies quoting the economy growth rates in order to restrain the all too fast periphery development (on the possibilities of such quoting see Grinin, 2009d). If the limitations cannot be achieved through direct quotes on economic growth, they can still be pursued under the mask of struggling for the global ecology, for the rights of wage earners, against dishonest rivalry etc.

pected shocks, such as the modern crisis, can wake Western politicians and societies up. Surges of national, civilizational, and hegemonic pride are also useful.

3. Will the Deficit of Global Governance and World Fragmentation Increase?

We have already mentioned in our previous works that economic and financial globalization greatly advances the development of international law and political globalization (Grinin, 2008b, 2008c, 2009e, 2009f, 2009h, 2012). Will the political component of the World System lag behind the economic one even more in the decades to come? The answer to a great extent depends on what the economic development will be in the near future. Numerous economists and social scientists, presenting various arguments (sometimes basing on the dynamics of the famous Kondratieff cycles), suppose that in the next 15–20 years the world economic development will most probably proceed at a slower rate than in the preceding period.[42] We support this point of view (see Grinin & Korotayev, 2010). However, if this forecast comes true, will not the political component of the World System be able to catch up slightly? Besides, the weakening of the US leadership and the absence of an alternative (in any case changing a leader is a long and complicated process) must obviously lead to the international system transforming faster and in a more substantial way. Consequently, we are entering a period of searching for new structural and systemic solutions within the World System, which means a considerably complicated period awaiting us in the near future. Working out and stabilizing the model of a new political order within the World System will be a complicated, lengthy, and rather contentious process.

This way or another, the global governance deficit is present, and in the forthcoming decades it obviously will not disappear. Supposedly, it will become more ideologically sensible, while the project of eliminating this deficit will become relatively feasible. However, global governance requires great effort and substantial sacrifice. To what extent will the states and non-state subjects wish or be able to endure the growing burden of global governance? A refusal to divide this burden will aggravate the situation of increasing institutional lack (National Intelligence Council 2008). Seemingly, there will be quite a few countries ready to take some burden of international regulation by themselves, in the same way as nowadays only a few states take obligations to make large contributions to international organizations including the UN. That is why for some time many countries will still be interested in the USA leadership

---

42. However, much depends on which methods are chosen for GDP calculation. Many changes actually cheapen the final product, as a result of which the impression of GDP decrease arises.

even though, as has been mentioned above, the USA itself in certain conditions may not be interested or capable of continuing to be a leader. Certain large states rivaling the USA leadership will be most probably incapable of global governing as well.

Such a situation may reveal the most important spheres the regulation of which will be profitable, as well as certain important fields where it will be compulsory to participate according to international obligations. This should strengthen the trend to various collective activities, formation of associations, and developing different types of cooperation. This will also transform the global governance towards new technologies.

American analysts suppose that: a) in the nearest future politicians and the public will have to cope with the growing demand for multilateral cooperation; b) current trends are leading to the emergence in 15–20 years of a fragmented and contradictory world; c) multipolarity and structurelessness are the main features of the future system (National Intelligence Council, 2008).

As regards the demand for multilateral cooperation, which is already high nowadays, it will continue to grow. It seems, however, that the growth of this demand a) gives an opportunity for certain regional states and unions to strengthen their positions; b) will contribute to faster emergence of various formats of multilateral cooperation. It should be noted that the new international order would best emerge with the formation of a sufficient number of supranational unions, coalitions, coordination centers, multilateral agreements, as well as influential NGOs and networks varying in type and scale, on the one hand, and with presence of de jure (or at least de facto) accepted institutionalized leading center of the World System.

Multipolarity (though this term is interpreted differently) has become a geopolitical motto for some states and it seems to be forming (see also Nye, 2002). New centers of power (first of all, economic, but also military and political) are established, causing new configurations within the World System. However, in any case multipolarity in the context of peaceful coexistence implies the presence of a certain order, so multipolarity and structurelessness are opposites to each other.

Fragmentation increase would imply the World System disintegration (at least temporary). To what extent is it possible? We consider this to be unlikely due to some realities of a certain quasi-unity being customary for us. Even the crisis did not lead to disintegration; on the contrary, it united the world to some extent. Certain global conscience seems to be formed. Let us bring just one example of unexpected metamorphoses of modern economic psychology. During the period of a particularly low fall of production indexes in 2008–2010 many economists were circulating the idea that the Chinese economy would pull the whole world out, and so things are not too disastrous.

4. The Epoch of New Coalitions and Sovereignty Transformation

We have already written on the transformation process with respect to national sovereignty, i.e., on the decrease of the real volume of state sovereign powers, which is to a great extent voluntary (for more details see: Grinin, 2008b, 2009d, 2009e, 2009f, 2009h). The necessity to pull up the political component of the World System and to strengthen the global regulation of financial and other agents contributes to the sovereignty transformation process, as the states must voluntarily limit themselves in some spheres, and sometimes undertake additional functions (for more details see Grinin, 2008c, 2009c, 2009d, 2009e, 2009f, 2009g, 2012; Grinin & Korotayev, 2009a). The global crisis has revealed the sovereignty limitations more clearly and showed that even the USA cannot act without real support of other countries.

'By 2025 a single "international community" composed of nation-states will no longer exist. Power will be more dispersed with the newer players bringing new rules of the game while risks will increase that the traditional Western alliances will weaken' (National Intelligence Council, 2008: iv). Indeed, the real composition of 'international community' will most likely be more complicated in the next decades due to the addition of some supranational unions, official or informal councils of leaders of states and unions, temporary or constant coalitions, and, possibly, the NGOs.

However, sovereignty transformation within the new world order creation is not a unidirectional and unilinear process. Firstly, national state will for a long time remain the leading player in the world arena, as in the foreseeable future only the state will be capable of solving a number of questions. Secondly, sovereignty may even increase in some aspects, as the modern crisis shows once more that the fate of national economies to a great extent depends on the state strength. Thus, it is quite probable that the nearest future may reveal a certain 'renaissance' of the state role and activity in the world arena. In some countries sovereign powers that had previously been (sometimes thoughtlessly) given away to supranational organizations, unions, and global capital may possibly be returned. In long-term trends such ebbs and fluctuations are not only possible but unavoidable. Thus, the seemingly steady movement towards democracy in the early 20th century suddenly made a swerve toward totalitarianism; the development of free market trade in the late 19th century was turned to protection-ism. Thus, a return to etatism can be both rather lengthy and rather useful.

Nevertheless, it should be mentioned that such a return to the increasing role of state cannot be performed on the former bases, when the benefits of a state (even within the fulfillment of undertaken obligations and the observation of common in-ternational norms) were accepted in international relations as the highest cause of its activity on the world arena. We suppose that the return of the state role cannot be successful without a substantial change in the state foreign policy ideology. In other words, we can suggest that purely egoistic interests of states will to a much lesser extent underlie the foreign policy concept and performance.[43] Naturally, na-tional egoism will not disappear altogether for a long time (if ever at all), but it will be more disguised by supranational interests and necessities than it is now. To put it more exactly, every action may require not only a real interest, but also an ideological grounding. Viewing the global arena as a 'great chess board' (Brzezinski, 1997) where the strongest wins, while small pieces may be exchanged or sacrificed will possibly not be in demand any more. The world arena will rather be viewed as a common field of interests where rules advantageous for everybody must be stated and some-how supported. The countries will more and more remarkably define not only their own security in such categories which would accord with interests of the others, as Brzezinski advises to the USA (see Brzezinski, 1997), but also all their large-scale ac-tions. That is why it is sensed that gradually the mottos of common (regional, global, group) good will strengthen in foreign policy, though the 'who-represents-the-global-interests-better' formula may, as always, disguise egoistic causes.[44] However, this will lead to substantial changes, generally positive. In any case, the countries continuing to roughly stand up for their national egoistic interests will eventually lose, sooner or later. Radical changes will be unavoidable in the policy of large states aimed at direct and rough domineering in global or regional scale (including the most independent and egoistic sovereign, the USA).

---

43. One of the numerous examples of such egoistic approach in the position of the USA on the ques-tion of greenhouse gases emission into the atmosphere. Outright declaration of a certain region as a zone of its special interests, intervention into the business of other states under the pretext of weak-ening somebody (e.g., Russia), as well as supporting the undisguised corrupted regimes etc. are all examples of undisguised egoistic policy. Let us also note that foreign policy of such major countries as India, China, and Japan does not essentially possess any special ideology altogether.

44. For sure, in modern history different actions have been and are still carried out under the aegis of common interests, in particular, intervention into dependent countries (on the part of both the USSR and the USA), pressure upon certain countries under the mottos of defending the human rights, de-mocracy, etc. Ideological aspect will presumably substantially increase.

In this case national interests assertion, rivalry forms on the world arena, conflicts and litigations will acquire a different form from now. Rivalry will increase on directing the process of new world order formation. Rival forces will perform under the mottos of a new, more honest world arrangement, of fair and crisis-less global development, against national (especially American) egoism, etc. For conducting such a policy, allies and blocks are obviously necessary. Thus, regrouping of forces on the world and regional arenas will invariably start. In the struggle for a place of honour in globalization and coalitions, in organization and functioning of a new world order the phenomenon occurs which we named the epoch of new coalitions (see Grinin, 2009d, 2009h). As a result, new force disposition may be outlined for quite a lengthy period.

Forms, particular aims and activity directions of the new coalitions will depend on numerous factors, in particular, on how far the process of making common decisions will go and what means and forms of common decision-making will be realized. Thus, it seems that the system of simple democracy (one state/participant—one vote) on the World System level will hardly be viable.[45] China and India cannot be equaled to Lesotho (less than 2,000,000 inhabitants) or all the more Tuvalu with the population less than 15,000.

Probably, for some time the mobility of partnerships within the World System will increase, the arising coalitions may turn out to be chimerical, ephemeral, or fantastic. In the course of search for most stable, advantageous, and adequate organizational supranational forms various and even rapidly changing intermediary forms may occur, where the players of the world and regional political arenas will be searching for most advantageous and convenient blocks and agreements. For example, if population number and other parameters will be taken into account at decision-making (and quota distribution),[46] countries and participants may block with each other basing on the relative advantages of everyone in order to accept a decision advantageous to them (similar to political parties). However, some of the new unions and associations may eventually turn from temporary into constant ones and accept specific supranational forms.

Some new imperatives of global law will start being worked out in the same process. This idea is quite supported by the events connected with the modern global crisis, in particular the G-20 meetings. Direction towards such supranational regulation

---

45. The EU experience shows that this substantially restrains the development process, while such rules in the global representations will simply block it up.

46. There exists an index of national power measurement which integrates GDP measurements, defense expenses, population and the state of technology. Some indexes may serve as a basis for institution at the counting of quotes and votes.

forms is obvious, though it is unclear whether namely G-20 will become a constant organ, as 20 is possibly too great a number. However, as it has been stated above, another variant of leading players' club expansion is possible. Bringing the number of 'G club' members up to at least 11, i.e., 7 plus BRIC countries could already make this organ more influential than it is today. However, presidential meetings once a year or even more rarely, and even ministerial meetings do not suffice. Such meetings bear more of a ritual than practical character. In order to make such an organ not just influential, but a real global one, at least de facto, it is necessary to arrange the formats of negotiations, consultations, private agreements etc. on various levels and in different combinations.

One more form, much less likely but, in our opinion, much more desirable, would be the form of certain union representation. It would be reasonable to create a certain organ representing 10–15 leading establishments of the world (EU, OAS, LAS, CIS, etc.). Its sessions could allow representing the whole world through a limited number of representatives, while within the frames of representatives' powers the unions and coalitions would have an opportunity for better understanding of their common interests. Even such a dialogue in itself could be useful.

The stability of new geopolitical and geoeconomic forms will depend on numerous factors. However, historical experience shows the most stable ones to be those with not only particular advantages and objective necessity, but also with certain nonpolitical bases for uniting (i.e., geographical, cultural, economic, ideological etc.).

As regards the particular reasons for the convergence of certain societies, it should be taken into account that in the condition of a certain bifurcation which the world is currently going through, new lines and vectors contributing to the countries uniting into supranational establishments, unions, groups, blocs and clusters depend on a variety of reasons, among which a certain feature of proximity may turn out to be critical. Beside geographical proximity, economic relations and common political (geopolitical) interests, the proximity of culture and mentality (i.e., civilizational affinity and similarity) can be the strongest in many cases. These political, cultural, and religious specific features may lead to the creation of some special regional or even interregional supranational approximation models. For example, Chilean political scientist Talavera asks whether there is a special Latin way of action in the globalized world? He states there is such a way and quite particular indicators exist which confirm this. Further on, he points at the formation of a development variant implying the coexistence of socioeconomic order based on openness and free market relations (i.e., a purely

Western phenomenon—L. G., A. K.), with conservative socioreligious regime (Talavera, 2002)., i.e., let us add, a cultural form typical for a part of the old Europe.

Naturally, the movement towards the new world order will proceed at different levels. The regional level is very important. Regional leaders gather power very quickly; consequently, they will probably play a more significant role than now.[47] Besides, some regional states will start playing the key role in the whole geopolitical disposition in huge territories. Nowadays, according to some opinions, Iran is starting to play such a key role (see, e.g., Bykov *et al.*, 2009: 101–102). The eminent role of regional states will be revealed not only in geopolitical and geoeconomic aspects, but also, so to say, in geocultural aspect, which would be in no way less important than the first two.[48] It is not improbable that, responding to the probable deficit of global governance, non-governmental actors will form networks concentrated on particular problems.[49] However, neither the role of NGOs nor networks should be exaggerated. The main part in the formation of a new world order will most probably be played by states, while supranational unions of all formats and forms will be gathering strength.

A coordination centre is desperately needed, without it the net world will become an uncontrolled conglomerate. Besides, there is a prevailing stereotype on the necessity of some global institutions, so they will be aspired to, more or less successfully. Thus, the question of the coordinating political center of the World System remains exceedingly important. If some collective political (coordination) center (with limited rights) could be created, the coexistence of other functional centers could become more possible and systemic, interactive.[50] Namely states and especially supranational

---

47. We have already mentioned in our previous works that the level of economy and economic relations development in certain peripheral countries most probably belongs to industrial type than to postindustrial one. Accordingly, the level of nationalism is higher there, in large regional states it is just suitable for playing the hegemonic role in their regions (Grinin, 2008a). In some Asian and Latin American states which are economically rising and ideologically consolidating, 'nationalism' frequently emerges as a state ideology shared by the population. Along with that, an aspiration arises to support one's own sovereign rights, including the right for nuclear weapons etc. (It is also a convenient way of attracting the attention of international community to the country for a long time like, e.g., Iran has been doing.) Thus, a successful nuclear test became a subject of national pride for the Indians, though it caused strong anxiety in the USA and Western countries (Srinivas, 2002: 94).

48. In Bernstein's (Bernstein, 2002: 245) opinion, countries lying beyond the West and sufficiently powerful in economy, such as Japan (on the penetration of the Japanese mass culture into Asian countries see Aoki, 2002), or in culture, such as India, are capable of influencing the global culture.

49. One could mention as examples of such networks the Financial Stability Forum, the Carbon Sequestration Leadership Forum, and the International Partnership for the Hydrogen Economy (National Intelligence Council, 2008: 85).

50. Such a centre could rally the separate centers of the World Systems differing in their innovativity, industriality, financial capabilities, etc. Even though the rivalry between them would not disappear, it would become more productive.

---

unions are most likely to be capable of moving towards the creation of such a center.

Search for global responses to major challenges will lead to various types of solutions at the highest political level, from the ones aimed at forming an order capable of functioning for decades (huge experience of the 20th century proves this to be quite possible) to a mass of non-systemic, pragmatic, and palliative ones. However, even impulsive decisions allow starting the formation of a new system of decisions and institutions, coalitions and unions. On the other hand, the success of certain institutions will allow creating some projections on world restructuring.

Thus, we regard a wide range of decisions, institutionally and juristically formalized, aimed at systemic building of a new world; global, but for narrow problems, which, along with important but less global decisions will gradually create the outlines of a new world order. In 15–25 years our world will be both similar to the present and already substantially different from it. Global changes are forthcoming, but not all of them will take a distinct shape. Contrary to that, new contents may be covered by old outdated surfaces (as in the Late Middle Ages the emerging centralized state was not quite distinctly seen behind the traditional system of relationships between the crown, major seniors, and cities). One may say that these will be such changes that could prepare the world to the transition to a new phase of globalization (it will be very fortunate if there are grounds to call it the phase of sustainable globalization) whose contours are not clear yet.

Finally, the future epoch will be an epoch of not only new coalitions, but also the one of new global institutions, and new international technologies of multilateral (diplomatic, social, and cultural) cooperation, on which much will depend. For example, the format of international congresses and multilateral agreements that originated after the Napoleonic Wars and reached its apogee in the 20th century is likely to be pressed by other formats which most probably will be based on modern communication technologies. Thus, some standing commission that work not at a bargaining table, but through the video conference format could become a convenient and rather low-cost organ which could work permanently on solving certain problems. As Charles de Gaulle said, politics is too serious a matter to entrust politicians with it (Belmis, 2009: 238). The same may be said about the diplomacy in the globalizing world.

REFERENCES

Amin, S., Arrighi, G., Frank, A. G., and Wallerstein, I. 2006. Transforming the Revolution: Social Movements and the World-System. Delhi: Aakar.

Antolin P. 2008. Pension Fund Performance. OECD Working Papers on Insurance and Private Pensions No 20. Paris: OECD.

Antolin P., Stewart F. 2009. Private Pensions and Policy Responses to the Financial and Economic Crisis. OECD Working Paper on Insurance and Private Pensions No 36. Paris: OECD Publishing.

Aoki, T. 2002. Aspects of Globalization in Contemporary Japan. In Berger and Huntington 2002: 68–89.

Arrighi, G., and Silver, B. J. 1999. Chaos and Governance in the Modern World System. Minneapolis: University of Minnesota Press.

Bacevich, A. 2002. American Empire. The Realities and Consequences of U.S. Diplomacy. Cambridge, MA: Harvard University Press.

Beglaryan, G. 2011. The Summary of the PRC National Assambly Congress: To Retain Power, One should Sacrifice Booming Growth of Economy. URL: http://www.bloom-boom.ru/blog/bloomboom/10270.html.

Berger, P. L., and Huntington, S. P. (eds.) 2002. Many Globalizations: Cultural Diversity in the Contemporary World. New York, NY: Oxford University.

Belmis, Ye. V. (ed.) 2009. Thoughts of Great Politicians. Saint-Petersburg: Paritet. In Russian (Бельмис, Е. В. (сост.). Мысли великих политиков. СПб.: Паритет).

Bernstein, A. 2002. Globalization, Culture, and Development. Can South Africa be More Than an Offshoot of the West? In Berger and Huntington 2002: 185–250.

Braudel, F. 1973. Capitalism and Material Life, 1400–1800. New York: Harper and Row.

Brzezinski, Z. 1997. The Grand Chessboard: American Primacy and Its Geostrategic Imperatives. New York, NY: Basic Books.

Brzezinski, Z. 2004. The Choice: Global Domination or Global Leadership. New York, NY: Basic Books.

Buchanan, P. J. 2002. The Death of the West: How Dying Populations and Immigrant Invasions Imperil Our Country and Civilization. New York, NY: St. Martin's Griffin.

Bykov, P., Vlasova, O., Zavadsky, M., Koksharov, A., and Sumlennyi, S. 2009. The World after London. Expert 13(652): 100–103. In Russian (Быков, П. и др. Мир после Лондона. Эксперт 13: 100–103).

Callahan, G. 2002. Financial Economics for Real People. Paper presented at the 'Boom, Bust, and Future' Seminar at Ludwig von Mises Institute, January 18–19. Auburn.

Cassard, M. 1994. The Role of Offshore Centers in International Financial Intermediation. Washington, D.C.: IMF (IMF Working Paper 94/107).

Chase-Dunn, C., and Hall, T. D. 1994. The Historical Evolution of World-Systems. Sociological Inquiry 64: 257–280.

Chase-Dunn, C., and Hall, T. D. 1997. Rise and Demise: Comparing World-Systems. Boulder, CO: Westview Press.

Doronin, I. G. 2003. The World Stock Markets. In Korolyov, I. S. (ed.), The World Economy: Global Trends for One Hundred Years (pp. 101–133). Moscow: Economist. In Russian (Доро-нин, И. Г. Мировые фондовые рынки. Мировая экономика: глобальные тенденции за 100 лет / ред. И. С. Королев, с. 101–133. М.: Экономистъ, 2003).

Dynkin, A. A. (ed.) 2008. World Economy: Forecast till 2020. Moscow: Magistr. In Russian (Дынкин, А. А. (ред.). Мировая экономика: Прогноз до 2020 года. М.: Магистр).

The Economist 2008. Asset-backed insecurity. Economist. January 17. URL: http://www.economist.com/node/10533428?story_id=10533428.

Evans, M. D. 1969 [1859]. The History of the Commercial Crisis, 1857–58 and the Stock Exchange Panic of 1859. New York, NY: B. Franklin.

Fokin, V. 2010. The Hide-and-Seek at the Summit Level. RBK. The Final Issue 1: 72–75. In Russian (Фокин, В. Прятки на высшем уровне. РБК. Итоговый выпуск № 1: 72–75).

Frank, A. G. 1990. A Theoretical Introduction to 5,000 Years of World System History. Review 13/2: 155–248.

Frank, A. G. 1993. The Bronze Age World System and its Cycles. Current Anthropology 34: 383–413.

Frank, A. G. 1997. Asia Comes Full Circle—with China as the 'Middle Kingdom'. Humboldt Journal of Social Relations 76(2): 7–20.

Frank, A. G. 1998. ReORIENT: Global Economy in the Asian Age. Berkeley, CA: University of California Press.

Frank, A. G., and Gills, B. K. (eds.) 1993. The World System: Five Hundred Years of Five Thousand? London: Routledge.

Gelbras, V. G. 2007. The Costa of China`s Economic Success. Mirovaya ekonomika i mezhdunarodnye otnosheniya 9: 26–34. In Russian (Гельбрас, В. Г. Цена экономических успехов Китая. Мировая экономика и международные . отношения 9: 26–34).

Goldstone, J. 1988. East and West in the Seventeenth Century: political crises in Stuart England, Ottoman Turkey and Ming China. Comparative Studies in Society and History 30: 103–142.

Goldstone, J. 1991. Revolution and Rebellion in the Early Modern World. Berkeley, CA: University of California Press.

Greenspan, A. 2007. The Age of Turbulence: Adventures in a New World. London: Penguin.

Grigoriev, L., and Salikhov, M. 2008. Financial Crisis—2008: Entering the World Recession. Voprosy ekonomiki 12: 27–45. In Russian (Григорьев, Л., Салихов, М. 2008. Финансовый кризис—2008: вхождение в мировую рецессию. Вопросы экономики 12: 27–45).

Grinin, L. E. 2008a. Globalization and the Models of Transformation of Sovereignty in Western and Non-Western Countries. In Kulpin, E. S. (ed.), Person and Nature: 'Challenge and Response' (pp. 56–88). Moscow: IATs-Energiya. In Russian (Гринин, Л. Е. Глобализация и модели трансформации суверенности в западных и незападных странах. Человек и природа: «Вызов и ответ» / ред. Э. С. Кульпин, с. 56–88. М.: ИАЦ-Энергия).

Grinin, L. E. 2008b. Globalization and Sovereignty: Why do States Abandon their Sovereign Prerogatives? Age of Globalization 1: 22–32.

Grinin, L. E. 2008c. Unwanted Child of Globalization. The Remarks on the Crisis. Vek globalizatsii 2: 46–53. In Russian (Гринин, Л. Е. Нежеланное дитя глобализации. Заметки о кризисе. Век глобализации 2: 46–53).

Grinin, L. E. 2009a. The Global Crisis as a Crisis of Overproduction of Money. Filosofia i obschestvo 1: 5–32. In Russian (Гринин, Л. Е. Глобальный кризис как кризис перепроизводства денег. Философия и общество 1: 5–32).

Grinin, L. E. 2009b. The Psychology of Economic Crises. Istoricheskaya psikhologia i sotsiologia istorii 2: 75–99. In Russian (Гринин, Л. Е. Психология экономических кризисов. Историческая психология и социология истории 2: 75–99).

Grinin, L. E. 2009c. Modern Crisis: New Lines and Classics of Genre. Istoriya i sovremennost 1: 3–32. In Russian (Гринин, Л. Е. Современный кризис: новые черты и классика жанра. История и современность 1: 3–32).

Grinin, L. E. 2009d. State and Historical Process. The Political Cut of Historical Process. 2nd ed. Moscow: URSS. In Russian (Гринин, Л. Е. Государство и исторический процесс. Политический срез исторического процесса. 2-е изд. М: ЛИБРОКОМ).

Grinin, L. E. 2009e. Globalization and the Transformation of National Sovereignty. Auckland. In Sheffield, J. (eds.), Systemic Development: Local Solutions in a Global Environment (pp. 47–53). ISCE Publishing: Goodyear.

Grinin, L. E. 2009f. Transformation of Sovereignty and Globalization. In Grinin, L. E., Beliaev, D. D., and Korotayev, A. V. (eds.), Hierarchy and Power in the History of Civilizations: Political Aspects of Modernity (pp. 191–224). Moscow: Librocom/URSS.

Grinin, L. E. 2009g. Understanding the Crisis. Global Crisis as the Crisis of Overproduction of Money. Filosophiya i obschestvo 1: 5–32. in Russian (Гринин, Л. Е. Глобальный кризис как кризис перепроизводства денег. Философия и общество 1: 5–32).

Grinin, L. E. 2009h. The State in the Past and in the Future. Herald of the Russian Academy of Sciences 79(5): 480–486.

Grinin, L. E. 2011. Chinese Joker in the World Pack. Journal of Globalization Studies 2(2): 7–24.

Grinin, L. E. 2012. Macrohistory and Globalization. Volgograd: Uchitel.

Grinin, L. E., and Korotayev, A. V. 2009a. Social Macroevolution: The Genesis and Transformation of the World System. Moscow: LIBROCOM. In Russian (Гринин, Л. Е., Коротаев, А. В. Социальная макроэволюция: Генезис и трансформации Мир-Системы. М.: ЛИБРОКОМ).

Grinin, L. E., and Korotayev, A. V. 2009b. Social Macroevolution: Growth of the World System Integrity and a System of Phase Transitions. World Futures 65/7: 477–506.

Grinin, L. E., and Korotayev, A. V. 2010. Global Crisis in Retrospect: A Brief History of Rises and Crises; From Lycurgus to Alan Greenspan. Moscow: LIBROCOM. In Russian (Гринин, Л. Е., Коротаев, А. В. Глобальный кризис в ретроспективе: Краткая история подъемов и кризисов: от Ликурга до Алана Гринспена. М.: ЛИБРОКОМ).

Grinin, L. E., and Korotayev, A. V. 2011. The coming epoch of new coalitions: possible Global scenarios World Futures 67(8): 531–563.

Grinin, L. E., Korotayev, A. V., and Malkov, S. Yu. 2010. A Mathematical Model of Juglar Cycles and the Current Global Crisis. In Grinin, L., Korotayev, A.,and Tausch, A. (eds.), History & Mathematics: Processes and Models (pp. 138–187). Volgograd: Uchitel.

Haberler, G. 1964 [1937]. Prosperity and Depression. Theoretical Analysis of Cyclical Movements. Cambridge, MA: Harvard University Press.

Hayek, F. A. von 1931. Prices and Production. London: Routledge.

Hayek, F. A. von 1933. Monetary Theory and the Trade Cycle. London: Jonathan Cape.

Held, D., and McGrew, A. (eds.) 2003. The Global Transformation Reader: An Introduction to the Globalization Debate. 2nd ed. Cambridge, UK: Polity Press.

Held, D., McGrew, A., Goldblatt, D., and Perraton, J. 1999. Global Transformations. Politics, Economics and Culture. Stanford, CA: Stanford University Press.

Hicks, J. R. 1946 [1939]. Value and Capital: An Inquiry into Some Fundamental Principles of Economic Theory. Oxford: Clarendon Press.

Hilferding, R. 1981 [1910]. Finance Capital. A Study of the Latest Phase of Capitalist Development. London: Routledge.

Hinz R., Heinz P. Rudolph, Pablo Antolín, and Juan Yermo. 2010. Evaluating the Financial Performance of Pension Funds. Washington, DC: The International Bank for Reconstruction and Development / The World Bank.

Inozemtsev, V. 2008. Post-American World—Dilettantes' Dream and a Complex Reality. Mirovaya ekonomika i mezhdunarodnye otnosheniya 3: 3–15. In Russian (Иноземцев, В. «Постамериканский мир»: мечта дилетантов и непростая реальность. Мировая экономика и международные отношения 3: 3–15).

Ivanter, A. 2009. Why do Everybody Avoid Loans like the Plague? Expert. The Best Articles. The Anatomy of Recession: 97–99. In Russian (Ивантер, А. Все как от огня шарахаются от кредитов. Эксперт. Лучшие материалы. Анатомия рецессии: 97–99).

Jervis, R. 2005. American Foreign Policy in a New Era. New York, NY: Routledge.

Johnson, K. L., and de Graaf, F. J. 2009 Modernizing Pension Fund Legal Standards for the 21st Century. Network for Sustainable Financial Markets: Consultation Paper No. 2. URL: http://www.oecd.org/dataoecd/28/62/42670725.pdf

Juglar, C. 1862. Des Crises Commerciales et de leur retour périodique en France, en Angleterre et aux États-Unis. Paris: Guillaumin.

Juglar, C. 1889. Des Crises Commerciales et de leur retour périodique en France, en Angleterre et aux États-Unis. 2nd ed. Paris: Alcan.

Kagan, R. 2003. Of Paradise and Power. America and Europe in the New World Order. New York, NY: Knopf Publishers.

Kanayev, Ye., and Kurilko, A. 2010. South-Eastern Asia in the Conditions of World Financial-Economical Crisis. Mirovaya ekonomika i mezhdunarodnye otnosheniya 2: 38–46. In Russian (Канаев, Е., Курилко, А. Юго-восточная Азия в условиях мирового финансово-экономического кризиса. Мировая экономика и международные отношения 2: 38–46).

Karsbol, D. 2010. I have a 'Bear' View on China. RBK. Final Issue 1: 100–103. In Russian (Карсбол, Д. У меня «медвежий» взгляд на Китай. РБК. Итоговый выпуск 1: 100–103).

Keynes, J. M. 1936. The General Theory of Employment, Interest, and Money. London: Macmillan.

Khalturina, D. A., and Korotayev, A. V. 2009. System monitoring of global and regional development. In Khalturina, D. A., and Korotayev, A. V. (eds.), System monitoring: global and regional development (pp. 11–188). Moscow: Librokom/URSS. In Russian (Халтурина, Д. А., Коротаев, А. В. Системный мониторинг глобального и регионального развития. Системный мониторинг: Глобальное и региональное развитие / ред. Д. А. Халтурина, А. В. Коротаев, с. 11–188. М.: ЛИБРОКОМ/URSS).

Kirichenko, E. V., Martsinkevich, V. I., Vasilevsky, E. K., Zapadinskaya, L. I., Lebedeva, Ye. A., Nikolskaya, G. K., and Perova, M. K. 2008. The United States of America. In Dynkin 2008: 185–230. In Russian (Кириченко, Э. В., Марцинкевич, В. И., Василевский, Э. К., Западинская, Л. И, Лебедева, Е. А., Никольская, Г. К., Перова, М. К. Соединенные Штаты Америки. Мировая экономика: Прогноз до 2020 год / ред. А. А. Дынкин, с. 185–230. М.: Магистр).

Korotayev, A., Malkov, A., and Khaltourina, D. 2006. Introduction to Social Macrodynamics: Compact Macromodels of the World System Growth. Moscow: URSS.

Korotayev, A., Zinkina, J., and Bogevolnov, J. 2011. Kondratieff waves in global invention activity (1900–2008). Technological Forecasting and Social Change 78: 503–510.

Kudrin, A. 2009. The World Financial Crisis and its Impact on Russia. Voprosy ekonomiki 1: 9–10. In Russian (Кудрин, А. Мировой финансовый кризис и его влияние на Россию. Вопросы экономики 1: 9–10).

Lan, V. I. 1976. The USA: From World War I to World War II. Moscow: Nauka. In Russian (Лан, В. И. США: от первой мировой до второй мировой войны. М.: Наука).

Lescure, J. 1907. Des Crises Générales et Périodiques de Surproduction. Paris: L. Larose et Forcel.

Maddison, A. 2007. Contours of the World Economy, 1–2030. Oxford: Oxford University Press.

Maddison, A. 2010. World Population, GDP and Per Capita GDP, A.D. 1–2003. URL: www.ggdc.net/maddison

Mandelbaum, M. 2005. The Case for Goliath: How America Acts as the World's Government in the Twenty-First Century. New York, NY: Public Affairs.

Marx, K. 1993 [1893, 1894]. Capital. Vol. I–II / Transl. by D. Fernbach. Harmondsworth: Penguin Group.

Meliantsev, V. A. 2009. Developed and Developing Countries in the Age of Transformations. Moscow: ID 'Klyuch-C'. In Russian (Мельянцев, В. А. Развитые и развивающиеся страны в эпоху перемен. М.: ИД «Ключ-С»).

Mendelson, L. A. 1959–1964. Theory and History of Economic Crises and Cycles. Vols. 1–3. Moscow: Izdatelstvo sotsialno-economicheskoi literatury. In Russian (Мендельсон, Л. А. Теория и история экономических кризисов и циклов. Т. 1–3. М.: Издательство социально-экономической литературы).

Mikhailov, D. M. 2000. The World Financial Markets. Trends and Tools. Moscow: Ekzamen. In Russian (Михайлов, Д. М. Мировой финансовый рынок. Тенденции и инструменты. М: Экзамен).

Mikheev, V. V. 2008. China. In Dynkin 2008: 303–329. In Russian (Михеев, В. В. Китай. В: Дынкин 2008: 303–329).

Minsky, H. P. 2005. Induced Investment and Business Cycles. Cheltenham: Elgar.

Mises, L. von 1981 [1912]. The Theory of Money and Credit. Indianapolis, IN: Liberty Fund.

Movchan, A. 2010. The Gold Rush. Forbes (Russian Edition) January: 49. In Russian (Мовчан, А. Золотая лихорадка. Forbes январь: 49).

National Intelligence Council 2008. Global Trends 2025: A Transformed World. Washington, D.C.: National Intelligence Council.

Naumov, I. 2008. The Pension Capital Fund of Russia (PFR) in the Current Year Sustained 10 billion roubles of Losses from the Investment of Funds in the State Securities. Nezavisimaya gazeta 19.12. URL: http://www.ng.ru/economics/2008-12-19/4_pensia.html. In Russian (Наумов, И. Пенсионный фонд России [ПФР] в текущем году зафиксировал убытки в 10 млрд руб. от размещения средств в государственных бумагах. Независимая газета 19.12. URL: http://www.ng.ru/economics/2008-12-19/4_pensia.html).

Nye, J. S. Jr. 2002. The Paradox of American Power. Why the World's Only Superpower Can't Go It Alone. New York, NY: Oxford University Press.

Pantin, V. I., and Lapkin, V. V. 2006. Philosophy of Historical Forecasting: Rhythms of History and the Perspectives of the World Development in the First Half of the 21st Century. Dubna: Feniks +. In Russian (Пантин, В. И., Лапкин, В. В. 2006. Философия исторического прогнозирования: ритмы истории и перспективы мирового развития в первой половине XXI века. Дубна: Феникс+).

Platonova, I. N., Nagovitsin, A. G., and Korotchenya, V. M. 2009. Rearrangement of the World Monetary System and the Position of Russia. Moscow: LIBROCOM/URSS. In Russian (Платонова, И. Н., Наговицин, А. Г., Коротченя, В. М. Перестройка мировой валютной системы и позиция России. М.: ЛИБРОКОМ/URSS).

Renwick, N. 2000. America's World Identity. The Politics of Exclusion. Basingstoke: Macmillan Press.

Samuelson, P. A., and Nordhaus, W. D. 2005. Economics. 18th ed. New York, NY: McGraw-Hill.

Samuelson, P. A., and Nordhaus, W. D. 2009. Macroeconomics. 19th ed. New York, NY: McGraw-Hill.

Schäfer, U. 2009. Der Crash des Kapitalismus. Frankfurt: Campus Verlag.

Shtefan, Ye. 2008. American Pension Capital Funds have lost 2 trillion dollars. Novyi region 2. 08.10. URL: http://www.nr2.ru/economy/199830.html. In Russian (Штефан, Е. Пенсионные фонды США потеряли два триллиона долларов. Новый регион 2. 08.10. Интернет-ресурс: http://www.nr2.ru/economy/199830.html).

Soros, G. 1998. The Crisis of Global Capitalism. Open Society Endangered. London: Public Affairs.

Srinivas, T. 2002. 'a Tryst with Destiny'. The Indian Case of Cultural Globalization. In Berger and Huntington 2002: 89–117.

Suetin, A. 2009. On the Causes of Current Financial Crisis. Voprosy ekonomiki 1: 40–51. In Russian (Суэтин, А. О причинах современного финансового кризиса. Вопросы экономики 1: 40–51).

Talavera, A. F. 2002. Trends toward Globalization in Chile. In Berger and Huntington 2002: 250–295.

Thurow, L. C. 1996. The Future of Capitalism: how today's economic forces shape tomorrow's world. New York, NY: Morrow.

Tinbergen, J. 1976. Reshaping the international order: A report to the Club of Rome. New York, NY: Dutton.

Tooke, T. A. 1838–1857. A History of Prices and the State of the Circulation. London: Longman.

Tugan-Baranovsky, M. I. 1954. Periodic Industrial Crises. Annals of the Ukranian Academy of Arts and Sciences in the United States 3/3: 745–802.

Tugan-Baranovsky, M. I. 2008 [1913]. The Periodical Industrial Crises. Moscow: Directmedia Publishing. In Russian (Туган-Барановский, М. И. 2008 [1913]. Периодические промышленные кризисы. М.: Директмедиа Паблишинг).

Van Der Wee, H. 1990. Histoire economique mondiale 1945–1990. Paris: Academia Duculot.

Wallerstein, I. 1987. World-Systems Analysis. Social Theory Today (pp. 309–324). Cambridge, UK: Polity Press.

Wang, H. H. 2010. Myth of China as a Superpower. URL: http://helenhwang.net/2010/04/myth-of-china-as-a-superpower/

Wilkinson, D. 1999. Unipolarity without Hegemony. International Studies Review 1(2): 141–172.

Wolf, M. 2005. The Paradox of Thrift: Excess Savings are Storing up Trouble for the World Economy. Financial Times 13.06.

World Bank 2012. World Development Indicators Online. Washington, D.C.: World Bank. URL: http://web.worldbank.org/WBSITE/EXTERNAL/DATASTATISTICS/0,,contentMDK: 20398986~pagePK:64133150~piPK:64133175~theSitePK:239419,00.html

Yan, Yunxiang 2002. Managed Globalization. State Power and Cultural Transition in China. In Berger and Huntington 2002: 19–47.

Zakaria, F. 2009. The Post-American World. New York, NY: Norton.

Zorome, A. 2007. Concept of Offshore Financial Centers: In Search of an Operational Definition. Washington, DC: IMF.

Zotin, A. 2010. The World Central Bank. As a Result of the Crisis the IMF can become a Global Financial Regulator. Who Makes Profit of it? RBK. The Final Issue 1: 46–49. In Russian (Зотин, А. Всемирный Центробанк. В результате кризиса МВФ может стать глобальным финансовым регулятором. Кому это выгодно? РБК. Итоговый выпуск № 1: 46–49).